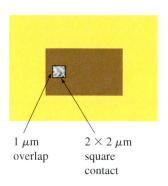 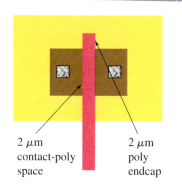 

1 μm overlap   2 × 2 μm square contact   2 μm contact-poly space   2 μm poly endcap   1 μm metal overlap of contact space

FIGURE 11-8

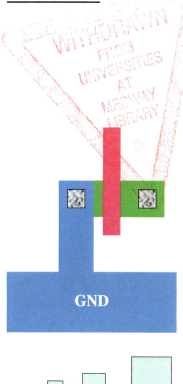

GND

Calibration squares 1, 2, 3, 5 units

FIGURE 11-9

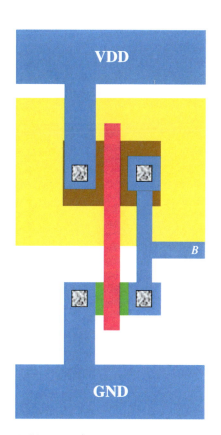

VDD

B

GND

FIGURE 11-10

CMOS Digital Integrated Circuits: A First Course

Charles Hawkins,
Jaume Segura,
and
Payman Zarkesh-Ha

University of Florida
University of Balearic Islands
University of New Mexico

SCITECH
PUBLISHING
an imprint of the IET

Edison, NJ
scitechpub.com

SciTech
PUBLISHING
an imprint of the IET

Published by SciTech Publishing, an imprint of the IET.
www.scitechpub.com
www.theiet.org

10 9 8 7 6 5 4 3 2 1

ISBN 978-1-61353-002-3

Typeset in India by MPS Ltd
Printed in the USA by Sheridan Books, Inc.

CMOS Digital Integrated Circuits: A First Course

CONTENTS

PREFACE

This book teaches introductory complementary metal-oxide-semiconductor (CMOS) digital electronics for electrical and computer engineering undergraduates. For many years the CMOS technology has dominated the method of designing and manufacturing digital (computing) integrated circuits. The selection of material here is not significantly different from the graduate texts by J. Rabaey et al. (Digital Integrated Circuits, 2003, Prentice Hall), N. Weste and D. Harris (CMOS VLSI Design, 2011, Addison-Wesley), or J. Baker (CMOS: Circuit Design, Layout, and Simulation, 2010, Wiley-IEEE Press), but the style is introductory with many examples, self-exercises, and end-of-chapter problems.

This book initially reviews material relevant to digital electronics that students learned in previous circuit and logic courses. The book then moves through chapters on basic physics of semiconductor materials and diodes; nMOS and pMOS field effect transistors circuit analysis; electronic properties of the metal interconnections; the CMOS inverter; the CMOS NAND, NOR, and transmission gates electronics; transformation from Boolean equations to CMOS transistor schematics and domino circuits; timing electronics; memory circuits; field-programmable gate arrays (FPGAs); CMOS layout; and CMOS fabrication basics. The emphasis is on transistor level electronics.

The principles of power dissipation are introduced with numerical examples. Lowering circuit power has special urgency today where total Internet power consumes about 10% of US electrical power generation.

Other features and objectives include:

- There are abundant examples, self-exercises with answers, and many problems at the end of chapters to give students reflexive skills in transistor circuit analysis.
- This course can be taught before or after a companion class in introductory analog electronics.
- The book strives for clarity and self-learning in an undergraduate presentation.
- The book doesn't overwhelm students with too much details; it defines teaching goals consistent with what they will take forward to the next level of electronics.
- Students are provided with an education that serves as a prerequisite for graduate or senior courses in digital electronics and allows entry level into the digital electronics industry.
- The book is light enough for students to carry to class.

Chapter Summaries

Chapter 1 reviews relevant logic theory that includes Boolean equation to logic gate schematics, DeMorgan's theorem, logic equivalence, and logic gate reduction. Basic circuit theory is next, with emphasis on analysis of terminal impedance, node voltages, and branch currents by inspection. Nonlinear circuit analysis techniques are introduced using the diode and its nonlinear current-voltage expression. Capacitor and inductor properties and circuits are reviewed, as is the power wasted in resistive and capacitive circuits. These topics are a few among many but are selected for their relevance in the digital circuit analysis that follows.

The second chapter introduces the semiconductor physics that underlie device operation. The goal is to impart a good visual model of the physics of materials and diodes, and to use basic equations for better understanding. Semiconductor physics is a complex subject that can involve more than one course at the graduate level, and Chapter 2 cannot replicate this. However, visual models of semiconductor materials and diodes are important because engineers often use qualitative language to communicate important properties of the physics of semiconductor diodes and transistors. Students should be able to answer the question, "How do diodes (and transistors) work and perform basic parameter calculations?". This chapter leads directly into Chapter 3 on field effect transistors.

CMOS circuits use two transistors types; the nMOS and pMOS field effect transistors. Chapter 3 describes how these transistors work, followed by numerical analysis of circuit node voltages and currents. Many examples, self-exercises, and end-of-chapter problems give students the reflexive response to analyze transistor digital circuits. Equal treatment is given to each transistor type.

Chapter 4 deals with metal properties, which are especially relevant in modern circuits since chip total metal length may be on the order of several miles, and minimum metal dimensions can be 22 nm or smaller. Metal properties are a major concern in attaining maximum IC frequency and minimum noise operation, and metal physics deserves as much study as does the transistor.

In Chapter 5, the CMOS inverter is discussed. The CMOS inverter is the most abundant logic gate in any digital integrated circuit (IC). It has one nMOS and one pMOS field effect transistor. This chapter introduces about a dozen important electronic properties, with numerical examples. Inverter properties are inherently important but are also the basis for electronic properties of NAND, NOR, and sequential logic circuits such as the master-slave flip-flop. Inverter power dissipation properties are emphasized.

Chapter 6 covers NAND and NOR gates, which build on the inverter by placing transistors in parallel and series to the inverter pair. These multi-input logic gates have all of the electronic properties of the inverter and a few that are unique. The electronic basis for the noncontrolling logic state is described in this chapter as it relates to circuit debugging, test engineering, and schematic reading. Pass transistor and CMOS transmission gate properties conclude the chapter. Transmission gates are abundant, comprising half of the logic gates in the master-slave flip-flop.

Chapter 7 develops design styles that assemble transistors into logic gates. It begins with a relatively simple technique to transform Boolean equations into a CMOS transistor schematic that performs the logic. Other design styles are presented along with the reasons for having different styles. Power dissipation is analyzed with a technique that allows a power comparison of different combinational logic configurations.

Chapter 8 discusses the accurate design and placement of timing signals, which may be the most challenging task for a designer. This often neglected undergraduate course topic is emphasized, giving it the importance it deserves. The edge-triggered flip-flop (FF) has a complexity that must be mastered. Timing parameters and rules must be exact otherwise circuits will fail. System-level timing builds on these foundations and introduces system timing parameters and constraints.

Chapter 9 covers memory circuits. They have always been embedded within the computing chips, but today microprocessor chips may dedicate more than 70% of the total transistor count to these memory circuits. Therefore, special emphasis is given on these static random access memory (SRAM) designs. Transistor sizing of SRAM cells is developed with numerical examples. Another high-volume memory design is dynamic random access memory (DRAM). This single transistor memory cell has different properties.

Chapter 10 looks at a unique and popular design style using field-programmable gate arrays (FPGA). This material follows from other design styles described in the preceding chapter. The electronics and method of operation are different, but FPGAs are common and abundant enough to devote a chapter.

In Chapter 11, the CMOS layout is discussed. A conversion occurs in the design process when transistor schematics are transformed to rectangular images on a photographic mask. The images represent transistor and metal line geometries. Masks are drawn for each of several layers in the buildup of the IC. Layout is not electronics but is the necessary first step in using photolithography to make the tiny transistors and metal interconnections. The mask layout step is introduced using manual layout of the inverter, NAND, and NOR gates. Several commercial layout tools exist, but cost and training time led us to consider the Microsoft PowerPoint program to draw the layouts. PowerPoint is typically available on all computers, training time is minimal, it appears to have long-term stability in the market, and students get a better grounding in design rules. PowerPoint has been successful as a teaching tool in the classroom for layout of simple logic gates circuits.

Chapter 12 describes the chemical, physical, and photolithography techniques that actually make the final circuit. This chapter is qualitative but sufficient enough to allow students to converse on the various sequenced fabrication techniques that achieve the end circuit result of the chip.

Comments for Instructors

The book uses long channel models for MOSFET analysis, even though short channel models are common in industry. The reasons are twofold. First, the short channel models are often simplified for undergraduate presentation where they lose accuracy. Also, full short channel models become too complex for hand calculations. Although the long channel models are also not accurate, they allow manual problem solving insight into the various bias regions of the transistors. We originally designed the book using short channel models, but found the simplified analytical expressions clumsy and inaccurate. A second observation is that modern industry electronic papers and oral presentations often refer to long channel models despite the use of short channel transistors. It is part of the language. The more accurate short channel models are best left to graduate courses and detailed computer models.

Other choices were made to avoid overly complex material at this undergraduate stage. The subjects and their depth were a trade-off between designing a one-term course and covering the important topics. For example, combinational logic power analysis uses the truth table analysis rather than logical effort. Chapter 9 on memory keeps the timing description simple but to the point. Memory design deserves a whole book for a full description.

The problems in this book most efficiently use the modern equation solving ability of scientific calculators. One great learning advantage is that time is spent on the problem itself and little on the grind of manually solving with quadratic equations or iteration. An unknown variable can be embedded anywhere in the equation, and the scientific calculator doesn't care. It solves for the unknown variable in seconds. Students and instructors can solve these problems any way they desire, but the scientific calculator is truly an advance in modern digital circuit teaching.

A Suggested Semester Chapter Order

CHAPTER	TITLE	TIME IN CLASS
Chapter 1	Basic Logic Gates and Circuit Theory	1 week
Chapter 2	Semiconductor Physics	1 week
Chapter 3	MOSFET Transistors	2 weeks
Chapter 5	CMOS Inverter	1.5 weeks
Chapter 6	CMOS NAND, NOR, and Transmission Gates	1 week
Chapter 7	CMOS Design Styles	1.5 weeks
Chapter 11	CMOS Circuit Layout	1 week
Chapter 12	How Chips are Made	1 week
Chapter 4	Metal Interconnection Properties	1 week
Chapter 8	Sequential Logic Gate Design and Timing	2 weeks
Chapter 9	Memory Circuits	1 week
Chapter 10	FPGAs	0.5 week

Chapter 4 on metal interconnects logically fits with device descriptions, but it interrupts the flow of electronic circuitry so it was put later. Chapters 11 and 12 continue the emphasis on circuitry and the IC before returning to Chapter 4.

Author Background

This book reflects the experience of the authors, who have taught this material at the graduate and undergraduate level and have worked closely with the digital electronic industry in their careers. Hawkins and Segura did sabbaticals with the Intel Corporation: Segura at the Intel campus in Portland Oregon, and Hawkins in Rio Rancho, New Mexico. Segura also did sabbatical work at Philips Semiconductor and received numerous research contracts from industry. Hawkins worked closely with the Sandia National Laboratory in New Mexico for over 20 years in its CMOS integrated circuits group. Both authors have long histories of committee work for the European DATE conference, the International Test Conference, and the VLSI Test Symposium. Hawkins was editor of the *Electron Device Failure Analysis* magazine.

Payman Zarkesh-Ha is professor in the ECE Department at the University of New Mexico (UNM). He teaches graduate and undergraduate VLSI, digital, and analog electronics. Prior to joining UNM, he worked for five years at LSI Logic Corp, where he worked on interconnect architecture design for the next ASIC generations. He has published more than 60 refereed papers and holds 12 issued patents. His research interests are statistical modeling of nanoelectronics devices and systems, and design for manufacturability, low power, and high performance VLSI designs. All of these activities outside of the classroom influenced our choice of material and style in the book. It is long overdue for electrical and computer engineering undergraduate students to rid themselves of outdated logic circuits and receive a course dedicated to digital CMOS electronics.

INTRODUCTION

Any sufficiently advanced technology is indistinguishable from magic.

Arthur C. Clarke's Third Law

The goal of this book is to prepare you to contribute to the computer evolution in the 21st century. It is about the electronics that propel the incredible surge in human communication and knowledge capability. The foundation of a computer is the transistor. Computer electronics deals with transistor-level behavior of circuits that perform all of the computer logic operations such as adding, multiplying, storing, comparing, and any operation described by Boolean equations. Billions of transistors and their wire connections are embedded in small, thin, rectangular silicon computer chips. The total wire connections on these tiny chips may be several miles in length, and power dissipation may range from a few microwatts to over 200 watts. The chip is also referred to as an integrated circuit (IC). Chips are complicated, and electrical and computer engineers must understand computing at this circuit level.

Engineers face challenges. How would you blend digital circuit knowledge with computer architecture to design a chip? How fast do we want to clock the computer, and where do we start? How do you interface chips on a circuit board? How much heat from chip power loss can you stand—how do you minimize it? As a customer, how do you talk to a chip designer? When your first chips are returned from the factory to evaluate and something is wrong, where do you begin to solve the problem? Failures may be temperature or power supply dependent and not simple static Boolean errors. What skills and knowledge do you need to identify and correct these failures? Whether you are an engineer at the chip level or you design at the higher board and system level, the solutions often reside with knowledge of chip properties at the transistor level.

A knowledge hierarchy exists in electronics. Semiconductor physics describes diode and transistor action using model equations that allow calculation of transistor circuit node voltages and path currents. Specific transistor configurations then form the different logic gates, such as the inverter, NAND, NOR, transmission gates, the D-flip-flop, and more complex combinational logics gates derived from arbitrary Boolean statements. These logic gates electronically perform the Boolean operations that define the computer, and we must understand their properties. What are the voltage, current, temperature, power drain, propagation delay time, and noise margins properties?

A master clock oscillator drives sequential circuits with pulses that synchronize data movement of the Boolean operations in the computer. Clock speed is an important parameter and often the first specification that a buyer looks at when shopping for a computer. The amount of computer memory may be the next question. Memory subcircuits are extensively built into the computer chips. So, what is a standard memory cell and how are memories organized? Modern computer chips may dedicate over 70% of the total transistor count to embedded memory. Memory embedded in the chip allows faster computing as opposed to sending signals back and forth to external memory chips mounted on a circuit board.

We might take for granted our computer-based miracles, such as the Internet, cell phone magic, email, Google, automobile electronics, biomedical instrumentation, GPS, YouTube, instant news, weather, and sports, automobile electronics, e-books, Facebook, and, yes, video games. You might ask, "Hasn't it always been like this?" The answer is no—the applications didn't really get rolling until the early 1990s, and all of these modern products depended on fast, cheap, and small computer chips.

Transistors and Computers—Until Death Do They Part

To get a better sense of our subject, let us track electronic progress in digital computer development and then its role in the Internet. We see not only the march of computers to smaller, faster, and cheaper but also the fascinating interplay of diverse forces. The Internet did not grow in a vacuum, and neither did computers.

The first computer circuit we are aware of was called the flip-flop by its English inventors, Eccles and Jordan, almost 100 years ago. A flip-flop remains stable in one of two voltage states until triggered to the other state by an external electrical pulse. The flip-flop stores a voltage state. Computers were not thought of at that time so the flip-flop remained dormant for many years. But today up to millions of flip-flops exist in every computer chip from the advanced Internet server chip to the chips in modern coffee makers or dishwashers. Flip-flops are at the heart of synchronizing data transfer.

In the late 1930s primitive computers combined Boolean algebra with mechanical switches to demonstrate simple computing machines. The Second World War sparked an interest in using computers for scientific calculations. The first vacuum tube computer was the ENIAC at the University of Pennsylvania in 1946. By the standards of its day, the 100 kHz clock was fast. It weighed 30 tons, was $80 \times 8.5 \times 3.5$ feet, and dissipated 150 kW of power. The old flip-flop was now an integral part of computer electronics. But the vacuum tube was a relatively large device requiring a glass enclosed vacuum and a heated metal filament. Tubes had poor reliability and were a challenge to cool. Something better was needed.

Bell Labs had a vision in the 1930s that a small, switching device could be constructed in a pure solid material. Bell Labs was thinking of replacing the slow, clunky mechanical relays in their telephone switching centers and not about computer development. In 1947, they struck gold with demonstration of a small, solid-state device called the transistor.

Approximately five years later, transistor computers emerged in production from several companies. Transistors were a giant step toward smaller, cooler, and more reliable computers. These computers used discrete (individual) transistors that were mounted in small metal cans and were not the small, integrated circuit chips with billions of transistors that were to follow. These mainframe computers as they were called still required a cooled, dedicated room, but steady progress was made into the 1970s when another revolution occurred.

Actually several things happened at the transistor level. The first was a rapid transition from the original Bell Labs transistor called a bipolar junction transistor to a newer device called a metal oxide field effect transistor (MOSFET) transistor. The MOSFET was blended in a unique design style called CMOS that was markedly cooler. The cooler CMOS allowed more transistors to be placed on a single chip without overheating thus increasing the computer functionality. CMOS also had the unusual property that if the transistor size were shrunk, the transistor would operate faster.

A third feature was that the smaller size of a CMOS transistor allowed more chips to be manufactured in a single operation than before because the total chip size could now be reduced. More chips could be accommodated per process run, and that drives the cost down. Often industry left the chip the same size and just added more transistors to increase functionality.

A final feature is that if the small particle defects that kill the chips in a production run remain the same density, then packing more chips in the same area will increase the fraction of good chips, (i.e., the yield). This gives a marked cost savings. CMOS has dominated computing chip design since around 1980, and CMOS technology today remains the focus of intense development.

It is a manufacturing miracle that next-generation chips could be sold for a lower price if the next-generation transistor was smaller. That was huge, and today you still pay about the same price (or less) for a personal computer as one that is a few years older. And these newer chips go faster and give more functionality while keeping the chip temperature under control. These CMOS features really fueled the development of computing chips. The reader should pause and dwell on the significance. What other product offers more dramatic performance each year for the same price or less?

Transistors and Computers—How Deep Can the Friendship Go?

In the early 1970s, Intel brought out the first microprocessors, first the 4-bit and then the 8-bit. Product innovation leaped on these transistor level advances. In 1974, the MITS Corporation in Albuquerque, New Mexico, offered the first personal computer, the PC. The MITS Altair 8800 was a primitive PC requiring code to be entered by toggle switches, but it had a video monitor and was the size of a typewriter. It had a 2 MHz clock and cost $498 assembled. It was also the first computer to be personally owned. It used a single microprocessor chip, the Intel 8080, to perform the computer function, and many engineers bought the Altair out of curiosity. Interestingly, Bill Gates and Paul Allen of the new Microsoft Corporation in Albuquerque wrote BASIC for the MITS Altair PC. In 1977, Apple launched the Apple II PC for $1200. No one had ever had a computer at that price, size, and capability—especially one they could call their own. But the IBM PC launched in 1980 had more impact because it brought in the business sector. There was no looking back. Businesses were being freed from the tedium of the big, central computer room, and later travelers found they could do work on the road with the coming of laptops. The ubiquitous typewriter was on the way out.

The PC launched a revolution in information accessibility that could not have been imagined. Technology and novel business enterprises were beginning to move together. A partnering of technology, business enterprise, and government support at crucial points drove this revolution. But a monster enterprise called the Internet lay quietly awaiting its entrance.

In the 1960s and 1970s, Internet development was marching in the background to its own beat driven by engineers and scientists who wanted to use each other's specialty computers across the country. It was government funding through the Advanced Research Projects Agency (ARPA) that allowed a mainframe computer from the University of California at Los Angeles (UCLA) to use an interface unit, called the Interface Message Processor (IMP), to talk to a similar hookup at Stanford University in October 1969. Long distance sharing of computer resources had happened. Messages, later called email, were exchanged, but the ability to do this was regarded then as a secondary feature and not a big deal. In fact, the first Internet exchange was not widely publicized. The response was sort of, "Isn't that nice that scientists and engineers can use each other's computers, but that won't affect my life." What an understatement.

The next necessary development occurred in 1989 when PC manufacturers began bundling internal modems in the PC. The Internet was now open to anyone. Email grew at a tremendous rate as users found it a good business tool, and as true today it was just plain fun to use. The mouse and graphical displays were huge steps toward friendly computers. And computer chips doubled their speed and transistor density about every two years following what is called Moore's law. Then spam, viruses, and hackers showed their ugly heads. Spam is expensive in system bandwidth and the required electrical energy generation to support its Internet hunger.

The Internet went global with introduction of the World Wide Web. "www" was a concept from CERN in Europe that was demonstrated in 1991. We now see "www" in our URL addresses. Browsers quickly followed with MOSAIC from the University of Illinois and the NETSCAPE browser from Netscape Corporation. Yahoo and Microsoft entered the competition, and the famous browser wars were on. Two students from Stanford introduced Google in 1998 with a novel concept in searches that became so successful that Google is now a verb.

Although clearly visible in these early applications, it was the special talents of the business entrepreneurs that carried the World Wide Web into its most recent surge. The incredible innovations now seem endless. The list of Amazon, eBay, PayPal, Google, Wikipedia, YouTube, bloggers, Refdesk, Facebook, email delivery each morning of your favorite newspaper, instant check on stocks and weather, instant Google satellite maps of the earth, online business carried on across the globe, tweeters, e-books, and many more brings us to our present state of information availability. These applications required computer chips that were faster, smaller, and cheaper. The miracle applications needed the base technologies.

Computers—Is There a Limit?

Computer chips depend on many disciplines. Electrical and computer engineers, computer scientists, mathematicians, physicists, chemical engineers, chemists, mechanical engineers, statisticians, manufacturing engineers, and marketing people work in harmony to achieve these miracle products. Technology products typically develop from an idea and a prototype demonstration. If the idea is sound, the product undergoes continuous improvement until performance limits are reached. How far can we push performance? Let's look at three other technologies to see their performance trajectory and how that might provide clues to our electronic future.

Train development spanned an increasing performance era from around 1820 to the 1950s. Then, except for the bullet train, it was over. Automobile development spanned from the 1890s to about the 1960s. Speed, comfort, and engine power peaked for trains and automobiles. Commercial aircraft basically spanned from 1903 to the 1960s when the Boeing 747 was produced. Speed and passenger carrying capability maxed out. Later in all three areas, the integrated circuit caused a second revolution in the 1980s, but if you lived in the height of these technology rushes, there was a feeling that "progress" would never end. The basic speed, power, and transport capability did end, though. What does this imply for computers?

Will the CMOS computer technology rush end? Will our computers provide dramatically more functionality each year? For several reasons, we believe that CMOS technology will hit a performance limit. If history is a guide, we will then squeeze every last design and manufacturing detail from our chips, and improvements will lie in more efficient manufacturing and using multiple processors on one chip. When will we see that soft end point? We see signs of reaching some of the limits now, so we hazard an educated guess that the CMOS technology limit may be reached by 2025. That is a guess. The exact date is immaterial to the thought that a performance limit exists within our professional lifetimes.

When CMOS performance development ends, we expect research will continue seeking another manufacturable electronic technology. The urge is strong to build faster, smaller, cheaper computers, and it will be novel transistor or transistor like devices that pushes us even further.

One significant challenge deals with electrical power. Tom Friedman in his book "Hot, Flat, and Crowded" (p. 31) quoted a Sun Microsystem engineer who put future power demand in perspective. He observed that the earth will add about one billion persons in the next 12 years. If each person were given a 60-watt light bulb then 60 billion watts of power generation would be required. If each person used that bulb for an average of four hours per day, then that average 10 billion watts would require about 20 coal-fired power plants. If each of the billion persons were also allowed to use a 120-watt computer for four hours then we would need an additional 40 power plants. The earth is power limited, and the pressure on low power computer chip design is huge.

The concept of technology is little different from early mankind using a wheel to support a cart to carry heavy weights. Technology is the use of materials and natural laws to ease our burdens. Electronic technology is no different, but we know that all technology has a downside. The Internet has done miracles, but that hasn't stopped hackers, spammers, and swindlers from peddling their dark objectives. The Internet can bring instant and accurate news, but it can also bring instant and inaccurate propaganda. These are issues to deal with as it has always been with technology. We should keep our eye on the benefits and continue the historic human battle of fighting the misuse of technology.

Future

We won't speculate much on the future other than that there is one. The Internet brought rapid changes in business and technology that we now take for granted. Startling new products will appear, and some old product names will disappear. This book addresses the education of the next generation of engineers who will continue to move this historic epic in information accessibility.

CHAPTER 1

Basic Logic Gate and Circuit Theory

Ben Franklin had a 50–50 chance of getting the current convention right – he didn't.

Jim Lloyd

This chapter reviews and emphasizes certain topics covered in prior courses such as Introduction to Logic Design and Introduction to Circuit Analysis. Relevant topics for digital circuits include logic circuit conversion from Boolean statements to logic gate circuits, DeMorgan equivalence, and Boolean minimization. Kirchhoff's laws are used repeatedly in transistor circuit analysis, so they are reviewed here with emphasis on circuit analysis by inspection. Capacitor networks, power relations, and voltage divider relations are relevant to integrated circuits (ICs). Inductance in the interconnection lines is developed later. Diode analysis in resistive networks provides experience with circuits containing nonlinear elements. The intention is to work sufficient problems in this chapter to develop circuit intuition and prepare for future chapters. This material was selected from an abundance of circuit topics as being more relevant to later chapters, which discuss how to analyze CMOS integrated circuits and how they work.

1.1. Logic Gates and Boolean Algebra

Combinational Logic

Boolean equations define what a digital circuit will do. Boolean algebra assumes two signal states: (1) a logic high, also called a logic-1; and (2) a logic low, called a logic-0. An output function is a logic-1 when the variables on the right-hand side of the Boolean equation are true. For example, if

$$F = AB + C\overline{D} \tag{1-1}$$

then F is a logic-1 if A and B are logic-1, or if C is logic-1 and D is logic-0. F is a *combinational logic* function since its value at any instant depends on the values of the

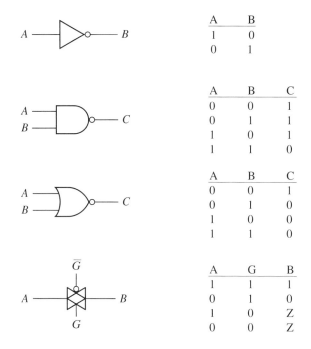

A	B
1	0
0	1

A	B	C
0	0	1
0	1	1
1	0	1
1	1	0

A	B	C
0	0	1
0	1	0
1	0	0
1	1	0

A	G	B
1	1	1
0	1	0
1	0	Z
0	0	Z

FIGURE 1-1.

Logic gate symbols inverter, NAND, NOR, and transmission gate and their truth tables.

circuit inputs, *ABCD.* Your introductory course in logic design described how these equations can specify computer operations such as addition, complement, comparator, and multiplication operations. Our next step in transforming Boolean equations to an actual computer uses the concept of logic gates.

We use logic gate symbols for logic statements in the Boolean equation. The greater importance is that we can build standard electronic circuits to perform the operation specified by a logic gate symbol. Figure 1-1 shows common logic gates and their truth tables. The transmission gate has a third state called the high-Z state, which stands for high impedance. A high-Z state circuit node has no direct current (DC) path to ground or to the power supply, so the node floats. The Z-state output of the transmission gate is essentially off and has no influence on other logic signals. There are times when parallel circuits are tied to a common node and must be isolated from one another. The high-Z state does that.

This book is about the electronic properties of circuits that perform logic functions; emphasis is placed on simple logic gates. There is a simplistic saying that, given inverters, NAND gates, and transmission gates, a designer can build a large complex computer. It is true, and the study of these basic circuits is essential.

EXAMPLE 1-1

Draw a logic gate schematic to solve $F = \overline{AB + C\overline{D}}$. There are two AND operations, one OR operation, and an invert. The output of the NOR gate is the function F. The NOR gate is equivalent to an OR gate and an inverter.

Solution

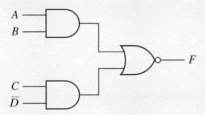

Self-Exercise 1-1
Draw logic gate circuits to solve the following Boolean statements.

$$F = (A + B)\, CD$$
$$F = \left(A + B\overline{C}\right) D$$
$$F = A\left(\overline{B} + D\right) + EG$$

   Logic gates initiate an action through their output voltage. Shapes appear on a video monitor screen, traffic lights respond to digital pulses, automobile brakes and engines are automatically optimized for performance, kitchen ovens bake on instructions, and computers and their monitors interact with us through Boolean algebra. Electronic designs typically start with Boolean statements and then electronic implementation of the gates. The transform of a Boolean equation to a logic gate is significant since we build standard electronic circuits for each of the logic gates and store these designs in a computer.

   DeMorgan's theorem shows the equivalent logic of a NOR and a NAND logic gate when inverted signals are used (Eqs. 1-2, 1-3). These conversions are useful since the electronic versions of NAND and NOR gates can have different timing properties and transistor layouts from DeMorgan AND and OR gates. DeMorgan's theorem may also reduce the number of logic gates to perform a given logic function.

$$\overline{X + Y} = \overline{X}\,\overline{Y} \tag{1-2}$$

$$\overline{XY} = \overline{X} + \overline{Y} \tag{1-3}$$

Truth tables provide a simple way to verify the equivalence of DeMorgan conversions. An example will illustrate.

EXAMPLE 1-2

Verify Eq. (1-2) using a truth table.

X	Y	$X+Y$	$\overline{X+Y}$	$\overline{X}$	$\overline{Y}$	$\overline{X}\,\overline{Y}$
0	0	0	**1**	1	1	**1**
0	1	1	**0**	1	0	**0**
1	0	1	**0**	0	1	**0**
1	1	1	**0**	0	0	**0**

The fourth and the last signal columns show the DeMorgan theorem equivalency.

Self-Exercise 1-2
Verify Eq. (1-3) using a truth table.

The DeMorgan theorem is applicable to more than two logic variables. The following is a guide to transforming the DeMorgan equivalent for any number of variables.

1. Product terms (AND) in the original function transform to sum (OR) terms in the DeMorgan equivalence.
2. Sum (OR) terms in the original function transform to product (AND) terms in the DeMorgan equivalence.
3. All variables are inverted when transforming to and from a DeMorgan equivalence.
4. An overbar on the original function transforms to no overbar in the DeMorgan equivalence, and vice versa.

EXAMPLE 1-3

Write the DeMorgan equivalent statement for

(a) $F = XY + \overline{Z}$
(b) $F = \overline{(\overline{X} + Y)\,(\overline{W}Z)}$

Solution

(a) $F = \overline{(\overline{X} + \overline{Y})\,Z}$
(b) $F = X\overline{Y} + W + \overline{Z}$

Self-Exercise 1-3

Write the DeMorgan equivalent statement for the Exclusive-Or (XOR) function.

$$F = X\overline{Y} + \overline{X}Y$$

Answer: $F = \overline{(\overline{X} + Y)\ (X + \overline{Y})}$

Self-Exercise 1-4

Write the DeMorgan equivalent statement for

$$F = \overline{(X + Y)\ Z + XY\overline{Z}}$$

Answer: $F = (\overline{X}\,\overline{Y} + \overline{Z})\ (\overline{X} + \overline{Y} + Z)$

Self-Exercise 1-5

Write the DeMorgan equivalent statement for

$$F = W\ (\overline{X} + \overline{Y}Z) + Z\ (W\overline{X}Y + \overline{W})$$

Answer: $F_{DM} = \overline{(\overline{W} + X\ (Y + \overline{Z}))\ (\overline{Z} + W\ (X + \overline{Y}))}$

1.2. Boolean and Logic Gate Reduction

Designers seek to minimize the number of logic gates. Each nonessential logic gate occupies only a small area, but when that area is multiplied by the possible millions of production chips then minimum logic gate count is an economic necessity. Chips with larger area cost more and may run slower.

This section reviews Boolean logic identities presenting examples and exercises to illustrate logic gate reduction. Table 1-1 lists basic Boolean identities where AND and OR operations are XY and $X + Y$, respectively.

TABLE 1-1 Basic Boolean Identities

$X + Y = Y + X$	$XY = YX$
$X + 0 = X$	$X + 1 = 1$
$X0 = 0$	$X1 = X$
$X + X = X$	$XX = X$
$X\overline{X} = 0$	$X + \overline{X} = 1$
$(X + Y) + Z = X + (Y + Z)$	$(XY)Z = X(YZ)$
$X(Y + Z) = XY + XZ$	$X + (YZ) = (X + Y)(X + Z)$
$\overline{X + Y} = \overline{X}\,\overline{Y}$	$\overline{XY} = \overline{X} + \overline{Y}$

EXAMPLE 1-4

Reduce the logic circuit to its minimum number of gates.

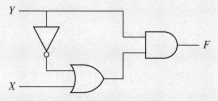

The original logic function is $F = Y(X + \overline{Y})$.
The reduction is

$$F = Y(X + \overline{Y}) = YX + Y\overline{Y} = XY$$

whose logic gate circuit is

EXAMPLE 1-5

A logic function is $F = (X + Y) + \overline{X}$. Reduce it to its minimum form.
The reduction is

$$F = (X + Y) + \overline{X} = (X + \overline{X}) + Y = 1 + Y = 1$$

The output is always a logic-1 no matter what the input values are. A designer should consider removing these gates as irrelevant.

Self-Exercise 1-6
Show in a series of equations that

$$(X + Y)(X + Z) = X + YZ$$

Self-Exercise 1-7
Minimize the logic gate function and draw its logic gate circuit.

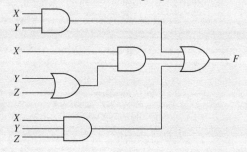

Answer:

The logic gate circuit reduction is

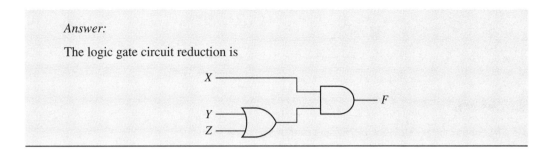

Self-Exercise 1-8
Find the minimum function for the gate circuit.

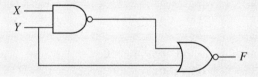

Answer:

The reduction is

$$F = 0$$

The logic design performs no useful logic function and should be deleted.

1.3. Sequential Circuits

Sequential circuits store data, and they have several forms. One form temporarily stores bits of data generated by combinational logic on one clock pulse and then sends that data to another combinational logic block on the next clock pulse (Chapter 8). Another type of sequential circuit stores data in longer-term memory (Chapter 9).

The basic sequential circuit for temporary data storage is the flip-flop, which is a term from the original Eccles-Jordan paper in 1918. The word flip-flop refers to the memory circuit's two logic states and its ability to hold either logic state. The output data are said to "flip into one logic state and then flop into the other when pulsed." Figure 1-2 shows a

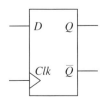

FIGURE 1-2.

D flip-flop circuit symbol.

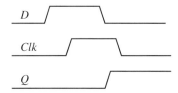

FIGURE 1-3.

Flip-flop input data (D), clock (Clk), and output (Q) waveforms.

flip-flop symbol with data input D, clock input Clk, and outputs Q and $\overline{Q}$. The incoming data are from an upstream combinational logic block of circuits. The D flip-flop stores a bit at the end of one clock cycle. The next clock pulse takes that data and transfers it to Q where it is delivered to another combinational logic circuit. The small triangle next to the Clk denotes a circuit that changes state on the positive edge of the Clk pulse. The D flip-flop is the dominant flip-flop in computer ICs.

Figure 1-3 shows flip-flop operational waveforms of the D, Clk, and output Q. The input data precede the Clk pulse, and the output data will be electronically delayed from the Clk pulse. In most cases it is the edge, or transition, of the clock that is the time mark about which other waveforms edges are referenced. The early arrival of D is measured from the clock edge, and the output Q event arrives after a delay from the clock edge. Designers must include these delays when making a digital chip work at high frequency, or at any frequency.

A collection of flip-flops can form a parallel register described as parallel data in and parallel data out. Figure 1-4 shows a 4-bit register. Each bit is a D flip-flop. Four bits of data form a 4-bit word. These data move as a unit. The clock controls when the word moves to a next stage for further Boolean processing. We see the synchrony between the data and the clock. Systems that operate with a master clock controlling the data are called *synchronous designs*. It is the most common timing design form in digital electronics. Chapter 8 will analyze the electronics of data registers and emphasize IC timing requirements.

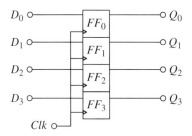

FIGURE 1-4.

A 4-bit register showing flip-flops and signal and clock connections.

Computer architecture

There is a design level above the combinational and sequential logic gates called design architecture. It is the design of an assembly of gates from functional blocks such as binary adders, subtractors, multiplexers, word bit shifters, and arithmetic logic units. These blocks are made from many small combinational or sequential logic gates, but we can design more efficiently if we define a larger electronic function as a block labeling its input and output signal lines. The blocks are predesigned therefore we don't have to redesign each time that we use a functional block. The designs are stored in computers. Functional blocks are treated by a designer with the same ease that a single inverter might be. This style is a *hierarchical design* method that goes even higher. Complete microprocessors can be treated as a functional predesigned block and integrated into the chip with other big blocks such as memories. We must be aware of the timing properties of these blocks and their voltage input-output (I/O) properties. How are these predesigned circuit blocks stored, and why are they so flexible? The answer lies in the ability to store a design as a software file. Chapter 11 will give insight when we discuss IC layout.

1.4. Voltage and Current Laws

Your introductory course in linear circuits is a base to build on analysis of transistor circuits. This section will first develop an inspection technique to analyze the equivalent resistance between two nodes. These techniques rely on Kirchhoff's circuit laws with solutions written as a single equation. The technique is rapid, visual, and allows speedy result checking. It is useful for estimating values as a manual or "back of the envelope" method. We will learn the shorthand technique by example.

1.4.1. Terminal Resistance Analysis by Inspection

EXAMPLE 1-6

Write the shorthand equation to calculate terminal resistance. Series resistors add so

$$R_{eq} = 10\,k\Omega + 20\,k\Omega + 5\,k\Omega + 2\,k\Omega = 37\,k\Omega$$

While this is a simple example, it does illustrate a rapid, single equation statement as a solution.

EXAMPLE 1-7

Write the shorthand equation to calculate terminal resistance.

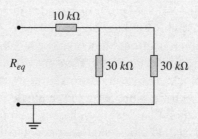

$$R_{eq} = 10\,k\Omega + 30\,k\Omega//30\,k\Omega = 25\ k\Omega$$

EXAMPLE 1-8

Write the shorthand equation to calculate terminal resistance.

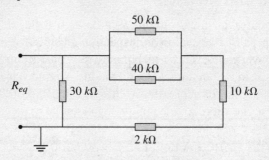

$$R_{eq} = 30\,k\Omega//(50\ k\Omega//40\,k\Omega + 10\,k\Omega + 2\ k) = 15.99\ k\Omega$$

Self-Exercise 1-9
Write the shorthand notation and solve for the terminal resistance.

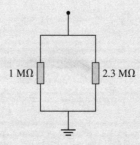

Answer: $R_{eq} = 697\ k\Omega$

Self-Exercise 1-10
Write the shorthand notation and solve for the terminal resistance.

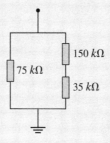

Answer: $R_{eq} = 53.4 \ k\Omega$

Self-Exercise 1-11
Write the single equation notation for the terminal resistance of the three circuits.
    $R_1 = 20 \ k\Omega$, $R_2 = 15 \ k\Omega$, $R_3 = 25 \ k\Omega$, $R_4 = 8 \ k\Omega$, $R_5 = 5 \ k\Omega$. Calculate R_{eq} for these three circuits using your single equations in (a).

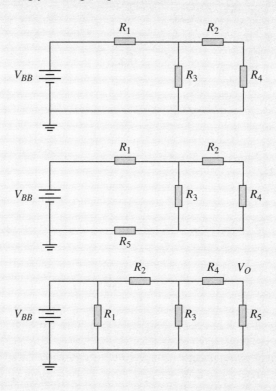

Answer:

(a) $R_{eq} = 31.98\ k\Omega$

(b) $R_{eq} = 36.98\ k\Omega$

(c) $R_{eq} = 10.82\ k\Omega$

We will next move to voltage and current dividers. Dividers allow circuit intuitions that are invaluable in manual estimations of answers. Circuit simulations are essential, but we must always check results for reasonableness.

1.4.2. Kirchhoff's Voltage Law and Analysis by Inspection

Two major inspection analysis techniques use *voltage divider* and *current divider* concepts to rapidly calculate circuit node voltages and branch currents. These techniques use Kirchhoff's voltage and current laws. Kirchhoff's voltage law (KVL) is a conservation of energy law stating, "*The voltage applied to a network is equal to the voltages dropped by the elements in the network.*" Figure 1-5 shows an analytical circuit with good visual voltage divider properties that we will illustrate in calculating V_3.

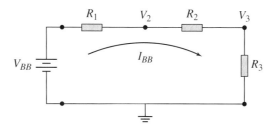

FIGURE 1-5.

Voltage divider circuit.

The KVL equation is

$$V_{BB} = I_{BB}(R_1) + I_{BB}(R_2) + I_{BB}(R_3) = I_{BB}(R_1 + R_2 + R_3)$$

$$I_{BB} = \frac{V_{BB}}{R_1 + R_2 + R_3} = \frac{V_3}{R_3} \tag{1-4}$$

then

$$V_3 = \frac{R_3}{R_1 + R_2 + R_3} V_{BB} \tag{1-5}$$

Eq. (1-5) is a shorthand statement of the voltage divider. It is written by inspection and calculations follow. The voltage dropped by each resistor is proportional to their fraction of the whole series resistance. Figure 1-5 is very visual, and you should be able to instinctively write the voltage divider expression by inspection for any voltage drop. For example, the voltage from node V_2 to ground is written as a single equation by inspection.

$$V_2 = \frac{R_2 + R_3}{R_1 + R_2 + R_3} V_{BB} \tag{1-6}$$

EXAMPLE 1-9

Use inspection to calculate the input resistance R_{in}, the battery current I_{BB}, and the voltage V_O. Verify that the sum of the voltage drops is equal to the applied voltage.

$$R_{in} = 4\,k\Omega + 12\,k\Omega // 20\,k\Omega = 11.5\ k\Omega$$

$$I_{BB} = \frac{1\text{ V}}{4\,k\Omega + 12\,k\Omega // 20\,k\Omega} = 87.0\ \mu A$$

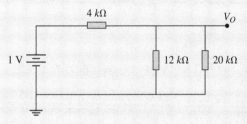

A voltage divider gives V_O.

$$V_O = \left(\frac{12\,k\Omega // 20\,k\Omega}{4\,k\Omega + 12\,k\Omega // 20\,k\Omega} \right) 1\text{ V} = 0.652 \text{ V}$$

and

$$V_{4k\Omega} = \left(\frac{4\,k\Omega}{4\,k\Omega + 12\,k\Omega // 20\,k\Omega} \right) 1\text{ V} = 0.348 \text{ V}$$

The sum of the resistive voltages is 1.000 V and equals the driving voltage.

Self-Exercise 1-12
Write the expression for R_{in}, V_O, and the power supply current. Solve for R_{in}, V_O, and I_{BB}.

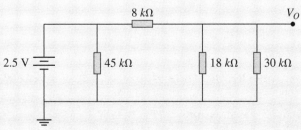

Answer: **(a)** $R_{in} = 13.48\,k\Omega$. **(b)** $V_O = 1.459$ V. **(c)** $I_{BB} = 185.4\,\mu A$

Self-Exercise 1-13

Write the shorthand expression for R_{in} between the terminals.

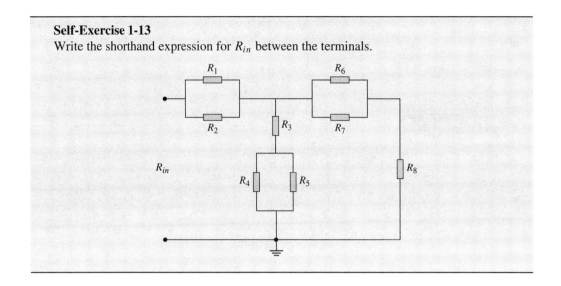

1.4.3. Kirchhoff's Current Law and Analysis by Inspection

Current dividers are based on the Kirchhoff's current law (KCL), which states, "*The sum of the currents at a circuit node is zero.*" Current is a mass flow of charge; therefore, the mass entering the node must equal the mass exiting. Current divider expressions are visual, allowing you to see the splitting of current as it enters branches. Figure 1-6 shows two resistors that share a total current I_{BB}.

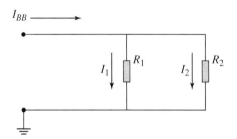

FIGURE 1-6.

Current divider.

The KVL gives

$$V_{BB} = (R_1 /\!/ R_2) I_{BB} = \frac{R_1 \times R_2}{R_1 + R_2} I_{BB} = I_1 R_1 = I_2 R_2 \qquad (1\text{-}7)$$

Giving the current divider expression

$$I_1 = \frac{R_2}{R_1 + R_2} I_{BB} \quad \text{and} \quad I_2 = \frac{R_1}{R_1 + R_2} I_{BB} \qquad (1\text{-}8)$$

Currents divide in two parallel branches by an amount proportional to the opposite leg resistance divided by the sum of the two resistors. This relation should be memorized as we did for the voltage divider. It is intuitively reasonable that the parallel leg of the lowest resistance will draw the most current.

EXAMPLE 1-10

Calculate V_O, $I_{2\Omega}$, and $I_{9\Omega}$.

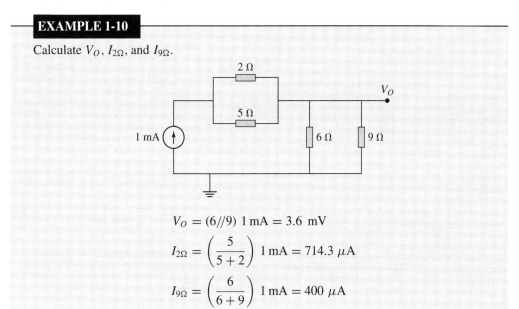

$$V_O = (6//9)\ 1\ \text{mA} = 3.6\ \text{mV}$$

$$I_{2\Omega} = \left(\frac{5}{5+2}\right)\ 1\ \text{mA} = 714.3\ \mu\text{A}$$

$$I_{9\Omega} = \left(\frac{6}{6+9}\right)\ 1\ \text{mA} = 400\ \mu\text{A}$$

EXAMPLE 1-11

Write the equation for I_{15k} by inspection and solve for the value.

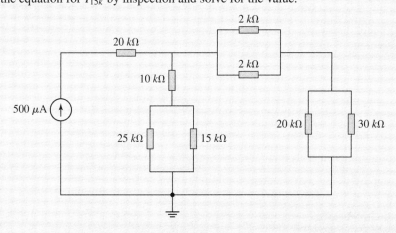

$$I_{15k} = \left[\frac{(2\,k//2\,k) + (20\,k//30\,k)}{10\,k + (25\,k//15\,k) + (2\,k//2\,k) + (20\,k//30\,k)}\right] \frac{25\,k}{25\,k + 15\,k} 500\ \mu\text{A} = 125.5\ \mu\text{A}$$

Self-Exercise 1-14

(a) Solve for current in all resistive paths using the technique of inspection.
(b) Calculate a new value for the $20\,k\Omega$ resistor so that its current is $5\,\mu$A.

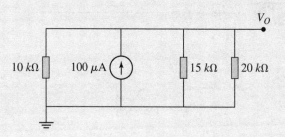

Answer: (a) $I_{10k} = 46.15\,\mu$A, $I_{15k} = 30.77\,\mu$A, $I_{20k} = 23.08\,\mu$A. (b) $R_{20k} = 114\,k\Omega$

Self-Exercise 1-15
Solve for I_{40k} using the technique of inspection.

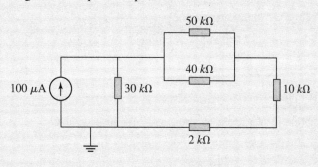

Answer: $25.95\,\mu$A

1.4.4. Mixing Voltage and Current Divider Analysis by Inspection

A common practice mixes divider concepts. If a circuit has a voltage source and a branch current is desired, then the first term in the equation might use the voltage source and input resistance to convert to a power supply current. That current can then be manipulated to the target branch using current dividers.

EXAMPLE 1-12

Solve for currents in the $12\,k\Omega$ and $20\ k\Omega$ paths by the method of inspection. *Hint*: first write the expression for current I_{BB} and then multiply that by the current divider.

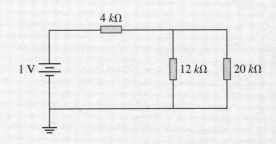

$$I_{12k} = \left[\frac{1\,V}{4\,k + 12\,k//20\,k}\right]\left[\frac{20\,k}{12\,k + 20\,k}\right] = 54.348\,\mu A$$

$$I_{20k} = \left[\frac{1\,V}{4\,k + 12\,k//20\,k}\right]\left[\frac{12\,k}{12\,k + 20\,k}\right] = 32.609\,\mu A$$

The sum of the currents is $86.957\,\mu A$, which is equal to the current in the battery.

$$I_{BB} = \left[\frac{1\,V}{4\,k + 12\,k//20\,k}\right] = 86.957\,\mu A$$

Self-Exercise 1-16

(a) Write the current expression by inspection and solve for currents in all resistors where $I_{BB} = 185.4\,\mu A$.

(b) Calculate V_{BB}.

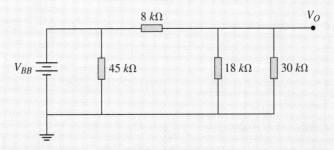

Answer: **(a)** $I_{45k} = 55.55\,\mu A$, $I_{8k} = 129.9\,\mu A$, $I_{18k} = 81.16\,\mu A$, $I_{30k} = 48.69\,\mu A$.
(b) $V_{BB} = 2.50\,V$

Self-Exercise 1-17

(a) Write R_{in} between the battery terminals by inspection and solve.
(b) Write the $I_{1.5k}$ expression by inspection and solve. This is a larger circuit, but it presents
no problem if we adhere to the shorthand style. Write R_{in} between battery terminals
by inspection, and calculate $I_{1.5k}$ by current divider inspection.

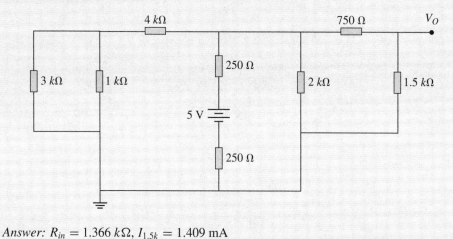

Answer: $R_{in} = 1.366\ k\Omega$, $I_{1.5k} = 1.409$ mA

Self-Exercise 1-18
If the voltage across the current source is 10 V, what is R_1? Watch your voltage polarity.

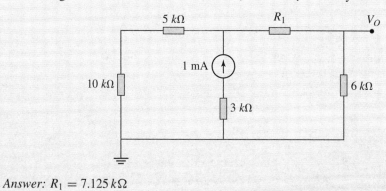

Answer: $R_1 = 7.125\ k\Omega$

1.5. Power Loss in Resistors

Power is desirable if it performs useful work such as heating the coils of a toaster. All
of the heat emitted from an electronic circuit is wasted, and it loads the room air cooler
pulling more electrical power. A single circuit may not have large impact, but integrated
circuit designs are replicated in production by millions and billions. Power loss in an

integrated circuit is mainly due to parasitic resistance and capacitance elements and the transistor properties. Circuits must minimize heat generated, since none of that waste goes into computing.

The product of voltage and current in an element is the power dissipated. The unit analysis is

$$P = V \left(\frac{\text{joule}}{\text{coulomb}} \right) \; I \left(\frac{\text{coulomb}}{\text{second}} \right) = P \left(\frac{\text{joule}}{\text{second}} \right) = P \; (\text{watts})$$

Ohms law allows equivalent expressions for power as

$$P = V I = I^2 R = \frac{V^2}{R} \tag{1-9}$$

EXAMPLE 1-13

(a) Calculate the power in each resistor.
(b) Calculate the total power delivered by the battery.
(c) Show if the battery power equals the resistor power loss.

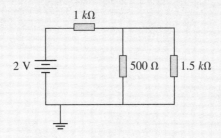

(a) First calculate resistor currents

$$I_{1k} = I_{BB} = \frac{2\,\text{V}}{1\,k + (500//1.5\,k)} = 1.455\ \text{mA} \quad P_{1k} = (1.455\ \text{mA})^2 \; 1\,k = 2.117\ \text{mW}$$

$$I_{500} = \left[\frac{1.5\,k}{1.5\,k + 500} \right] 1.455\ \text{mA} = 1.091\ \text{mA} \quad P_{500} = (1.091\ \text{mA})^2 \; 500 = 0.5954\ \text{mW}$$

$$I_{1.5k} = \left[\frac{500}{1.5\,k + 500} \right] 1.455\ \text{mA} = 363.8\ \mu\text{A} \quad P_{1.5k} = (363.9\ \mu\text{A})^2 \; 1.5\,k = 0.1985\ \text{mW}$$

(b) The battery power is

$$P_{BB} = V_{BB} I_{BB} = 2\,\text{V}\ (1.4545\,\text{mA}) = 2.909\ \text{mW}$$

(c) The sum of the resistive powers is 2.911 mW, so within decimal accuracy the applied power of the battery is equal to the dissipated power in the resistors.

EXAMPLE 1-14

(a) Calculate the power loss for $V_{BB} = 1\,V,\ 2\,V,\ 3\,V,\ 4\,V,\ 5\,V$.
(b) Sketch power versus V_{BB}.
(c) What do you conclude?

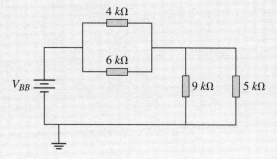

(a) The terminal resistance is

$$R_{eq} = (4\,k//6\,k) + (9\,k//5\,k) = 5.614\,k\Omega$$

$$P_{BB}(1\,V) = \frac{(1\,V)^2}{5.614\,k\Omega} = 0.1781\,\text{mW}$$

$$P_{BB}(2\,V) = \frac{(2\,V)^2}{5.614\,k\Omega} = 0.7125\,\text{mW}$$

$$P_{BB}(3\,V) = \frac{(3\,V)^2}{5.614\,k\Omega} = 1.603\,\text{mW}$$

$$P_{BB}(4\,V) = \frac{(4\,V)^2}{5.614\,k\Omega} = 2.850\,\text{mW}$$

$$P_{BB}(5\,V) = \frac{(5\,V)^2}{5.614\,k\Omega} = 4.453\,\text{mW}$$

(b)

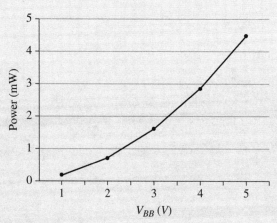

(c) Power dissipation increases with the square of the power supply voltage; therefore, voltage reduction is a first consideration in reducing circuit power.

Self-Exercise 1-19

(a) Calculate the heat generated in each resistor.
(b) Show that power dissipated is equal to power applied.

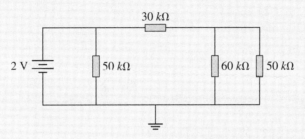

Answer: (a) $P_{50k} = 80\,\mu W$, $P_{30k} = 36.56\,\mu W$, $P_{60k} = 15.11\,\mu W$, $P_{50k} = 18.13\,\mu W$.
(b) $P_{BB} = (2\,V)(74.91\,\mu A) = 149.8\,\mu W$, which is the sum of the resistor powers.

Self-Exercise 1-20
Show with equations what happens to the power in the circuit when $R_2 \Rightarrow 0$.

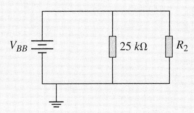

1.6. Capacitance

A parallel plate capacitor and symbol are shown in Figure 1-7. Capacitors are characterized by a parameter called *capacitance* (C), which is measured in farads. Capacitor properties appear in CMOS digital circuits as parasitic elements intrinsic to transistors or from the metals used for interconnections. These parasitic elements are often referred to by their location and capacitance to distinguish them from discrete capacitors.

Capacitance has an important effect on the time for a transistor to switch between the *ON* and *OFF* states, and it also contributes to propagation delay between gates due to interconnection capacitance. Intrametal capacitance also causes a type of noise appearing in high-speed circuits called *cross-talk*, in which the voltage on one interconnection line is

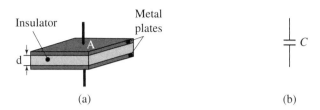

(a)   (b)

FIGURE 1-7.

(a) Parallel plate capacitor. (b) Circuit symbol.

affected by a change of voltage on another interconnection line that is isolated but located close to it. Cross-talk and capacitance calculations are discussed later.

Introductory courses in circuits teach equivalent terminal capacitive values for series and parallel networks. This section will use that information to describe capacitor power analysis and review capacitive voltage dividers. Both topics are relevant to digital ICs.

1.6.1. Capacitor Energy and Power

Figure 1-8 shows a capacitor circuit. When the switch closes at $t = 0$, electrons flow and charge the capacitor to V_B.

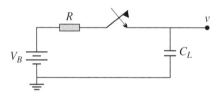

FIGURE 1-8.

RC time-constant circuit.

The capacitor charge–voltage relation is

$$q = C_L v$$

and the general energy relation is

$$dw \text{ (joules)} = v \left(\frac{\text{joules}}{\text{coulomb}} \right) dq \text{ (coulombs)}$$

The total energy W to charge the capacitor with Q coulombs is

$$W = \int_0^Q dw = \int_0^Q v \, dq = \int_0^Q \frac{q}{C_L} dq = \frac{Q^2}{2 C_L} \text{ joules}$$

(1-10)

$$W = \frac{1}{2} C_L V_B^2 \text{ joules}$$

The ideal capacitor doesn't heat or cool during the charge and discharge. The heat occurs in the power supply and its internal resistance. Compare this to a pendulum. It takes a certain amount of energy to move the pendulum from its low point to its high point. The driving machinery expends energy and heats up as it lifts the pendulum. The pendulum does not. When the pendulum is released from its high point, the potential energy converts to kinetic energy and not to heat. If the pendulum was slowly lowered from its high point to its low point, then heat would be generated in the braking mechanism. Another view is that if you lifted the pendulum weight to its high position, your muscles would generate heat but not the pendulum. Capacitor thermal properties are similar. Heat is generated in the power supply charge and discharge circuits.

Digital circuits have two capacitive waveform situations: (1) a single charge or a single discharge occurs during one clock period T; or (2) charge and discharge both happen in one clock period. If only a single event (charge or a discharge) occurs during a cycle, then $W = \frac{1}{2}C_L V_B^2$. If both a charge and a discharge occur during a cycle, then $W = C_L V_B^2$ or twice that of a single charge event. Both situations occur in digital ICs.

Since joule/s = watt, we can calculate the power in watts (heat) expended at particular driving frequencies. If a capacitor is charged and discharged at a frequency f, then the heat generated in the driving circuits in that capacitor is $P = C_L V_B^2 f_{Clk}$.

EXAMPLE 1-15

(a) Calculate the energy stored in a 10 pF capacitor if it is charged with 2 V in 1 second.
(b) What is the heat generated if the capacitor is driven by a square wave at 100 MHz?

Solution

(a) $W = \dfrac{1}{2}(10\,\text{pF})(2\,\text{V})^2 = 20 \times 10^{-12}$ joules $= 20$ pJ

(b) $P = (10\,\text{pF})(2\,\text{V})^2(100 \times 10^6) = 4\,\text{mW}$

Notice that the factor of one-half disappears since a charge and a discharge occurs during one *Clk* period of $T = f^{-1} = 10$ ns.

EXAMPLE 1-16

A capacitance C_L is 200 fF at a logic node and the driving voltage is 1.5 V. If C_L changes logic states once per clock pulse at a rate of 20 MHz, what is the power generated as heat in the driving circuit?

The node voltage changes its state only once per clock cycle. Therefore, the heat generated is

$$P = \frac{1}{2}(200\,\text{fF})(1.5\,\text{V})^2(20 \times 10^6) = 4.5\,\mu\text{W}$$

Self-Exercise 1-21
A microprocessor has hundreds of millions of transistors, and it is driven by a clock at 1 GHz. Assume a single logic transition per clock cycle. If the effective total microprocessor internal circuit load is 40 nF and the power supply voltage is 1.5 V, what is the heat generated?

Answer: $P = 45$ W

Imagine the heat generated by a 45 W light bulb in an object the size of a coin.

Self-Exercise 1-22
A video game chip is driven by a clock at 3 GHz assuming a single logic transition per clock cycle. If the effective load in 12 nF, what is the maximum power supply voltage if the heat generated must be kept at 50 W?

Answer: $V_{DD} = 1.667$ V

1.6.2. Capacitive Voltage Dividers

There are open-circuit defect situations in CMOS circuits where capacitance couples voltages to otherwise unconnected nodes. Figure 1-9 shows a simple connection of a capacitance voltage divider circuit. The voltage across each capacitance is a fraction of the total voltage V_{DD} across both terminals.

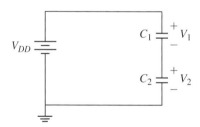

FIGURE 1-9.

Capacitive voltage divider.

EXAMPLE 1-17

Derive the relation between the voltage across each capacitance C_1 and C_2 in Figure 1-9 to the terminal voltage V_{DD}.

The charge across the plates of the series capacitance is equal so that $Q_1 = Q_2$. The capacitance relation $C = Q/V$ allows us to write

$$Q_1 = Q_2 \Rightarrow C_1 V_1 = C_2 V_2$$

or

$$V_2 = \frac{C_1}{C_2} V_1$$

since

$$V_{DD} = V_1 + V_2$$

then

$$V_2 = V_{DD} - V_1 = \frac{C_1}{C_2} V_1$$

Solve for

$$V_1 = \frac{C_2}{C_1 + C_2} V_{DD}$$

and

$$V_2 = \frac{C_1}{C_1 + C_2} V_{DD}$$

The form of the capacitor divider is similar to the resistor voltage divider except the numerator term differs.

EXAMPLE 1-18

Solve for V_{C1} and V_{C2} across C_1 and C_2.

$$V_{C1} = \frac{75\,\text{nF}}{45\,\text{nF} + 75\,\text{nF}} (100\,\text{mV}) = 62.5\,\text{mV}$$

$$V_{C2} = \frac{45\,\text{nF}}{45\,\text{nF} + 75\,\text{nF}} (100\,\text{mV}) = 37.5\,\text{mV}$$

Self-Exercise 1-23

(a) Calculate the power in each capacitor if the voltage source generates a square wave at $f_{Clk} = 1$ GHz and $V_B = 2$ V peak.

(b) Calculate the equivalent capacitance of the two series capacitors and compute the power to charge and discharge this capacitance at 500 MHz and 2 V peak.

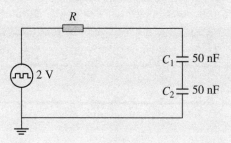

Answers: (a) $P_1 = P_2 = 50$ W. (b) $C_{eq} = 25$ nF, $P_{eq} = 100$ W

Self-Exercise 1-24

(a) If $V_2 = 700$ mV, what is the driving terminal voltage V_D?

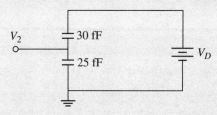

Answer: $V_D = 1.283$ V

1.7. Inductance

An inductor is formed by a conducting structure whose current is linked to a magnetic field around the inductor. The inductor can be a straight wire, or it can be a wire shaped into a coil. *Inductance* is a unit that is a measure of the inductor magnetic flux properties. Inductance is a constant of proportionality between the magnetic flux generated by the inductor and the current change in the inductor. The inductance unit is the henry, and in ICs we often see the unit pico henry (pH $= 10^{-12}$ H). The parasitic inductance of the IC metal wires is an important parameter to control in high-speed circuits. IC inductance relations are presented in Chapter 4.

1.8. Diode Nonlinear Circuit Analysis

A circuit analysis of the semiconductor diode is presented, while the next chapter discusses its physics and role in transistor construction. Diodes do not act like resistors; they are nonlinear. Diodes pass significant current in one voltage polarity and near zero current for the opposite polarity. A typical diode nonlinear current–voltage response and their sign conventions are shown in Figure 1-10a and its circuit symbol in Figure 1-10b. The positive terminal is called the anode, and the negative one is the cathode. The diode is a dominant element in ICs since every transistor has two diodes in its structure. This section will use the diode to show a method of solving nonlinear circuit equations.

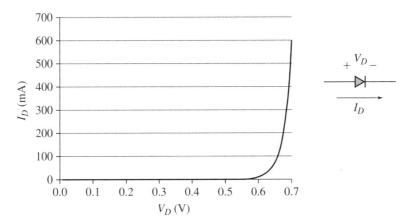

FIGURE 1-10.

Diode I-V characteristics and symbol.

The diode equation is

$$I_D = I_S \left(e^{q V_D / kT} - 1 \right) \tag{1-11}$$

where k is the Boltzmann constant ($k = 1.38 \times 10^{-23}$ J/K), q is the charge of the electron ($q = 1.6 \times 10^{-19}$ C), and I_S is the reverse bias saturation current. The quantity kT/q is called the *thermal voltage* (V_T), whose value is 0.0257 V at a room temperature of $T = 298$ K. We typically approximate $V_T = 26$ mV at room temperature. When the diode applied voltage is positive and well beyond the thermal voltage ($V_D \gg V_T = kT/q$), then the diode Eq. (1-11) approximates to

$$I_D = I_S e^{q V_D / kT} = I_s e^{V_D / V_T} \tag{1-12}$$

Most of our forward bias circuits will have V_D large enough to neglect the 1 term. The voltage across the diode can be derived from Eq. (1-11)

$$V_D = \frac{kT}{q} \ln \left(\frac{I_D}{I_S} + 1 \right) \tag{1-13}$$

For forward bias applications, this reduces to

$$V_D = \frac{kT}{q} \ln\left(\frac{I_D}{I_S}\right) = V_T \ln\left(\frac{I_D}{I_S}\right) \tag{1-14}$$

EXAMPLE 1-19

(a) Calculate the forward diode voltage if $T = 25°C$, $I_D = 200$ nA, and $I_S = 1$ nA. Compute from Eq. (1-13).

(b) At what current will the voltage drop be 400 mV?

(a) $V_D = 26\,\text{mV} \ln\left(\frac{200\,\text{nA}}{1\,\text{nA}}\right) = 137.8$ mV. (b) $I_D = 1\,\text{nA}\ e^{\frac{400\,mV}{26\,mV}} = 4.80$ mA

Diode Eqs. (11-14) are useful only at the temperature at which I_S was measured. These equations predict that I_D will exponentially drop as temperature rises, which is not true in practice. I_S is more temperature sensitive than the temperature exponential and doubles for about every 10°C rise. The result is that diode current markedly increases as temperature rises.

1.8.1. Diode Resistor Analysis

Example 1-20 shows a circuit that can be solved for the current and node voltages if we know the reverse bias saturation current I_S.

EXAMPLE 1-20

If $I_S = 10$ nA at room temperature, what is the voltage across the diode and what is I_D? Let $kT/q = 26$ mV.

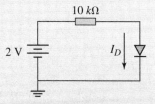

Write KVL using the diode voltage expression

$$2\,\text{V} = I_D(10\,k\Omega) + (26\,\text{mV}) \ln\left(\frac{I_D}{10\,\text{nA}} + 1\right)$$

This equation has one unknown (I_D), but is difficult to solve analytically, so two methods work. One is the historic way of solving by the iterative method. Values of I_D are substituted into the equation and the value that balances the left-hand side and right-hand side is a close approximation. An iterative solution is $I_D = 175\,\mu\text{A}$. Modern scientific calculators have a numerical method for solving these types of equations avoiding the manual plug and chug labor.

The diode voltage is

$$V_D = \frac{kT}{q} \ln\left(\frac{I_D}{I_S}\right)$$

$$= 26 \text{ mV} \times \ln\left(\frac{175 \ \mu A}{10 \text{ nA}}\right) = 254 \text{ mV}$$

EXAMPLE 1-21

Two circuits are shown with the diode cathode connected to the positive terminal of a power supply ($I_s = 100$ nA). What is V_O in both circuits?

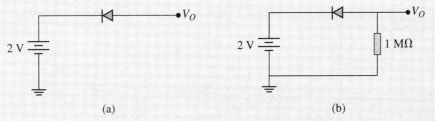

(a)   (b)

(a) There is a floating node at V_O so there is no current in the diode. Since $I_D = 0$, the diode voltage drop $V_D = 0$ and

$$V_O = V_D + 2 \text{ V} = 2 \text{ V}$$

(b) There is now a current path to ground. The diode is reversed biased and $I_{BB} = -I_D = 100$ nA. Then

$$V_O = I_{BB} \times 1 \text{ M}\Omega = 100 \text{ nA} \times 1 \text{ M}\Omega = 100 \text{ mV}$$

Self-Exercise 1-25
Estimate V_D and I_D for $I_S = 1$ nA.

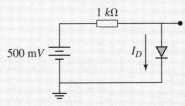

Answer: $I_D = 185 \ \mu A$, $V_D = 315.3$ V

These problems can be analyzed using Eqs. (11-14) or observing the process in the $I - V$ curve of Figure 1-10. In Example 1-21a the operating point is at the origin. In Self-Exercise 1-21b, it has moved to the right of the origin.

Self-Exercise 1-26

Calculate V_{D1} and V_O if $I_S = 100\,\mu\text{A}$ and $T = 25°\text{C}$.

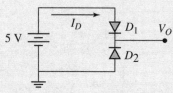

Answer: $V_{D1} = 18.02$ mV, $V_O = 4.982$ V

EXAMPLE 1-22

(a) Calculate I_{D1}, I_{D2}, and V_O. Assume the diode forward voltage is 0.7 V. Specify whether D_1 and D_2 are on or off.

(b) If $I_S = 10$ nA for each diode, what is V_B?

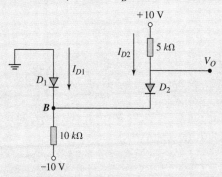

(a) Consider the diode states. If D_1 is on, then the cathode of D_1 at node-B is at -0.7 V. But if the $+10$ V supply pulls that node above -0.7 V then D_1 is off. Let's consider that D_1 is off and calculate the cathode node voltage at node-B.

$$10\,\text{V} = I_{D2}(5\,k\Omega) + 0.7\,\text{V} + I_{D2}(10\,k\Omega) - 10\,\text{V}$$

$$I_{D2} = 1.287 \text{ mA}$$

and

$$V_B = 1.287\,\text{mA}(10\,k\Omega) - 10\,\text{V} = 2.287 \text{ V}$$

If we were to now attach D_1 to node-B, there would be no effect since D_1 is reverse biased. D_1 is off and $I_{D1} = 0$.

$$V_O = 10\,\text{V} - (1.287\,\text{mA})(5\ k\Omega) = 3.565 \text{ V}$$

(b) We ignore the reverse bias saturation current of diode D_1 (10 nA). Diode D_1 is still off so

$$10\,\text{V} = I_{D2}(5\,k\Omega) + 0.026\ln\left(\frac{I_{D2}}{10\,\text{nA}}\right) + I_{D2}(10\,k\Omega) - 10\,\text{V}$$

$$I_{D2} = 1.313 \text{ mA}$$

$$V_B = (1.313\,\text{mA})(10\,k\Omega) - 10 = 3.13 \text{ V}$$

EXAMPLE 1-23

$I_{s1} = 1\,\mu A$ for diode D_1, and $I_{s2} = 20$ nA for the diode D_2. Calculate the voltage at the node connecting the diodes.

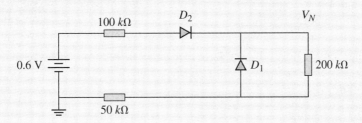

$$0.6 = I_D(150\,k) + 0.026 \ln \left[\frac{I_D}{20\,\mathrm{nA}} + 1 \right] + (I_D - 1\,\mu A)\,200\,k$$

Solve $I_D = 1.945\,\mu A$

$$V_N = [1.945\,\mu A - 1\,\mu A](200\,k\Omega) + (50\,k\Omega)(1.945\,\mu A) = 0.2863\text{ V}$$

1.9. Some Words about Power

Personal computers typically run quiet and clean with no smoke curling up, indicating a small power plant inside driving the logic gates. For many years, designers did not excessively concern themselves with wasted power in electronics. But ICs attracted attention when chip power dissipation grew as high as 150–200 watts. Think of holding a 150 W lightbulb constructed on the size of a postage stamp.

Let's examine personal computers and estimate the actual power expended at the electric generation plant. These calculations were taken from www.wikipedia.org under "coal." Assume that an average PC power dissipation is about 100 watts, and coal power plants deliver in an hour about 2.0 kW of electricity per kg of coal burned. In one year the PC burns 100 watt × 8760 hours/year × 1 year = 876 kW hours. The weight of coal burned for the power wasted in that PC in a year is

$$\frac{876\,kW \cdot hr}{2.0\,\frac{kW \cdot hr}{kg\ coal}} = 438 \text{ kg coal} = 964 \text{ pounds}$$

This is a large amount of coal to power one PC for a year. Visually, it is a cube of coal about 2.5 feet on a side. In addition, the power plant emits toxic gases including carbon dioxide, sulfur dioxide, mercury, and uranium. Coal contains about 50% carbon so that when carbon combines with two oxygen atoms in the air, the amount of coal burned to CO_2 emitted is about 1 kg of coal converts to 1.83 kg of CO_2. Therefore, one PC powered for a year is responsible for 801.5 kg CO_2 or 1763 pounds of CO_2. Internet servers waste more heat since they are designed for very high speed to expedite the digital traffic. Server system powers run from low end of about 200 W to over 15 kW for high-volume applications.

The point is that every fraction of a watt should be challenged. A design that burns 1 W less than another design and has one million ICs manufactured per year will save one million watt-hours of user's energy. Integrated circuit designs are manufactured in the millions and billions during the product life. The impact of lower power design is huge. This book will not detail the many tricks and theory of low power design but will later include some theory and problems to sensitize us to a paramount specification. By the way, the 100-watt PC example gives the same calculation for the yearly coal burned for a 100 W light bulb.

1.10. Summary

This chapter reviewed basic logic theory and analysis of circuits with power supplies, resistors, capacitors, and diodes. Kirchhoff's current and voltage laws were combined with Ohm's law to calculate node voltages and element currents for a variety of circuits. The technique of solving for currents and voltages by inspection is powerful for rapid insight into the nature of circuits. Finally, the diode illustrated circuit analysis with a nonlinear element. The problems at the end of the chapter should provide sufficient drill to prepare for subsequent chapters that will introduce the MOSFET transistor and its simple logic configurations.

Exercises

Logic Gates and Boolean Functions

When signal invert symbols appear in the following figures, do not draw an inverter; treat an inverted signal as it is.

1-1. (a) Draw the logic gate implementation for the Boolean function $G = A(B + C) \times (D\,\overline{E}\,F)$.

    (b) Draw the logic gate implementation for the Boolean function $G = (A + B) \times (C + D\,\overline{E}\,F)$.

1-2. (a) Draw the logic gate implementation for the Boolean function $F = \left(W\,\overline{X}\right)(Y + Z)$.

    (b) Write the DeMorgan equivalent Boolean statement in part (a) and draw its logic gate schematic.

1-3. Write the DeMorgan equivalent Boolean statement and draw its logic gate schematic for

    (a) $F = A\left(\overline{B} + D\right) + E\,F$

    (b) $F = (A + BC)\,D$

1-4. Draw the DeMorgan equivalent logic gate circuit for

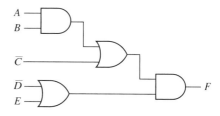

Boolean and Logic Gate Reduction

1-5. Minimize the function $F = X + \overline{X + Y} + XY$.

1-6. Minimize the function $F = \overline{\overline{XY} + \overline{ZY}}$.

1-7. Minimize the function $F = \overline{XY} + \overline{ZY}$ putting it in a logical inverter form.

1-8. Reduce the logic circuit to its minimum function.

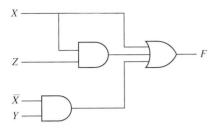

1-9. Reduce this logic gate circuit to its minimum function using Boolean reduction.

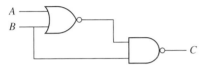

Terminal Resistance by Inspection

1-10. Write the shorthand expression for R_{eq} between the terminals.

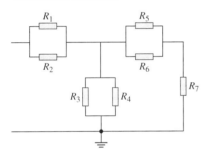

1-11. Write the shorthand expression for R_{eq} between the terminals.

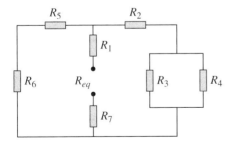

Voltage Dividers by Inspection

1-12. **(a)** Write the V_O expression by inspection and solve for V_O using a voltage divider.
   (b) Write the V_O expression by inspection and solve for V_O using a current divider.

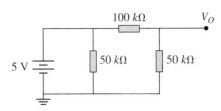

1-13. Calculate V_O by first writing a voltage divider expression and then numerically solving for V_O for both circuits.

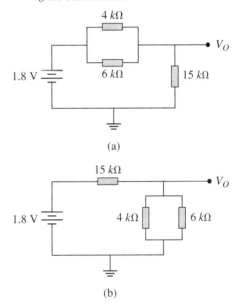

(a)

(b)

1-14. **(a)** Write the V_{O1} expression by inspection and solve for V_{O1} using a voltage divider.
   (b) Write the V_{O2} expression by inspection and solve for V_{O2} using a voltage divider.

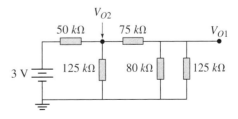

Current Dividers by Inspection

1-15. Repeat Problem 1-14, but solve by including a current divider.

1-16. Write the general expressions and solve for all resistor currents.

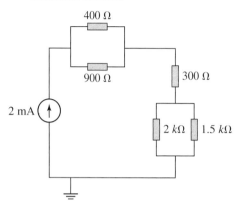

1-17. Given the circuit

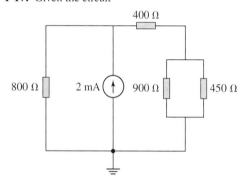

(a) Write the expression for I_{450} and solve.
(b) Write the expression for V_{800} and solve.
(c) Show that $I_{800} + I_{400} = 2$ mA.

1-18. (a) Calculate V_O using a voltage divider written by inspection.
(b) Calculate I_{2M} using a current divider written by inspection.

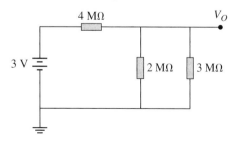

1-19. For the circuit
(a) Solve for V_O using a voltage divider expression.
(b) Solve for I_{2k}.
(c) Solve for I_{900}.

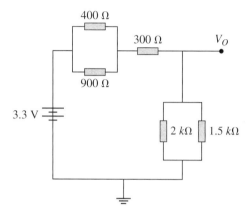

1-20. Calculate I_{2k} using the circuit analysis technique by inspection.

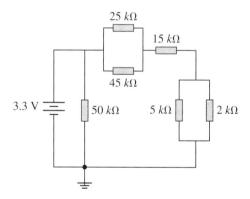

1-21. Using analysis by inspection, write the expression for the voltage across the 2 $k\Omega$ resistor and solve for its value.

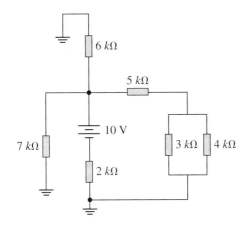

Mixing Voltage and Current Divider Analysis

1-22. Find I_{6k}.

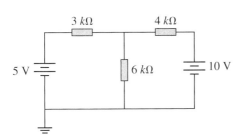

Hint: When we have two power supplies and a linear (resistive) network, we solve in three steps.

(1) Set one power supply to 0 V and calculate current in the 6 $k\Omega$ resistor from the nonzero power supply.

(2) Reverse the power supply roles and recalculate I_{6k}.

(3) The final answer is the sum of the two currents.

This is known as the superposition theorem and can be applied only for linear elements.

1-23. Solve for V_O using a method of inspection (current divider, voltage-divider, or both).

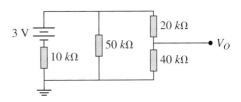

Capacitors

1-24. Find the equivalent capacitance at the input nodes and calculate the charge–discharge energy W for the parallel capacitors.

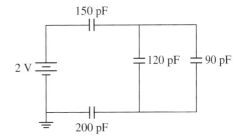

1-25. **(a)** What is the energy W needed to charge the circuit?

(b) Write the capacitance voltage divider expression for V_O and solve for the value.

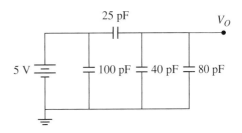

1-26. Find C_1 and the energy to charge C_1.

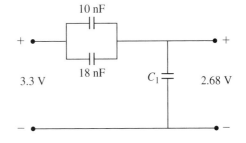

1-27. The 2 nF capacitors are precharged to 3 V, and the 5 nF capacitor is precharged to 1.2 V. At $t = 0$, switch S1 closes. What is the final voltage?

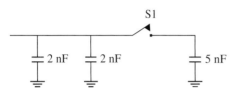

Diodes

1-28. Calculate V_O and the current through each resistor. Assume that the forward bias diode voltage is 0.7 V.

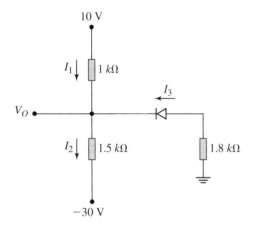

1-29. Given that $I_S = 10$ nA. Calculate I_D and V_D for (a) $V_{BB} = 1$ V and (b) $V_{BB} = 10$ V.

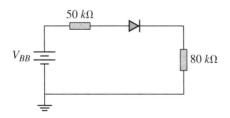

1-30. Calculate V_O given that the reverse bias saturation current $I_S = 1$ nA and you are at room temperature.

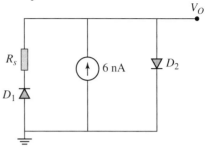

1-31. Diode D_1 has a reverse bias saturation current of $I_{s1} = 1$ nA, and diode D_2 has $I_{s2} = 4$ nA. At room temperature, what is V_O?

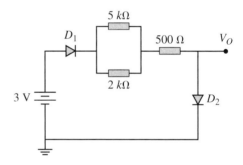

1-32. Calculate the voltage across the diodes given that the reverse bias saturation current in D_1 is $I_{s1} = 175$ nA, and $I_{s2} = 100$ nA.

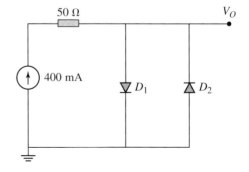

1-33. Given that $I_{sD1} = I_{sD1} = 100$ pA. Calculate I_{D1} and V_{D1}. Calculate I_{5k}.

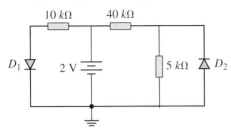

1-34. Calculate the diode current and voltage.

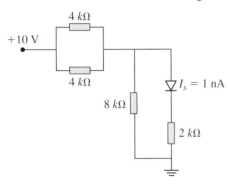

1-35. $I_s = 2\mu A$ for the diode. Calculate V_D and I_D.

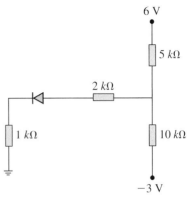

CHAPTER 2

Semiconductor Physics

> *If ... all of scientific knowledge were to be destroyed, and only one sentence passed on to the next generations of creatures ... it is ... all things are made of atoms.*

> Richard Feynman

2.1. Material Fundamentals

2.1.1. Metals, Insulators, and Semiconductors

Transistor and diode operation is understood through semiconductor physics. Emphasis here will be on introducing these principles and setting a base for how devices work. The language and mathematical models of semiconductor physics permeate CMOS integrated circuit (IC) design and manufacturing. Our integrated circuits would never have happened without a physics-based model of transistor operation.

Materials are classified by their physical properties. Conductivity is one such physical property that measures current through a material when a voltage is applied. Conductivity allows three material classifications: metals, insulators, and semiconductors. Metals have high conductivity and present little resistance to current, while insulators allow virtually no current under an applied voltage. Semiconductors are unique acting as conductors or insulators.

The classic explanation of conduction differences between these materials uses the energy band model of solids from quantum mechanics. Neils Bohr found that electrons in an atom could not have arbitrary energies, but had specific (quantum) energy values. A basic principle is that energy is distributed in quantum packets and cannot take continuous values. Electrons orbit at discrete distances from the nucleus. Figure 2-1a shows the allowed energy levels of a single hydrogen atom electron. These energies that the electrons can take are discrete. The *s* and *p* energy level symbols are taken from quantum mechanical convention.

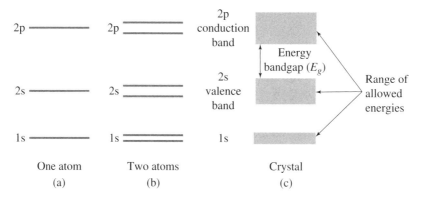

FIGURE 2-1.

(a) Energy levels in a single atom. (b) Two atoms. (c) A solid.

In a two-atom system each energy level in the single atom system splits into two sublevels as shown in Figure 2-1b since the quantum mechanical solution shows that an energy level can be occupied by only one electron at a time. When many more atoms are added to construct a crystalline solid, the energy levels successively split merging into the picture of Figure 2-1c, where energy bands separated by gaps of forbidden energies (called band gaps) replace single energy levels. The band gap energy magnitude depends on the type of atom, and it influences the conductive properties of the material.

Energy bands have different properties. The outermost energy band is called the *conduction band*, and the next lower one is called the *valence band* or outer shell. An electron having an energy corresponding to the conduction band is not tied to any atom and can move "freely" through the solid. A conduction band electron contributes to current when a voltage is applied. But an electron in the valence band is tightly bound to an atom of the semiconductor, and it is not free to move within the solid when a voltage is applied. There is a second charge mechanism in the valence band called hole conduction that will be described shortly.

Energy bands allow an interpretation of the conductive properties of materials. Figure 2-2a shows the energy bands of a *metal* where the lowest energy value of the conduction band is below the maximum energy of the valence band. The bands merge so that the metal conduction band has an abundance of electrons taken from the valence band that are available for conduction. It takes virtually no additional energy to move an electron from the valence to the conduction band in a metal. The metal energy band model predicts that metals will have a low resistance.

Figure 2-2b shows an energy band structure of an *insulator* with its large energy gap between the valence and conduction bands. The energy for an electron to go from the valence to the conduction band is so high that at room temperature only a few electrons within the material can acquire sufficient thermal energy and jump over the gap. A voltage

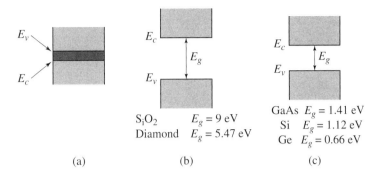

FIGURE 2-2.

Energy bands in solids: (a) Metal. (b) Insulator. (c) Semiconductor.

applied to the material will cause almost no current since virtually all electrons are tied to atoms in the valence band.

Semiconductors are the third class of conducting material and show an intermediate behavior (Figure 2-2c). The valence and conduction bands are not merged, but the energy gap is small enough so that more electrons are energetic enough to jump across it. The energy for electrons to jump across the gap can come from either ambient temperature or photon energy. We will deal with thermal energy since most integrated circuits are sealed and admit no light from the environment. The process of gaining enough thermal energy and jumping from the valence to the conduction band is inherently statistical. Electrons also move from the conduction band back to the valence band. When this happens, a photon is created. Electrons in a solid are continuously moving up and down between the valence and conduction bands. At a given temperature, there is a small average population of carriers that contribute to current.

Since the energy taken by an electron to jump the gap is thermal, the population of electrons in the conduction band depends on the temperature. At absolute zero temperature there is no thermal energy in a pure semiconductor, and no electron has energy to jump across the gap. The conduction band is empty. As temperature increases, the number of conducting electrons increases exponentially.

The different gap energies between insulators and semiconductors are related to how electrons are arranged within atoms. Electrons group into layers around the atomic nucleus, and electrons in the internal layers cannot be separated from the nucleus. Only electrons from the outside valence layer may jump from their bounded valence state to the conducting state (relatively free from the attractive forces of the atomic nucleus).

2.1.2. Carriers in Semiconductors: Electrons and Holes

The most common IC semiconductor material is silicon with four electrons in its outer energy valence band (Figure 2-3a), making silicon a Group IV atom in the Periodic Table.

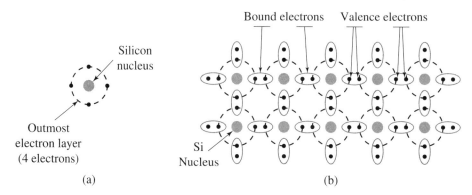

FIGURE 2-3.

(a) Representation of a silicon atom with its four electrons at the outmost layer. (b) Silicon structure in a crystalline solid.

When silicon is crystalline, each electron of the valence layer is shared with one electron of a neighbor atom so that by sharing with four electrons, each Si atom has eight outer shell electrons (Figure 2-3b). Eight valence band electrons are one of nature's mysteries that provide stable crystalline solids.

If a valence electron gains enough thermal energy to jump to the conduction band, it leaves an electron vacancy. The vacancy position has a positive charge site that is available for another valence electron to move into and leave a vacancy in its original site. This process can be repeated now for a third valence electron moving to this last vacancy and so on. Valence band electrons are driven by thermal energy and move randomly or can be driven in a particular direction by an applied electric field.

We could track the hopping electrons as a current, but a better way views the electron vacancy as a "particle" of positive charge that moves in the opposite direction to the valence electron movement. It is a virtual particle with a positive charge q and is called a *hole*. Holes have a net charge $+q$ (q being the charge value of an electron $q = -1.6 \times 10^{-19}$ coulomb). Holes can move and become a current. Hole motion is treated as a physical positive charge motion even though it is electron hopping that causes the appearance of positive charge motion. This model works, and, as we will see, it is better than treating moving electrons in the valence band.

The only contribution to current in a metal is from electrons in the conducting band. In contrast, semiconductor current has two contributions. One from electrons in the conduction band and the other from holes in the valence band caused by electrons that jumped into the conduction band. It is significant that this process does not require the moving valence electron to go to the conduction band.

TABLE 2-1 Semiconductor Materials Constants

Semiconductor	E_g (eV)	B (cm^{-3}) K$^{-3/2}$
Silicon (Si)	1.12	5.23×10^{15}
Germanium (Ge)	0.66	1.66×10^{15}
Gallium arsenide (GaAs)	1.41	2.10×10^{14}

2.1.3. Determining Carrier Concentrations

We can calculate n_i, the density of conduction band electrons of intrinsic material from

$$n_i = BT^{3/2} e^{\frac{-E_g}{2kT}} \tag{2-1}$$

where T is the absolute value of temperature in Kelvin (K), k is the Boltzmann constant 1.38066×10^{-23} J/K, and B is a constant dependent on the material (Table 2-1). Since the Boltzmann constant is typically given in J/K, and the semiconductor bandgap is expressed in electron volts (eV), knowing that $1\text{eV} = 1.60219 \times 10^{-19}$ J, the Boltzmann constant may be expressed as

$$k = \frac{1.38066 \times 10^{-23}\,\text{J/K}}{1.60219 \times 10^{-19}\,\text{J/eV}} = 86.17\,\frac{\mu\text{eV}}{\text{K}}$$

The Boltzmann constant is a statistical value giving the mean energy value of particles in thermal equilibrium. Material constants are given in Table 2-1.

EXAMPLE 2-1

Calculate n_i for Si at $T = 300$ K.

$$n_i = 5.23 \times 10^{15} \left(300^{3/2}\right) \exp\left[\frac{-1.12}{2 \times 86.17 \times 10^{-6} \times 300}\right]$$

$$n_i = 1.062 \times 10^{10} \left(\frac{\text{electrons}}{\text{cm}^3}\right)$$

The room temperature electron density calculation of intrinsic Si is 1.062×10^{10} (electrons/cm^3), and it is a small percentage of the total number of silicon atoms whose density is 5×10^{22} (atoms/cm^3). This means that only a small number of atoms are ionized at a room temperature of 300 K. Many textbooks use $n_i = 1.5 \times 10^{10}$ at room temperature, and that is a measured value. However, we will use the calculated value since subsequent numerical problems are done at different temperatures. Also, room temperature is a slightly vague term.

EXAMPLE 2-2

If the measured density of carriers is 1.5×10^{10} cm^{-3}, what is the calculated temperature from Eq. (9-1)?

$$1.5 \times 10^{10} = 5.23 \times 10^{15} \left(T^{3/2}\right) \exp\left[\frac{-1.12\,\text{eV}}{2 \times 86.17 \times 10^{-6} \times T}\right]$$

Your calculator solution is $T = 304.5$ K.

The density of holes is equal to the density of electrons ($p_0 = n_0$) in pure silicon. When the semiconductor is in equilibrium, electrons and holes move randomly in space, and no net current is observed. When an electric field is applied, the charge movement is not random, since holes and electrons drift in the field. The net current contribution is from holes in the valence band plus electrons in the conduction band. Electrically, a negative charge moving in one direction is equivalent to a positive charge moving in the opposite direction. These dual hole–electron conduction mechanisms in solids are detailed in the next section.

In a pure silicon material, electrons and holes are created in pairs. The "creation" of a free electron jumping into the conduction band creates a hole in the valence band, while an electron dropping from the conduction to valence band implies that a hole disappears. When an electron jumps from the valence to the conduction band, the process is called *electron–hole pair generation*, and when an electron jumps back from the conduction to the valence band, the process is referred to as *electron–hole recombination*, or simply *recombination*. Energy is needed for electron–hole pair formation, but in the opposite process, energy is released when an electron recombines with a hole. The recombination event releases a photon of energy. This band gap light emission property is used in semiconductor lasers and light-emitting diodes (LED).

Figure 2-4 illustrates the band energy gap model of a semiconductor and its solid-state physical representation. It emphasizes that electrons are mobile carriers in the conduction band and holes are mobile carriers in the valence band. Electron–hole pair creation and recombination is inherently statistical. Electron–hole pairs are continuously created and recombined at a given temperature.

Self-Exercise 2-1

What is the relative increase in intrinsic carrier concentration if the temperature of a silicon crystal is raised from 310 K to 410 K?

Answer: The relative electron density increase is $= 252.8$.

Self-Exercise 2-2

If $n_i = 4.773 \times 10^{13}$ what is the semiconductor temperature?

Answer: $T = 467$ K

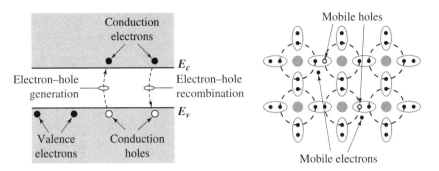

FIGURE 2-4.

Electron–hole pair creation and recombination in the band gap model, and its representation in the solid.

2.2. **Intrinsic and Extrinsic Semiconductors**

The previous concepts apply to "pure" semiconductors where all atoms are of the same type such as pure silicon or pure germanium. A pure crystalline silicon is referred to as an *intrinsic semiconductor* where the number of conduction band electrons is equal to the number of valence band holes since they are generated in pairs. *Extrinsic semiconductors* intentionally add impurities[†] to the semiconductor to increase the concentration of one carrier type (electrons or holes) without increasing the concentration of the other, thus breaking the symmetry between the number of electrons and holes. Impurities refer to atoms with different numbers of electrons in their outer shell, such as boron (B), arsenic (As), or phosphorus (P). The intentional substitution of a silicon atom by another element is called *doping*.

Doping properties depend on the specific impurity atom and the number of atoms added. There are *n*-type impurities that increase electron concentration and *p*-type impurities that increase hole concentration. The number of dopant atoms (atoms/unit volume) injected into the solid is much less than the number of silicon atoms, so the crystalline structure of the semiconductor is not globally disturbed.

2.2.1. *n*-Type Semiconductors

Silicon has four electrons in its outer layer that form bonds with neighbor atoms (Figure 2-5). We can increase the electron concentration by replacing some silicon atoms with the Periodic Table Group V atoms that have five electrons in their external layer (arsenic, phosphorus). Four of the five external electrons of the substituting atom create bonds with the neighboring silicon atoms, while the fifth electron is free to move within the solid. Room temperature thermal energy is enough to activate the fifth electron into the

---

[†]Any material, no matter its quality, always has unintended impurities. The effect of such unintended impurities can be neglected, if they are kept down to a minimum. But the word impurity refers here to a different atom in the Periodic Table intentionally added to the silicon crystal to alter electrical properties.

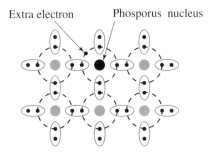

FIGURE 2-5.

Adding a donor atom creates a mobile electron without creating a hole.

conduction band. The electron jumps into the conduction band from the impurity energy level without creating a hole, so no electron–hole pairs are generated, only free electrons. Each impurity atom is called a *donor*, since it implies an extra electron donated to the conduction band. Silicon is a Group IV atom in the Periodic Table, and donor atoms come from the Group V atoms.

When silicon is doped with donors at room temperature and higher, the number of conducting electrons is equal to the number of donor atoms injected (N_D) plus a few electrons from thermal electron–hole pair creation. When donor concentration greatly exceeds the normal intrinsic population of carriers, then the conducting electron concentration is essentially that of the donor concentration. By adding a specific concentration of donors, the population of electrons can be made significantly higher than that of holes. An extrinsic semiconductor doped with donor impurities is called an *n*-type semiconductor.

When a considerable number of *n*-doping atoms are added (10^{15}–10^{17} atoms per cm^3), the band gap model predicts an impurity atom energy level within the gap close to the conduction band energy E_c (Figure 2-6). At 0 Kelvin temperature all electrons are at this energy level and are not mobile within the solid. At room temperature, the energy required for donor atoms to jump into the conduction band is small, so that essentially all donor atoms are ionized and in the conduction band.

We can calculate the density of electrons or holes in the conduction band from the important relation

$$n_o p_o = n_i^2 \qquad (2\text{-}2)$$

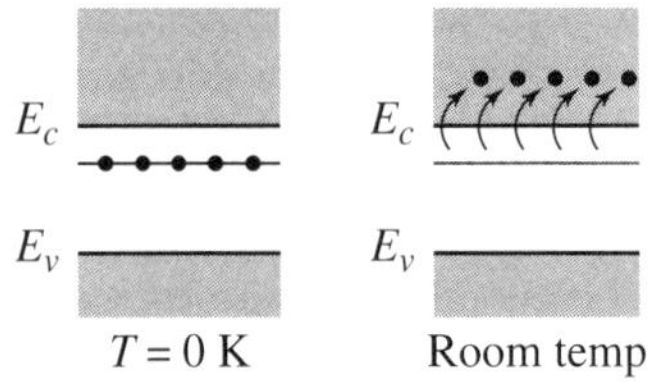

FIGURE 2-6.

Donor effect on conduction band carriers with temperature increase.

where n_o is the free electron concentration, p_o is the free hole concentration, and n_i is called the concentration of free carriers in an intrinsic material. $n_o = p_o$ in an intrinsic semiconductor, but $n_o \neq p_o$ when doping occurs.

Eq. (2-2) can be extended to estimate the carrier concentrations in a doped semiconductor. For example, the density of holes in n-doped silicon can be estimated by letting N_D be the concentration of donor atoms and n_o be the concentration of electrons in the conduction band. Since virtually all donor atoms are ionized and contribute many more electrons than do the thermal electron–hole pairs, then $n_o \approx N_D$ and the minority concentration p_o is

$$p_o \approx \frac{n_i^2}{N_D} \tag{2-3}$$

Eq. (2-3) shows that as the electron majority concentration increases, the minority hole concentration decreases. Visually this because more electrons can shower down on the valence band recombining and eliminating holes.

EXAMPLE 2-3

If $N_D = 10^{16}$ (donor atoms/cm³), calculate the minority concentration at $T = 300$ K. $n_o \approx N_D = 10^{16}$ (electrons/cm³) and

$$p_o = \frac{(1.062 \times 10^{10})^2}{10^{16}} = 1.128 \times 10^4 \left(\frac{\text{holes}}{\text{cm}^3} \right)$$

EXAMPLE 2-4

Given that $T = 350$ K for a Si crystal

(a) Calculate n_i.

$$n_i = B(T)^{3/2} \exp \left[\frac{-E_g}{2 \left(\frac{k}{q} \right) T} \right] = 5.23 \times 10^{15} (350)^{3/2} \exp \left[\frac{-1.12}{2 \times 86.17 \times 10^{-6} \times 350} \right]$$

$$= 2.956 \times 10^{11} \left(\frac{\text{carriers}}{\text{cm}^3} \right)$$

(b) If $N_D = 10^{17}$ (cm⁻³), what is the minority carrier concentration?

$$N_D = 10^{17} \left(\frac{\text{electrons}}{\text{cm}^3} \right) \text{ then}$$

$$p_o = \frac{n_i^2}{N_D} = \frac{[2.965 \times 10^{11}]^2}{10^{17}} = 8.73 \times 10^5 \left(\frac{\text{holes}}{\text{cm}^3} \right)$$

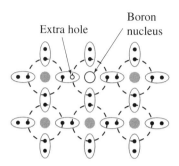

FIGURE 2-7.

Adding acceptor atoms creates a mobile hole without creating an electron.

Doping a semiconductor does not increase its net charge, since the excess negative q charge of the fifth electron with respect to the "replaced" silicon atom is balanced by the atomic number[†] of the donor impurity (5 instead of 4 for silicon). This positive charge compensates the electron charge.

2.2.2. p-Type Semiconductors

We can increase the number of holes without increasing the number of electrons by doping silicon atoms with elements from Group III in the Periodic Table that have three electrons in the outermost layer. The three electrons of the impurity atom bond with three of the neighbor silicon atoms, and the fourth bond is missing (Figure 2-7). The Group III atom ionizes when an electron from the valence band is thermally activated to the impurity energy level. Impurities with three electrons in the outermost layer are called *acceptors* since a vacancy or hole is created in the valence band by the impurity atom that can accept and hold electrons.

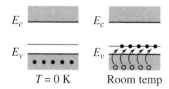

FIGURE 2-8.

Effect of acceptor doping in the band gap model.

Acceptor doping in the energy-band model creates an energy level close to the silicon valence band as illustrated in Figure 2-8 that shows boron-doping atoms. At room temperature, valence band electrons have enough energy to jump to this level creating a hole in the silicon without injecting electrons into the conduction band. At a given temperature

---

[†]The atomic number is the number of protons of a neutral (nonionized) atom.

the concentration of holes in a semiconductor in which a number N_A of acceptor impurities were introduced will be N_A plus a few of the holes contributed by the thermal electron–hole pairs created at that temperature. Boron is a small Group III atom and is the most popular of acceptor doping atoms.

Self-Exercise 2-3
If $N_A = 5 \times 10^{17}$ (acceptor atoms/cm^3), calculate the minority carrier concentration at $T = 300$ K.

Answer: $n_o = 226 \left(\dfrac{\text{electrons}}{\text{cm}^3} \right)$

Self-Exercise 2-4
The hole concentration in *n*-doped silicon is 1.5×10^5 cm^{-3} at 325 K. What is the doping concentration?

Answer: $N_D = 2.67 \times 10^{16} \left(\dfrac{\text{donor atoms}}{\text{cm}^3} \right)$

Self-Exercise 2-5
Calculate the temperature at which $n_i = 2.5 \times 10^{11}$ cm^{-3}.

Answer: $T = 347.1$ K

2.2.3. Carrier Concentration in *n*- and *p*-Doped Semiconductors

A fourth environment contains both *p*- and *n*-doping in the semiconductor. It is called *compensated doping*, and this situation exists in real diodes and transistors. If $N_D > N_A$ then the material will act as an *n*-type material. If $N_A > N_D$ then the material will act as a *p*-type material. If $N_D = N_A$ then the material will act as an intrinsic material.

If $N_D > N_A$ then the material acts as an *n*-type material, and $n_o = N_D - N_A$ and the minority concentration is

$$p_o = \frac{n_i^2}{n_o} = \frac{n_i^2}{N_D - N_A} \tag{2-4}$$

If $N_A > N_D$ then the material acts as a *p*-type material, and $p_o = N_A - N_D$ and the minority concentration is

$$n_o = \frac{n_i^2}{p_o} = \frac{n_i^2}{N_A - N_D} \tag{2-5}$$

EXAMPLE 2-5

If $N_A = 5 \times 10^{17}$ cm^{-3}, $N_D = 10^{16}$ cm^{-3}, and $T = 300$ K, what is the minority carrier concentration?

$$n_o = \frac{n_i^2}{N_A - N_D} = \left(\frac{(1.062 \times 10^{10})^2}{5 \times 10^{17} - 10^{16}} \right) = 230 \left(\frac{\text{electrons}}{\text{cm}^3} \right)$$

EXAMPLE 2-6

If $N_A = 5 \times 10^{16}$ cm^{-3}, $N_D = 10^{18}$ cm^{-3}, and $T = 380$ K, what is the minority carrier concentration?

$$n_i = BT^{3/2} \exp \left[\frac{-Eg}{2 \left(\frac{kT}{q} \right)} \right] = 5.23 \times 10^{15} (380)^{1.5} \exp \left[\frac{-1.12}{2(86.17 \, \mu\text{eV}) \, 380} \right]$$

$$= 1.448 \times 10^{12} \left(\frac{\text{electrons}}{\text{cm}^3} \right)$$

$$p_o = \frac{(1.448 \times 10^{12})^2}{10^{18} - 5 \times 10^{16}} = 2.208 \times 10^6 \left(\frac{\text{holes}}{\text{cm}^3} \right)$$

Self-Exercise 2-6

A compensated silicon crystal has an acceptor doping concentration of $N_A = 10^{16}$ cm^{-3}, $T = 375$ K, and a minority carrier hole concentration of 6×10^7 holes/cm^3. What is the n-doping concentration?

Answer: $N_D = 3.129 \times 10^{16} \left(\frac{\text{atoms}}{\text{cm}^3} \right)$

Self-Exercise 2-7

$T = 400$ K, $N_A = 10^{18}$ cm^{-3}, $N_D = 5 \times 10^{17}$ cm^{-3}. What is the minority carrier concentration?

Answer: $n_o = 2.706 \times 10^7 \left(\frac{\text{electrons}}{\text{cm}^3} \right)$

Self-Exercise 2-8

A compensated silicon structure has $N_D = 2 \times 10^{18}$ cm^{-3}, $N_A = 6 \times 10^{17}$ cm^{-3}, and hole minority carrier concentration $p_o = 3 \times 10^6$ (hole/cm^3). What is the temperature of the material?

Answer: $T = 387.2$ K

2.3. **Carrier Transport in Semiconductors**

Electron and hole net movement is called carrier transport. Computation of carrier transport requires the carrier concentration. It also needs the laws governing movement of carriers within the solid. Movement of electrons and holes within a semiconductor is different from carriers traveling in free space since collisions of carriers with the lattice impact their *mobility*. These carrier collisions induce *scattering*, and the more scattering that occurs the lower the mobility. Temperature is significant in carrier transport. It affects the carrier population and carrier movement within the solid, because atomic thermal vibration impedes carrier movement. Although carriers are in constant motion within a solid, an unbiased semiconductor will not register a net current since carrier movement is random. A force is required for net charge movement to occur. The two main carrier force mechanisms in solids are drift and diffusion.

2.3.1. Drift Current

Drift is carrier movement due to an electric or magnetic field. An electric field E applied to a semiconductor causes electrons in the conduction band to move in the direction opposite to the electric field, while holes in the valence band move in the same direction of this field. Carrier movement is illustrated in Figure 2-9. Electron and hole current densities (electron hopping in the valence band) have the same direction and both contribute to current.

The *flux* of charge carriers J_d $\left(\frac{A}{cm^2}\right)$ in an electric field is derived from a series of relations shown in the following equations where v is the velocity of moving charge, n_o is the moving electron concentration, p_0 is the moving hole concentration, q is carrier charge, and E is the electric field drawing the charge with a force F through a solid with mobility μ. For an n-doped material

$$J_{dn} = v_{dn}n_0 = (\mu_n F)n_0 = (\mu_n q E)n_0 \qquad (2\text{-}6)$$

and for p-doped material

$$J_{dp} = v_{dp}p_0 = (\mu_p F)p_0 = (\mu_p q E)p_0 \qquad (2\text{-}7)$$

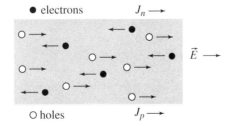

FIGURE 2-9.

Electron and hole direction in an external applied field E.

When holes and electrons are present in a semiconductor, the total drift flux is

$$J = \mu_n q E n_o + \mu_p q E p_o \qquad (2\text{-}8)$$

The two drift terms on the right are positive representing positive current convention. Both drift components sum to total current.

EXAMPLE 2-7

An n-type silicon has $n_o \approx N_D = 10^{17}$ cm^3 at 300 K, $\mu_n = 1350$ cm^2/V · s, $\mu_p = 480$ cm^2/V · s, and the electric field is 10 V/cm. (a) What is the minority hole carrier density p_o (b) What is the total current flux J?

$$\textbf{(a)} \quad p_o = \frac{n_i^2}{n_o} = \frac{(1.062 \times 10^{10})^2}{10^{17}} = 1.128 \times 10^3 \left(\frac{\text{holes}}{\text{cm}^3} \right)$$

$$\textbf{(b)} \quad J = (1350)(1.6 \times 10^{-19})(10)(10^{17}) + (480)(1.6 \times 10^{-19})(10)(1.128 \times 10^3)$$

$$= 216 + 8.7 \times 10^{-13} = 216 \left(\frac{\text{A}}{\text{cm}^2} \right)$$

The following calculation is useful when designing a semiconductor material cross-sectional area given the calculated current density, doping, temperature, mobility, and electric field conditions.

Self-Exercise 2-9

A device is to carry 2 mA. What cross-sectional area should it have if the current flux is $J = 216 \, (A/\text{cm}^2)$.

Answer: Area $= 9.3 \times 10^{-6}$ cm^2. This would lead to a device that is about 30 μm $\times$ 30 μm.

Drift flux is a function of the electric field E and proportionality constant σ called the *conductivity*. This is a form of Ohm's law:

$$J = \sigma E = \frac{1}{\rho} E \qquad (2\text{-}9)$$

where ρ is the material resistivity.

2.3.2. Diffusion Current

Diffusion is the motion of particles due to thermal energy. A high concentration of thermally active particles will experience a net movement (flux) toward a lower concentration of particles. A concentration difference, called a gradient, is necessary for net diffusion. Diffusing particles may be charged or neutral. Electrons and holes diffusing in a direction

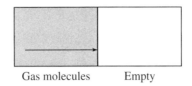

Gas molecules   Empty

FIGURE 2-10.

Diffusion and concentration gradient of a gas.

are moving charged particles that create a diffusion current. Semiconductors have many high gradient situations partly due to short micron or nanometer (nm) gradient dimensions.

Figure 2-10 shows two gas compartments separated by a permeable membrane. The gas molecules in the left-hand side have a higher concentration of particles than the right-hand side, and some move into the empty chamber. The atoms on both sides of the membrane move in random directions powered by thermal energy. More molecules are directed across the barrier from the left side, while fewer are directed from right to left. A net transfer of gas molecules happens from left to right. The net displacement of particles in time is called a flux that is measured in particles/cm$^2 \cdot$ s. The same diffusion phenomenon happens in semiconductors when holes and electrons have concentration gradients, but the diffusing medium is a solid. The force of diffusion is strong, and the flux of charged particles is a current. Ampere currents can be generated if high gradients and small distances exist.

Diffusion flux is a net movement of particles/cm^2-sec, and is described by Fick's law:

$$J_{diff} = -D\frac{dc}{dx} \qquad (2\text{-}10)$$

where dc/dx is the gradient of particles, and D is the *diffusion coefficient*. The equation states that the flux of particles is zero if there is a uniform distribution (i.e., dc/dx is zero).

We first consider hole diffusion. Figure 2-11 sketches a hypothetical profile of hole concentration versus distance x. The holes are positive charged so their direction is the same as positive current convention. Notice that the slope of the gradient is negative as we go from high to low concentration.

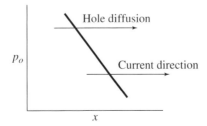

FIGURE 2-11.

Hole concentration p_o versus a distance x.

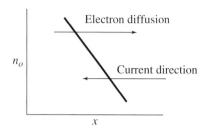

FIGURE 2-12.

Electron concentration n_o versus a distance x.

Electron diffusion also follows Fick's laws. Figure 2-12 sketches a profile of electron concentration n_o versus distance x. The electrons are negatively charged so their direction is opposite to the positive current convention.

Electrons and holes diffusing in the opposite direction have the same current direction since their charge sign is opposite. Expressions for absolute electron and hole diffusion fluxes are

$$J_{ndiff} = qD_n \frac{dn_0}{dx} \tag{2-11}$$

$$J_{pdiff} = qD_p \frac{dp_0}{dx} \tag{2-12}$$

D_n and D_p are electron and hole diffusion coefficients that differ since electron and hole mobilities are different. Electron mobility is about twice that of hole mobility.

The Einstein relation between the diffusion coefficient and mobility is

$$\frac{D_x}{\mu_x} = \frac{kT}{q} \tag{2-13}$$

where x must be substituted by n for electrons and by p for holes. Drift is the dominant charge transport mechanism in CMOS field effect transistors, while diffusion plays a secondary role.

EXAMPLE 2-8

Given the hole concentration profile in the figure and $D_p = 10 \left(\frac{cm^2}{s} \right)$, calculate the hole current density J.

$$J_p = qD_p \frac{dp_o}{dx} = qD_p \frac{\Delta y}{\Delta x}$$

$$J_p = (1.6 \times 10^{-19})(10)\left[\frac{10^{12} - 10^{15}}{10^{-4} - 0}\right]$$

$$J_p = -16.0 \left(\frac{A}{cm^2} \right)$$

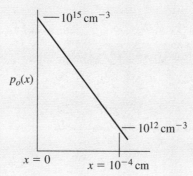

The holes move from the high to the low concentration, and since the concentration slope is negative, the equation carries a negative sign. The absolute current is

$$J_p = 16.0 \left(\frac{A}{cm^2} \right)$$

EXAMPLE 2-9

An initial electron concentration difference of 10^{17} and 10^{14} exists across a short separation of 1 μm. If $D_n = 35 \dfrac{cm^2}{s}$, what is the absolute initial flux surge?

$$J_{ndiff} = qD_n \frac{dn_o}{dx} = (1.6 \times 10^{-19} \text{ coulomb}) \left(35 \frac{cm^2}{s} \right) \left(\frac{10^{14} - 10^{17}}{10^{-4} - 0} \right) = 5.594 \times 10^3 \left(\frac{amp}{cm^2} \right)$$

Electron flux would have a negative sign, but conventional current is positive.

Self-Exercise 2-10

n_o is twice the p_o concentration gradient in a semiconductor. The flux is $J = 1$ mA/cm^2, $\mu_n = 1250$ cm^2/V · s, and, $\mu_p = 450$ cm^2/V · s. Neglect the effects of the electric field and determine the absolute concentration gradients of n_o and p_o.

Answer: $\dfrac{dp_o}{dx} = 2.119 \times 10^{12} \left(\dfrac{holes}{cm} \right)$ $\dfrac{dn_o}{dx} = 4.238 \times 10^{12} \left(\dfrac{electrons}{cm} \right)$

Self-Exercise 2-11

What is the current density in A/μm^2 for an electron flux in a material 100 μm long with 1 V across the length? $D_n = 20$ cm^2/s, $\mu_n = 1300$ cm^2/V · s, $dn/dx = -10^{19}$ atoms/cm, and $N_D = 10^{18}$ cm^{-3}.

Answer: $J = 208.3 \dfrac{\mu A}{\mu m^2}$

Self-Exercise 2-12

(a) A p-semiconductor has a gradient of 10^{20} holes/cm, and $D_p = 15$ cm^2/s. What is the current density in cm^2 and μm^2?

(b) What is the current if the material has a length of 200 μm, a thickness of 1 μm, and a width of 10 μm?

Answers: **(a)** $J = 240 \dfrac{A}{cm^2} = 2.4 \dfrac{\mu A}{\mu m^2}$. **(b)** $I = 24 \ \mu A$

2.4. The *pn* Junction

Diodes are the semiconductor building blocks of MOS transistors. Diodes form a junction by physically joining a p-type to an n-type semiconductor in what is called a *pn junction*. To visualize *pn* junction properties, assume an ideal case in which two pieces of semiconductors with opposite doping are initially separated (Figure 2-13a) and then joined (Figure 2-13b). The Si lattice structure is not lost at the joining surface, but a dynamic reaction occurs when the two materials touch.

Initially the n-type and p-type semiconductor bars are in equilibrium, so the total net charge in each bar is zero. Strong concentration gradients exist for electrons on the

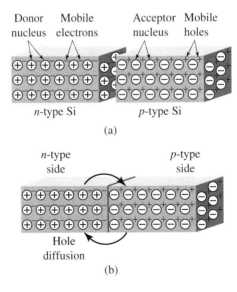

(a)

(b)

FIGURE 2-13.

Two pieces of semiconductor materials with opposite doping. (a) Separated. (b) Joined with arrows showing hole and electron diffusion currents across the *pn* junction.

n-side and holes on the *p*-side. At the instant when the semiconductor pieces join, the electron and hole gradients cause a current surge at the joining region. The subsequent nonequilibrium adjustment exists for a short time during which electrons at the junction rapidly diffuse from the *n*-type bar into the *p*-type, while holes close to the junction move away from the *p*-doped semiconductor into the *n*-type. The minority carriers that moved to the other side of the junction rapidly recombine and disappear in a sea of majority carriers.

If electrons and holes were not charged particles, then this diffusing process would continue until electron and hole concentrations were uniform along the whole area of joined semiconductors. But electrons and holes are charged particles, and their diffusion and subsequent recombination across the junction boundary create an electric field in the semiconductor at both sides of the junction. Carriers close to the junction immediately diffuse away and recombine when they meet opposite carrier types across the junction. Simplistically, mobile charge disappears after crossing the junction.

This carrier migration and recombination action ionizes all dopant atoms close to the joining surface creating a zone of net charge (Figure 2-14) around the junction (positive at the *n*-side and negative at the *p*-side). This induces an electric field pointing from the *n*-side with its fixed positive charges to the *p*-side with its fixed negative charges. Carriers moving by diffusion now "feel" the electric field as an opposing force when trying to move across the junction. As more carriers move by diffusion, more atoms are ionized, and the

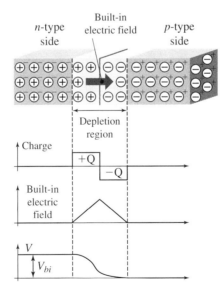

FIGURE 2-14.

A picture of a diode in equilibrium showing the charge, electric field, and potential internal distribution.

electric field strength increases. The net diffusion mechanism stops when the induced internal electric field (that increases with the number of carriers moving by diffusion) reaches a value such that its force exactly balances the force tending to diffuse carriers across the junction. The result is a narrow depletion region with all donors and acceptors ionized. A high electric field exists sweeping out any free charge. The thermally generated hole–electron pairs are instantly ejected from the depletion region.

Figure 2-14 illustrates a diode junction in equilibrium (no external electric field, light, or temperature gradients are present) and the charge, electric field, and potential distribution at each point within the diode. Notice that in the regions where atoms are not ionized (out side the depletion region) there is charge neutrality, so that no electric field or potential drop exists. The net charge in the depletion region is positive on the n-type side and negative on the p-type side. As a result, the electric field increases while moving from the neutral regions to the junction site. The electric field distribution causes a voltage difference between the two opposite doped regions that depends mainly on the doping levels. This voltage is called the *built-in junction potential* (V_{bi}) shown in the bottom of Figure 2-14.

The built-in junction potential (V_{bi}) can be calculated from the doping concentrations

$$V_{bi} = \frac{kT}{q} \ln \left(\frac{N_A N_D}{n_i^2} \right) = V_T \ln \left(\frac{N_A N_D}{n_i^2} \right) \tag{2-14}$$

where V_T is the thermal voltage, $V_T = \dfrac{kT}{q}$.

EXAMPLE 2-10

Calculate the built-in potential at 300 K if $\dfrac{kT}{q} \approx 0.026$ V, $n_i = 1.062 \times 10^{10}$ cm^{-3}, $N_A = 10^{16}$ cm^{-3}, and $N_D = 10^{17}$ cm^{-3}.

$$V_{bi} = 0.026 \ln \left[\frac{10^{16} \times 10^{17}}{\left(1.062 \times 10^{10} \right)^2} \right]$$

$$V_{bi} = 0.775 \text{ V}$$

The potential difference across the junction cannot be measured by a voltmeter because new potential barriers are formed between the probes. V_{bi} maintains equilibrium so this voltage produces no current.

EXAMPLE 2-11

What is the built-in potential of a diode if $T = 150°C$, $N_A = 10^{18}$ cm^{-3}, and $N_D = 10^{16}$ cm^{-3}?

$$T = 150 + 273 = 423 \text{ K}$$

$$n_i(423 \text{ K}) = 5.23 \times 10^{15}(423)^{3/2} \exp\left[\frac{-1.12}{2(86.16 \times 10^{-6})\,423}\right] = 9.676 \times 10^{12} \left(\frac{\text{carriers}}{\text{cm}^3}\right)$$

$$V_{bi} = (86.17 \times 10^{-6})(423) \ln\left[\frac{10^{18} \times 10^{16}}{(9.676 \times 10^{12})^2}\right] = 0.674 \text{ V}$$

Self-Exercise 2-13

This exercise will estimate initial surge currents when a p-semiconductor bar is joined to an n-semiconductor bar. Given $D_n = 35$ cm^2/s, $(n_o)_{n-side} = 5 \times 10^{18}$ cm^{-3}, and $(n_o)_{p-side} = 10^5$ cm^{-3}:

(a) At $t = 0$ the depletion width is 0 nm; what is the theoretical initial surge in J_D and I_D?
(b) At $t = 0^+$ the depletion width is 500 nm. What is the initial electron surge in J_D?
(c) At $t = 0^+$ the depletion width is 100 nm. What is the initial surge in I_D?

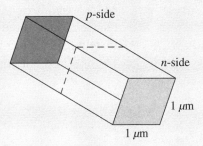

Answers: **(a)** Both are infinite. **(b)** $J_D = 560 \dfrac{\text{kA}}{\text{cm}^2}$. **(c)** $I_D = 28$ mA

2.5. Biasing the *pn* Junction

We showed that when two semiconductor bars of opposite doping are joined, a built-in electric field appears preventing electrons from diffusing too far away from the n-side and holes from diffusing away from the p-side. Once the system is in equilibrium, no net current exists since the junction is isolated with no external conducting path. We will now analyze the typical behavior of the junction when an external voltage is applied.

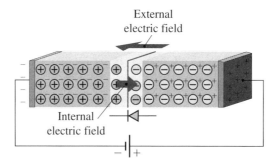

FIGURE 2-15.

Forward biased diode showing internal junction depletion field reduction.

2.5.1. The *pn* Junction under Forward Bias

Assume that an external voltage is connected to a diode with a positive voltage at the *p*-side with respect to the *n*-side (Figure 2-15). The external voltage creates an electric field opposed to the built-in one reducing its strength. The built-in electric field is a barrier for electrons diffusing from the *n*-side to the *p*-side and for holes diffusing from the *p*-side to the *n*-side, but an external reduction of the built-in field favors the diffusion process. Since many conducting electrons are on the *n*-side and many holes are on the *p*-side, then many electrons will diffuse into the *p*-side and many holes diffuse into the *n*-side. This process causes a permanent current since electrons must be replaced at the *n*-type terminal by the voltage source and holes are required at the *p*-type terminal. The larger the applied forward bias voltage the higher the diode current. *pn* junction current increases exponentially with forward bias.

2.5.2. The *pn* Junction under Reverse Biasing

An applied positive voltage at the *n*-side with respect to the *p*-side induces an external electric field that increases the built-in junction electric field (Figure 2-16). This reduces electron diffusion from the *n*-side silicon into the *p*-side and hole diffusion in the opposite direction. Only a few thermally generated electrons or holes within the junction region itself are accelerated by the junction electric field and contribute to overall leakage current. Since few thermally generated electrons are present in a *p*-type material and few holes are in the *n*-type side, the reverse bias current is small. This current is called the *reverse bias saturation current* and it arises from thermal generation of electron–hole pairs in the depletion region itself. The thermally generated free electron and hole carriers are rapidly swept out of the junction by the depletion region electric field forming the small current at the diode terminals.

The device level current–voltage characteristics of the diode were introduced in Chapter 1. The relation between the applied diode voltage V_D and the obtained current I_D

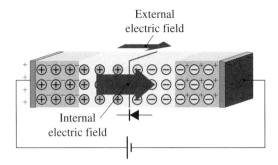

FIGURE 2-16.

Reverse biased diode showing internal junction depletion field increase.

is exponential and reproduced as

$$I_D = I_S \left(e^{\frac{q V_D}{kT}} - 1 \right) \tag{2-15}$$

where I_S is the reverse bias saturation current. When V_D is large and negative, then the exponential term goes to zero and $I_D = -I_S$. The I-V characteristic of the diode current rapidly increases for positive diode voltages, while very little current is obtained for negative diode voltages (Figure 2-17).

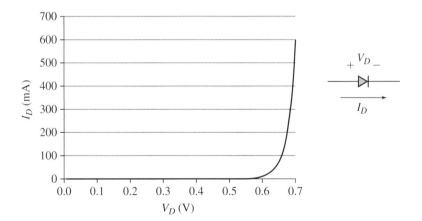

FIGURE 2-17.

Diode I_D versus V_D curve and diode symbol.

EXAMPLE 2-12

A diode has $I_S = 10^{-13}$ A and $T = 300$ K. Find I_D for $V_D = 0.6$ V and $V_D = 0.4$ V.

$$I_D = 10^{-13}(e^{\frac{600}{26}} - 1) = 1.05 \text{ mA}$$

and

$$I_D = 10^{-13}(e^{\frac{400}{26}} - 1) = 480 \text{ nA}$$

Notice that the -1 term is negligible for most forward bias voltages and can be neglected in the calculation.

Self-Exercise 2-14

If $I_S = 10$ fA, $T = 300$ K, what is the voltage that will bring a diode forward bias current to 625 μA?

Answer: $V_D = 646.3$ mV

2.6. Diode Junction Capacitance

All diodes deviate from ideality, and this deviation can be modeled by *parasitic elements*. Parasitic elements are unintended capacitance, inductance, or resistance. In CMOS technology two diodes exist in each transistor. These diodes are always reverse biased in ICs, and their main degradation effects at the circuit level are related to signal delay by parasitic junction capacitance in the diode and to reverse current leakage.

When a diode is reverse biased, the depletion region widens and capacitance decreases. Conversely, a forward bias narrows the depletion region and increases capacitance across the *pn* junction. A diode has an *internal parasitic capacitance* since there is a charge variation induced by a voltage variation. Capacitance is the charge variation due to voltage variation at its terminals.

$$C = \frac{\partial Q}{\partial V} \tag{2-16}$$

The parasitic capacitance inherent to the *pn* junction is different than the passive or parallel-plate capacitors studied in Chapter 1 because the charge-voltage ratio varies nonlinearly with the applied voltage. Since the Q/V quotient is not constant when the applied voltage changes, a fixed capacitor value cannot be assigned to the parasitic capacitance.

The junction capacitance with the reverse voltage V_R is

$$C_j = \frac{C_{j0}}{\left(1 + \dfrac{V_R}{V_{bi}}\right)^{1/2}} \tag{2-17}$$

where C_{j0} is a constant defined by the capacitance at $V_R = 0$. C_{j0} depends on the *p*- and *n*-doping concentration, fundamental constants of the silicon, the area of the surfaces

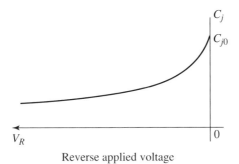

FIGURE 2-18.

Plot of C_j versus reverse applied voltage (V_R) for a diode.

being joined, and the built-in junction potential V_{bi}. V_R is the reverse applied voltage, and is a positive value in Eq. (2-17).

Figure 2-18 plots the capacitance normalized to C_{j0} with respect to the reverse applied voltage. Eq. (2-17) breaks down in the diode forward bias regions for large positive values of V_D. The effect of the diode capacitance at the circuit level is significant, since it must be charged and discharged when a gate switches. This contributes to circuit delay and heat.

EXAMPLE 2-13

Given that $T = 325$ K, $N_A = 10^{15}$ cm^{-3}, $N_D = 10^{14}$ cm^{-3}, and $C_{j0} = 40$ fF, calculate the diode C_j at $V_R = 1.5$ V.

$$C_j = \frac{C_{j0}}{\left(1 - \dfrac{V_A}{V_{bi}}\right)^{1/2}}. \text{ But we must find } V_{bi} = \frac{kT}{q} \ln\left(\frac{N_A N_D}{n_i^2}\right) \text{ and } n_i \text{ for this equation}$$

$$n_i = (5.23 \times 10^{15})(325)^{3/2} e^{\frac{-1.12}{2(kT/q)}} = 6.34 \times 10^{10} \left(\frac{\text{carriers}}{\text{cm}^3}\right)$$

$$V_{bi} = (86.17 \times 10^{-6})(325) \ln\left(\frac{10^{15} \times 10^{14}}{(6.34 \times 10^{10})^2}\right) = 0.477 \text{ V}$$

$$C_j = \frac{40 \text{ fF}}{\left(1 + \dfrac{1.5}{0.477}\right)^{1/2}} = 19.65 \text{ fF}$$

Self-Exercise 2-15

Given that $T = 400$ K, $C_j = 600$ aF, $N_A = 3 \times 10^{18}$ cm^{-3}, $N_D = 2 \times 10^{16}$ cm^{-3}, and reverse junction bias is $V_R = 1.2$ V, what is C_{j0}?

Answer: $C_{j0} = 961.4$ aF

Self-Exercise 2-16

A *pn* junction diode has $V_R = 1.5$ V, $C_j = 5$ fF, $C_{j0} = 9$ fF, $N_A = 10^{16}$ cm^{-3}, and $T = 300$ K. Calculate N_D.

Answer: $N_D = 1.73 \times 10^{15}$ cm^{-3}

2.7. Summary

The construction of integrated circuits depends on a strong physical theory. Chapter 2 introduced these essential concepts using a few modeling equations to deepen our thought processes. The purpose is qualitative understanding of the language and physical flow of semiconductor current, related doping properties, and diode characteristics. The diode depletion region is one of the essential properties of transistors. These ideas permeate later chapters. More information can be found in [1,2].

References

[1] D. A. Neamen, *Semiconductor Physics and Devices—Basic Principles*, 3rd Ed., McGraw Hill, 2010.

[2] Robert F. Pierret, *Semiconductor Device Fundamentals*, Addison-Wesley Pub. Co., 1996.

Exercises

Assume silicon material and use the constants given in this chapter.

Free Carrier Concentration

2-1. If $n_i = 1.67 \times 10^{11}$ (carriers/cm^3), what is the temperature?

2-2. What fraction of Si atoms is ionized at $T = 100°$C? Assume $5 \times 10^{22} \dfrac{\text{atoms}}{\text{cm}^3}$

2-3. The intrinsic carrier concentration ratio at two temperatures is 303. If the lower temperature is at room temperature (300 K), what is the other temperature?

2-4. Plot n_i versus temperature for $T = 250$ K, 300 K, 350 K, and for 400 K. Why would a log plot be preferred over a linear plot? What does the plot tell us?

2-5. A semiconductor chip is operating at $100°$C. What temperature would reduce the intrinsic carrier concentration by a factor of five?

Extrinsic Semiconductors

2-6. If $N_A = 10^{17}$ cm^{-3} at 300 K, what is the electron concentration?

2-7. The silicon to donor atom ratio is 10,000:1. What is the hole concentration at $T = 300$ K? Assume $5 \times 10^{22} \dfrac{\text{atoms}}{\text{cm}^3}$

2-8. If $T = 385$ K and acceptor doping $N_A = 6 \times 10^{18}$, calculate the minority carrier concentration.

2-9. The electron concentration in extrinsic silicon at 300 K is 5×10^4 cm^{-3}. The carrier intrinsic concentration of silicon at 300 K was measured at 1.5×10^{10} cm^{-3}. Determine the hole concentration and indicate if the material is *p*-type or *n*-type.

2-10. The minority concentration is $n_o = 4.5 \times 10^4$ and the doping concentration is $N_A = 10^{18}$. What is the temperature of the semiconductor?

Compensation Doping

2-11. If $N_D = 10^{18}$ cm^{-3} and $N_A = 10^{16}$ cm^3, calculate the minority carrier dopant concentration at $T = 300$ K.

2-12. Repeat Problem 2.11 if the temperature is elevated 50°C.

2-13. In a compensated semiconductor, $p_o = 4 \times 10^6$ cm^{-3} and $N_D - N_A = 4.99 \times 10^{18}$ cm^{-3}. What is the temperature for this condition?

2-14. If $T = 390$ K, $N_D = 5 \times 10^{18}$, and minority carrier concentration is 1.1×10^6 cm^{-3}, what is the minority carrier doping concentration in a compensated semiconductor?

Carrier Transport—Drift Current

2-15. A *p*-silicon material has $p_o = N_A = 10^{18}$ cm^3 at 280 K. $\mu_n = 1500$ (cm^2/V $\cdot$ s), and $\mu_p = 500$ (cm^2/V $\cdot$ s). The semiconductor has 1 V across a 20 μm dimension.
(a) What is the electric field in V/cm?
(b) What is the electron carrier density n_o?
(c) What is the current density J (A/cm^2)?
(d) What is the current density J (A/μm^2)?

2-16. (a) Neglect minority carrier current density in an *n*-doped semiconductor. If $\mu_n = 1200$ cm^2/V $\cdot$ s at $T = 325$ K, $N_D = 10^{18}$ and current density is $J_n = 10$ kA/cm^2, what is the electric field?
(b) If 2 V causes this E-field across the material, what is the material dimension in microns?

2-17. What are the conductivity and resistivity in Problem 2.16?

2-18. The current density is $J = 300$ A/cm^2, conductivity $\sigma = 0.5$ A/V $\cdot$ cm, and $\mu_n = 1350$ cm^2/V $\cdot$ s. What is the donor doping concentration?

Carrier Transport—Diffusion Current

2-19. If an electron concentration gradient is 4×10^{18} electrons/cm^3 and $D_n = 25$ cm^2/s, what is the diffusion current?

2-20. $D_n = 35$ cm^2/s, $D_p = 12$ cm^2/s, $J = 15$ mA/cm^2, the free electron concentration gradient is three times that of the free hole concentration. What are the free carrier concentration gradients?

2-21. At room temperature, $D_n = 35$ cm^2/s and $D_p = 10$ cm^2/s. What are μ_n and μ_p?

2-22. At room temperature $\mu_n = 1300$ (cm^2/V $\cdot$ s) and $\mu_p = 400$ (cm^2/V $\cdot$ s). If electron and hole concentration gradients are 10^{20} cm^{-1} and 10^{17} cm^{-1}, what is total current density?

pn Junction Diodes

2-23. A *pn* junction has $N_A = 10^{15}$ cm^{-3}, $N_D = 10^{16}$ cm^{-3}, and $T = 300$ K. Calculate V_{bi}.

2-24. A *pn* junction has $N_D = 10^{18}$ cm^{-3} and $N_A = 10^{16}$ cm^{-3}.
(a) Calculate V_{bi} at $T = 300$ K.
(b) Calculate V_{bi} at $T = 400$ K.

2-25. Calculate the built-in potential of a *pn* junction if $T = 345$ K, acceptor doping is 10^{18} cm^{-3}, and donor doping is 10^{15} cm^{-3}.

2-26. If $N_D = 10^{17}$, $T = 300$ K, and $V_{bi} = 0.725$ V, what must N_A be set at?

2-27. If $N_D = 10^{17}$, $T = 420$ K, and $V_{bi} = 0.725$ V, what must N_A be set at?

2-28. A diode has $I_S = 10$ pA, $T = 300$ K, and $V_D = 0.625$ V. What is the diode current I_D?

2-29. The diode equation is $I_D = I_S \left(e^{V_D/V_T} - 1 \right)$. The -1 term becomes negligible with respect to the exponential in most forward bias situations and can be neglected. At what value of V_D does the exponential become ten times greater than the one term? Assume room temperature.

2-30. A silicon *pn* junction is operating in the forward bias region. Determine the increase in forward bias voltage that will cause a factor of 100 increase in the diode current. Assume room temperature.

pn Junction Capacitance

2-31. A *pn* junction diode has $C_{j0} = 2$ pF and $V_{bi} = 0.65$ V. Calculate the reverse bias depletion capacitance for reverse bias voltages of 1 V, 2 V, and 3 V.

2-32. Calculate the capacitance of the *pn* junction diode where $C_{j0} = 100$ fF and the built-in potential is 0.72 V.

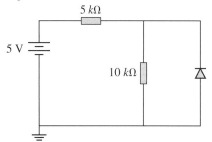

2-33. Given a *pn* junction diode with a reverse bias of 4 V. $C_{j0} = 50$ fF, $T = 350$ K, $N_A = 10^{15}$ cm^{-3}, and $N_D = 10^{16}$ cm^{-3}. Calculate C_j.

2-34. A *pn* junction diode has a reverse bias capacitance $C_j = 80$ fF, $C_{j0} = 150$ fF, and a reverse bias voltage of 2 V. What is V_{bi}?

2-35. A diode has $N_A = 10^{15}$ cm^{-3}, $N_D = 10^{18}$ cm^{-3}, $T = 390$ K, $C_{j0} = 100$ pF, and $V_R = 2$ V. Calculate C_j the depletion capacitance.

2-36. Calculate the junction capacitance of a reverse biased *pn* junction when given $T = 400$ K, $C_{j0} = 50$ fF, $N_A = 10^{14}$ cm^{-3}, $N_D = 10^{18}$ cm^{-3}, and reverse bias voltage is $V_R = 2$ V.

CHAPTER 3

MOSFET Transistors

Last year, more transistors were produced—and at a lower cost—than grains of rice.

SEMI Annual Report 2005

MOSFET transistors are the core of today's integrated circuits (ICs). Original computers used mechanical switches and then vacuum tubes to solve Boolean operations. But the smaller, faster, cooler MOSFET transistors allowed computers to evolve and dominate our lives. Our goal is to acquire the analytical ability and transistor insights that engineers need to design and troubleshoot digital ICs. Abundant examples and self-exercises will develop intuitive responses to transistor circuit operation. Chapter 3 is an important foundation for subsequent chapters.

3.1. Principles of Operation

The two major types of transistors are the metal-oxide semiconductor field-effect transistor (MOSFET), which will be our focus, and the bipolar junction transistor (BJT). Digital integrated circuits almost exclusively use the MOSFET while the BJT has application in analog electronics. Transistors differ from passive resistor, capacitor, inductor, and diode elements in that MOSFET transistor output current and voltage characteristics vary with the voltage on a control terminal. Transistors have three terminals concerned with signal transmission, while passive elements have two terminals.

Figure 3-1a sketches a MOSFET transistor. The bottom rectangular block of material is the silicon substrate often referred to as the bulk. There are four electronically active regions: (1) *gate* (G); (2) *source* (S); (3) *drain* (D); and (4) the *bulk* terminal (B) to which the gate, drain, and source voltages are typically referenced. The rectangular gate region lies on top of the bulk separated by a thin silicon oxide dielectric with thickness T_{OX}. Two other important dimensions are the transistor gate length and width. The drain

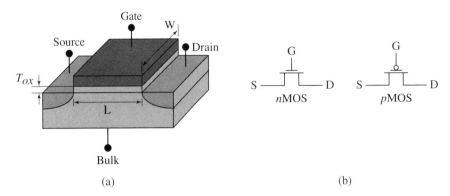

FIGURE 3-1.

(a) MOS structure. (b) Symbols used at the circuit level.

and source regions are embedded in the substrate but have an opposite doping to the substrate.

Two types of MOSFET transistors—the *n*MOS transistor and the *pMOS* transistor—differ in the polarity of carriers responsible for transistor current. The charged carriers are holes in *pMOS* transistors and electrons in *nMOS* transistors. Figure 3-1b shows symbols commonly used for MOSFETs where the bulk terminal is either labeled (B) or more typically implied (not drawn).

3.1.1. The MOSFET as a Digital Switch

A simple description treats the MOSFET transistor as a switch. The gate terminal is analogous to the light switch on the wall. When the gate has a high voltage, the transistor closes like a wall switch, and the drain and source terminals are electrically connected. Just as a light switch requires a certain force to activate, the transistor gate terminal needs a certain voltage level to switch and connect the drain and source terminals. This voltage is called the *transistor threshold voltage* V_t and is a fixed voltage for nMOS and for pMOS devices in a given fabrication process.

An ideal transistor has zero ohms between the drain and source when it is in the On-state and infinite resistance between these terminals in the Off-state. The ideal device should also switch between On- and Off-states with a zero delay time as soon as the control variable changes state. Figure 3-2 shows a switch, a resistor, and a 2 V battery. When the mechanical switch closes, $V_{logic} = 0$ V , which is logic-0. When the switch is open $V_{logic} = 2$ V, which is a logic-1 state. If we replace the mechanical switch with the transistor, then the same logic states appear when the gate terminal turns the transistor On and Off. A *pMOS* transistor acts similarly as a switch. This transistor principle exists in all logic gates we will study.

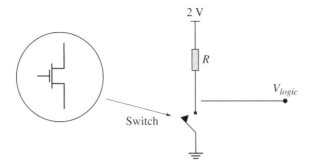

FIGURE 3-2.

A transistor modeled as a switch.

3.1.2. Physical Structure of MOSFETs

MOS transistor construction begins with a lightly doped host crystalline substrate structure. nMOS substrates are doped with p-doped silicon, while the pMOS substrates use n-doped silicon. Figure 3-1 shows the transistor thin oxide (T_{ox}) that electrically isolates the gate terminal from the semiconductor crystalline structures underneath. The gate oxide is made of oxidized silicon forming a noncrystalline, amorphous SiO_2. The gate oxide thickness (T_{ox}) typically ranges from near 15 Å to 100 Å (1 Å = 1 Angstrom = 10^{-10} m). SiO_2 molecules are about 3.2 Å in diameter so that this vital dimension is now a few molecular layers thick. The purpose of the SiO_2 dielectric is to support an electric field generated by the gate voltage to influence the amount of charge passing between the drain and source. A thinner gate oxide allows the electric field gate to better control the device state and allows faster transistors. The thin gate oxide has been referred to as the beating heart of the transistor.

Figure 3-3 shows nMOS and pMOS transistor structures. The nMOS transistor has a p-doped silicon substrate with n^+-doping for the drain and source. pMOS transistors have a complementary structure with an n-doped silicon bulk and p^+-doped drain and source regions. The gate region above the thin oxide dielectric is constructed with polysilicon in both transistors. Polysilicon is a conducting material, and is made of many small silicon crystals. The region between the drain and source just under the gate oxide is called the

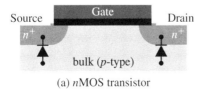
(a) nMOS transistor

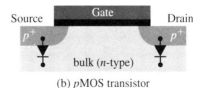
(b) pMOS transistor

FIGURE 3-3.

Relative doping and equivalent electrical connections between device terminals for (a) nMOS and (b) pMOS transistor.

channel and is where charge conduction takes place. The electronic distinction between the two transistors is that electrons are the channel current in the *n*MOS transistor, and holes are the channel current in the *p*MOS transistor. Since drain and source dopants are opposite to the substrate (bulk), they form *pn* junction diodes that are either reverse or zero biased in normal operation (Figure 3-3).

The distance from the drain to the source is a parameter called the *channel length* (L) and the lateral dimension is the transistor *channel width* (W) (Figure 3-1a). Transistor length and width are parameters set by the circuit designer and process engineer. The width to length ratio (W/L) is linearly related to the drain drive current capability of the transistor. A wider transistor will pass more current. The gate is the control terminal, while the source provides electron or hole charge carriers that empty into the channel and are collected by the drain.

Other parameters, such as the transistor oxide thickness, threshold voltage, and doping levels depend on the fabrication process and cannot be changed by design. These are technology parameters set by the process and device engineers.

3.1.3. MOS Transistor Operation: a Descriptive Approach

Transistors must have exact terminal voltage polarities to operate (Figure 3-4). The bulk or substrate of *n*MOS (*p*MOS) transistors must always be connected to the lower (higher) voltage that is the reference terminal. The bulk is usually connected to ground for an *n*MOSFET, and the bulk of a *p*MOSFET is connected to the power supply voltage. We will assume that the bulk and source terminals are shorted to simplify the description. The positive current convention in an *n*MOS (*p*MOS) device is from the drain (source) to the source (drain) and is referred to as I_{DS} or simply I_D since drain and source current

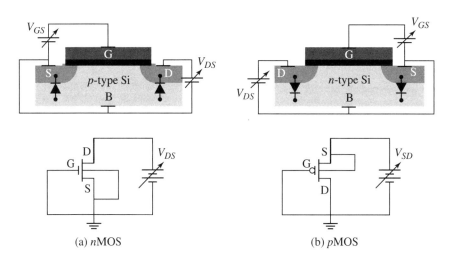

(a) *n*MOS   (b) *p*MOS

FIGURE 3-4.

Normal biasing: (a) *n*MOS. (b) *p*MOS.

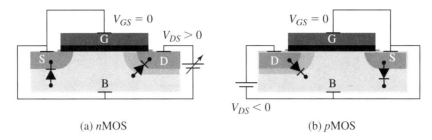

(a) *n*MOS   (b) *p*MOS

FIGURE 3-5.

When the gate-source voltage is zero, the drain-source voltage prevents current from the drain to the source.

are equal. When a voltage is applied to the drain terminal, the drain current depends on the voltage applied to the gate control terminal. V_{GS}, V_{DS}, and I_{DS} are negative values for *p*MOS transistors for reasons explained shortly.

A channel of free carriers requires a minimum gate to source threshold voltage to induce a sheet of mobile charge. If V_{GS} is zero in an *n*MOS device, then there are no appreciable free charges between the drain and source and an applied drain voltage reverse biases the drain-bulk diode (Figure 3-5a). As a result there is no current when $V_{GS} = 0$ for *n*MOS devices (the same for *p*MOS). This is the Off-state, or nonconducting state of the transistor.

We begin the analysis with the source, drain, and substrate all at 0 V. Then when the gate voltage of an *n*MOS (*p*MOS) transistor is slightly increased (decreased) a vertical electric field exists between the gate and the substrate across the oxide. In *n*MOS (*p*MOS) transistors, the holes (electrons) of the *p*-type (*n*-type) substrate close to the silicon-oxide interface initially "feel repelled" in this electrical field, and move away from the interface. This causes a depletion region beneath the oxide interface for this small gate voltage (Figure 3-6). The depletion region contains no mobile carriers, so the application of a drain voltage provides no drain current for this low V_{GS}.

When the gate voltage of the *n*MOS (*p*MOS) device is further increased (decreased), the strong vertical electric field attracts *minority carriers* (electrons in the *n*MOS and holes in the *p*MOS device) from the bulk toward the gate. These minority carriers are attracted to

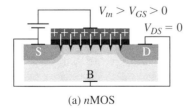
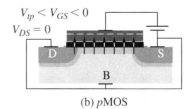

(a) *n*MOS   (b) *p*MOS

FIGURE 3-6.

(a) Deplete the *n*MOS channel of holes with small positive values of gate-source voltage. (b) Deplete the *p*MOS channel of electrons with small negative values of gate-source voltage.

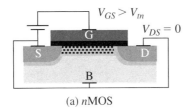
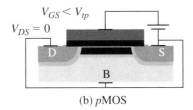

(a) nMOS   (b) pMOS

FIGURE 3-7.

Creating the conducting channel for (a) nMOS and (b) pMOS transistors.

the gate, but the silicon dioxide insulator stops them, and the electrons (holes) accumulate at the silicon to oxide dielectric interface. These carriers form a minority carrier inversion region or conducting channel that can be viewed as a "short-circuit" between the drain and source regions (Figure 3-7). When $V_D = V_S$ and $I_D = 0$, the transistor is on, and the channel carrier distribution is uniform along the device.

The threshold voltage of an nMOS transistor V_{tn} is positive, while V_{tp} is negative for a pMOS transistor. An nMOS (pMOS) transistor has a conducting channel when the gate-source voltage is greater than (less than) the threshold voltage, that is, $V_{GS} > V_{tn}$ ($V_{GS} < V_{tp}$). The gate of the pMOS transistor must be negative with respect to the source and substrate to attract a sheet of holes to the SiO_2 channel surface.

When the channel forms in the nMOS (pMOS) transistor, a positive (negative) drain voltage with respect to the source creates a horizontal electric field moving the electrons (holes) toward the drain forming a positive (negative) drain current coming into the transistor. The positive current convention is used for electron and hole current, but in both cases electrons are the actual charge carriers. If the channel horizontal electric field is of the same order or smaller than the vertical thin oxide field, then the inversion channel remains almost uniform along the device length. This continuous carrier profile from drain to source puts the transistor in a bias state that is equivalently called either the *nonsaturated*, *linear*, or *ohmic bias state*. The drain and source are effectively short-circuited. The ohmic state happens when

$$V_{GS} > V_{DS} + V_{tn} \quad n\text{MOS transistor}$$

$$(3\text{-}1)$$

$$V_{GS} < V_{DS} + V_{tp} \quad p\text{MOS transistor}$$

Drain current is linearly related to drain-source voltage over small intervals in the linear bias state. These equations state that when the gate to drain oxide voltage is itself greater than the threshold, then a continuous carrier profile exists across the whole channel. Accept these equations for now, and later Eq. (3-1) will be derived.

But if the nMOS drain voltage increases beyond the limit of Eq. (3-1) so that $V_{GS} < V_{DS} + V_{tn}$, then the horizontal electric field becomes stronger than the vertical field at the drain end, creating an asymmetry of the channel carrier inversion distribution shown in Figure 3-8. If the drain voltage rises while the gate voltage remains constant, then V_{GD} can

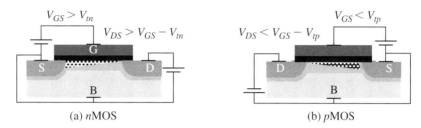

$V_{GS} > V_{tn}$

$V_{DS} > V_{GS} - V_{tn}$

$V_{GS} < V_{tp}$

$V_{DS} < V_{GS} - V_{tp}$

(a) nMOS (b) pMOS

FIGURE 3-8.

Channel pinch-off for (a) nMOS and (b) pMOS transistor devices.

go below the threshold voltage in the local drain region. There can be no carrier inversion at the drain-gate oxide region, so the inverted portion of the channel retracts from the drain, and no longer "touches" this terminal. The pinched-off portion of the channel forms a depletion region with a high electric field. The n-drain and p-bulk form a reverse bias pn junction. When this happens the inversion channel is said to be *pinched off* and the device is in the *saturation region*.

The asymmetric electron distribution in this saturation bias for the nMOS transistor is shown in Figure 3-8a. Channel inversion cannot take place in the drain region when the gate to drain voltage is less than threshold. Figure 3-8b shows the hole inversion channel profile for a pMOS transistor. Although there are no inversion charges at the drain end of the channel, the drain region is still electrically active. Carriers leave the source and move under the force of the horizontal field. Once they arrive at the pinch-off point of the channel, they travel from that point to the drain driven by the high electric field of the depletion region.

The *pinch-off point* is the channel location that varies with changes in bias voltages. The drain-channel region forms a reversed bias pn junction with a high electric field depletion region. Charges ejected from the high electric field depletion region enter the drain and are ejected from the drain terminal from a reverse bias pn junction (drain to substrate). The high impedance constant current ejection forms a current source, which is a property of the saturated bias state.

CMOS transistors use all three bias states described here: Off-state, saturated state, and linear state (ohmic, nonsaturated). We will next look at curves illustrating MOS transistor response to changes in parameters and learn the analytical equations that predict and analyze transistor behavior. It is important to work through all examples and exercises. It is instructive to return to this transistor description after acquiring skill in transistor circuit analysis.

3.2. MOSFET Input Characteristics

MOSFETs have four terminals and need two sets of current-voltage curves; the input characteristic and the output characteristic. The input characteristic relates drain current response to the input gate-source driving voltage. Since the gate terminal is electrically

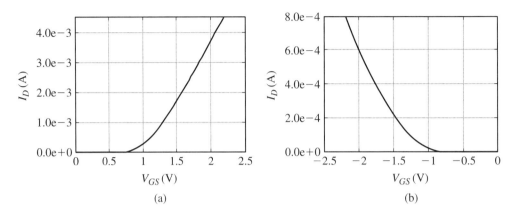

FIGURE 3-9.

Measured input characteristics (I_D versus V_{GS}) for (a) nMOS transistor and (b) pMOS transistor.

isolated from the remaining terminals (drain, source, and bulk), the gate current is essentially zero, so that gate current is not part of device characteristics. The input characteristic curve can locate the gate voltage at which the transistor passes current and leaves the Off-state. This is the device threshold voltage.

Figure 3-9 shows measured input characteristics for nMOS and pMOS transistors. A small 0.1 V potential across their drain to source terminals puts the transistors in their nonsaturated bias states. As V_{GS} increases for the nMOS transistor in Figure 3-9a, the threshold voltage is reached where drain current elevates. For V_{GS} between 0 V and 0.7 V, I_D is nearly zero indicating that the equivalent resistance between the drain and source terminals is extremely high. Once V_{GS} reaches 0.7 V, the current increases rapidly with V_{GS} indicating that the equivalent resistance at the drain decreases with increasing gate-source voltage. Therefore, the threshold voltage of this nMOS transistor is about $V_{tn} \approx 0.7$ V. When a transistor turns on and current moves through a load, then voltage changes occur in a circuit that translate into logic levels.

The pMOS transistor input characteristic in Figure 3-9b is analogous to the nMOS transistor except the I_D and V_{GS} polarities are reversed. V_{DS} is negative ($V_{DS} \approx -0.1$ V). Additionally, the gate is at a voltage lower than the source terminal voltage to attract holes to the channel surface. The threshold voltage of the pMOS device in Figure 3-9b is $V_{tp} \approx -0.8$ V.

3.3. nMOS Transistor Output Characteristics and Circuit Analysis

MOS transistor output characteristics plot I_D versus V_{DS} for several values of V_{GS}. Figure 3-10 shows this family of curves measurement for an nMOS transistor. Two conduction states are distinguished when the device is in the On-state: the saturated state and the

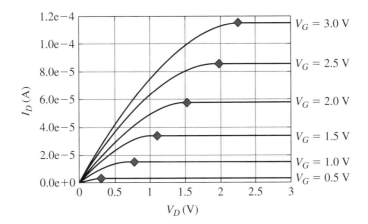

FIGURE 3-10.

nMOS transistor output characteristics as family of curves. The diamond symbol marks the pinch-off voltage V$_{DSAT}$.

nonsaturated state. The saturated curve is the flat portion and defines the *saturation region.*
For $V_{GS} < V_{DS} + V_{tn}$, the nMOS device is a current source, and I_D is independent of V_{DS}.
For $V_{GS} > V_{DS} + V_{tn}$, the transistor is in the *nonsaturation region* and the curve is a half parabola. When the transistor is off ($V_{GS} < V_{tn}$), then I_D is zero for any V_{DS} value.

The boundary of the saturation/nonsaturation bias states is marked as a diamond point for each curve in Figure 3-10. It is the intersection of the straight line of the saturated region with the quadratic curve of the nonsaturated region. This intersection point occurs at the channel pinch-off voltage called V_{DSAT}. The diamond symbol marks the pinch-off voltage V_{DSAT} for each value of V_{GS}. V_{DSAT} is defined as the minimum drain-source voltage that is required to keep the transistor in saturation for a given V_{GS}. In the nonsaturated state, the drain current initially increases almost linearly from the origin before bending in a parabolic response—thus the name ohmic or linear for the nonsaturated region.

The drain current in saturation is virtually independent of V_{DS}. The transistor acts as a current source because there is a depletion region and no carrier inversion at the drain region of the channel. Carriers are pulled into the high electric field of the drain/substrate pn junction and ejected out of the drain terminal. A near constant current is driven from the transistor no matter the drain to source voltage. pMOS transistor I_D versus V_{DS} curves have a shape similar to that in Figure 3-10, but the voltage and current polarities are negative to account for hole inversion and drain current that enters the transistor (pMOS device curves are shown later).

We next develop skills with equations that predict node voltages and currents in a transistor for any point in the family of curves in Figure 3-10. MOS equations can be derived by calculating the amount of charge in the channel at each point and integrating this expression from the drain to the source. This procedure is found in several books [1] and leads to drain current equations in the saturated and linear states:

$$I_D = \frac{\mu_n \varepsilon_{ox}}{2T_{ox}} \frac{W}{L} (V_{GS} - V_{tn})^2 \qquad \text{(Saturated State)} \qquad (3\text{-}2)$$

$$I_D = \frac{\mu_n \varepsilon_{ox}}{2T_{ox}} \frac{W}{L} [2(V_{GS} - V_{tn}) V_{DS} - V_{DS}^2] \quad \text{(Nonsaturated State)} \qquad (3\text{-}3)$$

where μ_n is the electron mobility, ε_{ox} is the thin oxide (SiO_2) dielectric constant, T_{ox} is transistor thin oxide thickness, and W and L are transistor effective gate width and length, respectively. A constant K is introduced to indicate the current drive strength of the transistor as

$$K = \frac{\mu \varepsilon_{ox}}{2T_{ox}} \qquad (3\text{-}4)$$

We often use K_n as a current drive symbol for nMOS transistors. If all constants are known, then Eq. (3-2) can predict I_D for any value of V_{GS} in the saturated region, and Eq. (3-3) can predict I_D in the nonsaturated region if V_{GS} and V_{DS} are specified. Eq. (3-2) is a square law relation between I_D and V_{GS} that is independent of V_{DS}. It is a flat line for a given V_{GS} since there is no V_{DS} term in Eq. (3-2). These two MOS equations allow calculations of transistor voltages and current, and their use is similar to what the linear Ohm's law does for resistors; that is, Eqs. (3-2) and (3-3) relate transistor current to applied voltage.

The saturated and nonsaturated states intersect at V_{DSAT} where either equation describes the current and voltage at that point. We can solve for this important bias condition where the saturated and nonsaturated states intersect (V_{DSAT}). This knowledge is essential for solving problems that follow. Figure 3-11 plots three parabolas of Eq. (3-3) at $V_{GS} = 2.0$ V, 1.6 V, and 1.2 V. Only the left-hand side of the parabolas is used to predict the curves in Figure 3-10, but the parabola also has a right-hand side. The dotted lines on the right-hand side of the curves are part of the continuous solution to the parabola but are electronically invalid, as examples will show.

The midpoint at zero slope defines the useful upper region of Eq. (3-3) and also defines the boundary between the saturated and nonsaturated bias states. We can define the boundary bias condition by differentiating Eq. (3-3) with respect to V_{DS}, setting the expression to zero and then solving for the conditions. Eq. (3-5) shows the derivative of Eq. (3-3) set to zero:

$$\frac{dI_D}{dV_{DS}} = \frac{\mu \varepsilon_{ox}}{2T_{ox}} \frac{W}{L} [2(V_{GS} - V_{tn}) - 2V_{DS}] = 0 \qquad (3\text{-}5)$$

Terms cancel giving the important bias condition at the transition between saturation and nonsaturation states as

$$V_{GS} = V_{DS} + V_{tn} \qquad (3\text{-}6)$$

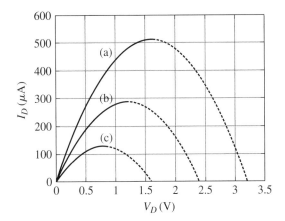

FIGURE 3-11.

Plot of parabola of nonsaturation state equation (Eq. (3-3)). $K_n = 100\,\mu\text{A/V}^2$, $V_{tn} = 0.4$ V, and $W/L = 2$. The solid line is the valid region, but dotted line is not. (a) $V_{GS} = 2.0$ V. (b) $V_{GS} = 1.6$ V. (c) $V_{GS} = 1.2$ V.

This equation holds for each of the intersection points in Figure 3-11 denoted at the peak of each curve. Eq. (3-6) can be extended to define the nMOS saturated bias condition

$$V_{GS} < V_{DS} + V_{tn} \quad \text{or} \quad V_{DS} > V_{GS} - V_{tn} \tag{3-7}$$

and the nMOS nonsaturated condition

$$V_{GS} > V_{DS} + V_{tn} \quad \text{or} \quad V_{DS} < V_{GS} - V_{tn} \tag{3-8}$$

These relations only hold for V_{GS} greater than the threshold voltage. The transistor must be on.

EXAMPLE 3-1

Determine the bias state for the three circuit conditions if $V_{tn} = 0.4$ V. The source voltage is always lower than the drain voltage in an nMOS transistor. First identify the correct terminals.

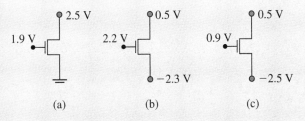

Answers: **(a)** $V_{GS} = 1.9$ V, $V_{DS} = 2.5$ V, $V_{tn} = 0.4$ V, therefore $V_{GS} = 1.9$ V < 2.5 V $+$ 0.4 V $= 2.9$ V. Eq. (3-7) is satisfied, and the transistor is in the saturated state described by Eq. (3-2). **(b)** $V_{GS} = V_G - V_S = 2.2$ V $- (-2.3$ V$) = 4.5$ V. $V_{DS} = V_D - V_S = 0.5$ V $- (-2.3) = 2.8$ V. Therefore, $V_{GS} = 4.5$ V > 2.8 V $+ 0.4$ V $= 3.2$ V. Eq. (3-8) is satisfied, and the transistor is in nonsaturation. **(c)** $V_{GS} = V_G - V_S = 0.9$ V $- (-2.5$ V$) = 3.4$ V. $V_{DS} = V_D - V_S = 0.5$ V $- (-2.5$ V$) = 3$ V. Therefore, $V_{GS} = 3.4$ V $= V_{DS} + V_{tn} = 3$ V $+ 0.4$ V $= 3.4$ V, and the transistor is at the boundary of the saturated and nonsaturated regions. Either Eq. (3-2) or (3.3) can be used to calculate I_D.

Self-Exercise 3-1
Determine the bias state for the three circuit conditions if $V_{tn} = 0.4$ V.

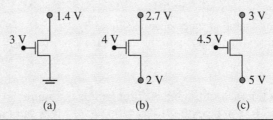

(a)   (b)   (c)

After determining the proper bias state equations in Example 3.1 and Self-Exercise 3.1, you may check your bias answer by referring to the *n*MOS transistor family of curves in Figure 3-10. Find the coordinates in the example and self-exercise, and verify that the bias state is correct. A series of examples and exercises with the *n*MOS transistor will reinforce these important relations and will show how to analyze circuits with transistors and resistors.

EXAMPLE 3-2

Calculate I_D and V_{DS} if $K_n = 100$ μA/V^2, $V_{tn} = 0.6$ V, and $W/L = 3$ for transistor M1.

The bias state of M1 is not known so we must initially assume one of the two states and then solve for bias voltages and check for consistency against that transistor bias assumption. Initially, assume that the transistor is in the saturated state so that

$$I_D = \frac{\mu \varepsilon_{ox}}{2T_{ox}} \frac{W}{L} (V_{GS} - V_{tn})^2 = K_n \frac{W}{L} (V_{GS} - V_{tn})^2$$

$$= (100 \ \mu A) \ (3) \ (1.2 - 0.6)^2$$

$$= 108 \ \mu A$$

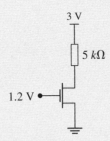

Using Kirchhoff's voltage law (KVL)

$$V_{DS} = V_{DD} - I_D R$$

$$= 3 - (108\ \mu A)(5\ k\Omega)$$

$$= 2.46\ V$$

We assumed that the transistor was in saturation, so we must check the result to see if that is true. For saturation

$$V_{GS} < V_{DS} + V_{tn}$$

$$1.2\ V < 2.46\ V + 0.6\ V$$

so the transistor is in saturation, and our assumption and answers are correct.

EXAMPLE 3-3

Repeat Example 3.2, finding I_D and V_{DS} if $V_G = 2$ V.

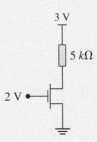

Assume a transistor saturated state and

$$I_D = (100\ \mu A)\ (3)\ (2 - 0.6)^2 = 588\ \mu A$$

$$V_{DS} = 3 - (588\ \mu A)\ (5\ k\Omega) = 60\ mV$$

The bias check gives

$$V_{GS} > V_{DS} + V_{tn}$$

$$2.0 \text{ V} > 0.06 \text{ V} + 0.6 \text{ V}$$

The initial saturated state assumption was wrong, so we repeat the analysis using the non-saturated state assumption

$$I_D = \frac{\mu \varepsilon_{ox}}{2T_{ox}} \frac{W}{L} \left[2(V_{GS} - V_{tn})V_{DS} - V_{DS}^2 \right]$$

$$I_D = K_n \frac{W}{L} \left[2(V_{GS} - V_{tn})V_{DS} - V_{DS}^2 \right]$$

$$= (100 \, \mu\text{A})(3) \left[2(2 - 0.6)V_{DS} - V_{DS}^2 \right]$$

$$= 300 \, \mu\text{A} \left[2.8 \, V_{DS} - V_{DS}^2 \right]$$

This equation has two unknowns, so another equation must be found. We will use the KVL

$$V_{DD} = I_D R + V_{DS}$$

$$I_D = \frac{(V_{DD} - V_{DS})}{R}$$

$$I_D = \frac{(3 - V_{DS})}{5 \, k\Omega}$$

The two equations can be equated to their I_D solution giving

$$\frac{(3 - V_{DS})}{5 \, k\Omega} = 300 \, \mu\text{A} \left[2.8 \, V_{DS} - V_{DS}^2 \right]$$

The solutions are $V_{DS} = 0.731$ V and 2.736 V.

The valid solution is $V_{DS} = 0.731$ V, since this alone satisfies the nonsaturation condition that was used in its solution.

$$V_{GS} > V_{DS} + V_{tn}$$

$$2 \text{ V} > 0.731 \text{ V} + 0.6 \text{ V}$$

and

$$I_D = \frac{(V_{DD} - V_{DS})}{5 \, k\Omega}$$

$$= \frac{(3 \text{ V} - 0.731 \text{ V})}{5 \, k\Omega} = 453.8 \, \mu\text{A}$$

EXAMPLE 3-4

What value of R_D will drive transistor M1 just at nonsaturation if $K_n = 50 \ \mu A/V^2$, $V_{tn} = 0.4$ V, and $W/L = 10$?

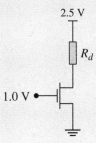

Since the bias state is at the boundary, either Eq. (3-2) or (3-3) can be used. Eq. (3-2) is simpler so

$$I_D = \frac{\mu \varepsilon_{ox}}{2T_{ox}} \frac{W}{L} (V_{GS} - V_{tn})^2 = K_n \frac{W}{L} (V_{GS} - V_{tn})^2$$

$$= (50 \ \mu A)(10)(1.0 - 0.4)^2$$

$$= 180 \ \mu A$$

The bias boundary condition

$$V_{GS} = V_{DS} + V_{tn}$$

becomes

$$V_G - V_S = V_D - V_S + V_{tn}$$

$$V_D = V_G - V_{tn}$$

$$V_D = 1.0 \text{ V} - 0.4 \text{ V} = 0.6 \text{ V}$$

Then

$$R_D = \frac{V_{DD} - V_D}{I_D}$$

$$= \frac{2.5 - 0.6}{180 \ \mu A}$$

$$= 10.56 \ k\Omega$$

Self-Exercise 3-2

Find I_D and V_D, and verify the bias state of your choice of MOS drain current model for $V_{tn} = 0.5$ V, $K_n = 120 \, \mu\text{A/V}^2$, and $W/L = 2$.

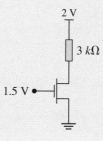

Answer: $I_D = 240 \, \mu\text{A}$, $V_D = 1.28$ V

Self-Exercise 3-3

Find I_D and V_D and verify the bias state consistency of your choice of MOS drain current model for $V_{tn} = 0.5$ V, $K_n = 25 \, \mu\text{A/V}^2$, and $W/L = 3$.

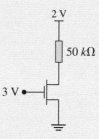

Answer: $I_D = 37.9 \, \mu\text{A}$, $V_D = 0.103$ V

Self-Exercise 3-4

Calculate V_{GS}, and give the correct bias state for transistor M1 where $V_{tn} = 0.5$ V.

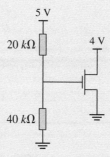

Answer: $V_{GS} = 3.333$ V, Saturated bias state

Self-Exercise 3-5

Find I_D and V_D, and verify the bias state consistency of your choice of MOS drain current model for $V_{tn} = 0.5$ V, $K_n = 50\,\mu\text{A/V}^2$, and $W/L = 2$.

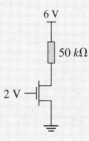

Answer: $I_D = 111.3\,\mu\text{A}$, $V_D = 0.434$ V

3.4. _p_MOS Transistor Output Characteristics and Circuit Analysis

_p_MOS transistor analysis is similar to the _n_MOS transistor with a major exception: care must be taken with the polarities of the drain current and node voltages. The _p_MOS transistor major carrier is the hole that emanates from the source into the channel and exits the drain terminal as a negative network convention current. The gate to source threshold voltage V_{tp} needed to invert an _n_-substrate is negative with respect to the source to attract holes to the channel surface. The equations to model the _p_MOS transistor in saturation and nonsaturation have a form similar to the _n_MOS device but have polarity considerations. We will choose a _p_MOS transistor equation form that is close to the _n_MOS transistor equations. Eqs. (3-9) and (3-10) describe the current _magnitude_ of a _p_MOS device.

$$I_D = \frac{\mu\varepsilon_{ox}}{2T_{ox}}\frac{W}{L}\left(V_{GS} - V_{tp}\right)^2 \qquad \text{(Saturated State)} \qquad (3\text{-}9)$$

$$I_D = \frac{\mu\varepsilon_{ox}}{2T_{ox}}\frac{W}{L}\left[2\left(V_{GS} - V_{tp}\right)V_{DS} - V_{DS}^2\right] \quad \text{(Nonsaturated State)} \qquad (3\text{-}10)$$

The form of the equations is identical to the _n_MOS transistor, but V_{GS}, V_{DS}, and V_{tp} have negative values for the _p_MOS transistor. Figure 3-12 shows a measured _p_MOS transistor family of curves with all voltages given with respect to the source. The plot is shown in quadrant-1 even though the drain current and voltage are negative. This choice is made here to retain similarity to the _n_MOS transistor family of curves and to minimize confusion with signal polarity. In order to ease the sign confusion in _p_MOS parameters, we suggest that the reader use absolute current values and treat the _p_MOS as an _n_MOS.

The boundary of the bias states is again found by differentiating Eq. (3-10), setting the result to zero and solving for the conditions to get

$$V_{GS} = V_{DS} + V_{tp} \qquad (3\text{-}11)$$

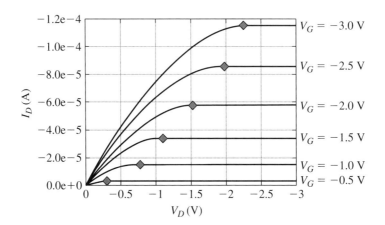

FIGURE 3-12.

*p*MOS transistor family of curves. The diamond symbol marks the pinch-off voltage V_{DSAT}.

The condition for *p*MOS transistor saturation is

$$V_{GS} > V_{DS} + V_{tp} \quad \text{or} \quad V_{DS} < V_{GS} - V_{tp} \tag{3-12}$$

and the condition for *p*MOS transistor nonsaturation is

$$V_{GS} < V_{DS} + V_{tp} \quad \text{or} \quad V_{DS} > V_{GS} - V_{tp} \tag{3-13}$$

An example will illustrate the polarity issues.

EXAMPLE 3-5

Determine the bias state for the *p*MOS transistors where $V_{tp} = -0.4$ V. It is helpful with *p*MOS transistors to first identify and label the source and drain terminals. The source terminal in a *p*MOS transistor has a higher voltage than the drain terminal.

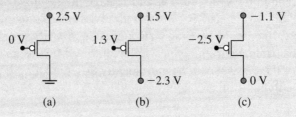

(a) $V_{GS} = -2.5$ V, $V_{DS} = -2.5$ V, therefore $V_{GS} > V_{DS} + V_{tp}$, or -2.5 V $> -2.5 + (-0.4)$ V, so the transistor is in saturation. **(b)** The gate voltage is not sufficiently more negative than either drain or source terminal to invert holes at the oxide interface, so the transistor is in the Off-state. **(c)** $V_{GS} = -2.5 - (-1.1) = -1.4$ V, $V_{DS} = 0 - (-1.1) = 1.1$ V. What is wrong? The gate voltage is sufficiently negative to turn on the transistor, but the source to drain voltage is negative. Holes must leave the source and flow to the drain, but they can't under this conclusion. The answer is that the drain terminal with a lower voltage is on the top and the source on the bottom so that $V_{GS} = -2.5 - 0 = -2.5$ V, $V_{DS} = -1.1 - 0 = -1.1$ V. Therefore, $V_{GS} > V_{DS} + V_{tp}$, or -2.5 V $< -1.1 + (-0.4)$ V, so the transistor is in nonsaturation. The *p*MOS source terminal always has a higher voltage than the drain terminal.

Self-Exercise 3-6
Give the correct bias state for the three *p*MOS transistors where $V_{tp} = -0.4$ V.

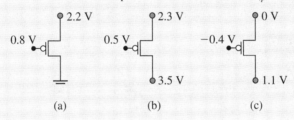

(a)   (b)   (c)

After determining the proper bias state equations in Example 3.5 and Self-Exercise 3.6, you may check your work by referring to the *p*MOS transistor family of curves in Figure 3-12. Find the coordinates in the example and exercise, and verify that the bias state is correct. A series of examples and exercises with the *p*MOS transistor will reinforce these important relations. We will use the current magnitude whose direction is from V_{DD} to GND. The *p*MOS source is identified as the node with a higher voltage than the drain.

EXAMPLE 3-6

Calculate I_D and V_{DS} for $V_{tp} = -1.0$ V, $K_p = 100 \ \mu A/V^2$, and $W/L = 4$. The source node is tied to V_{DD}.

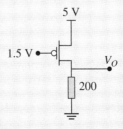

Assume a saturated bias state and a current convention from V_{DD} to GND.

$$I_D = \frac{\mu \varepsilon_{ox}}{2T_{ox}} \frac{W}{L} (V_{GS} - V_{tp})^2 = 100 \, \mu A(4)[1.5 - 5 - (-1)]^2 = 2.5 \text{ mA}$$

$$V_0 = I_D(200 \, \Omega) = (2.5 \text{ mA})(200 \, \Omega) = 0.5 \text{ V}$$

then

$$V_{DS} = 0.5 \text{ V} - 5 \text{ V} = -4.5 \text{ V}$$

The bias state check is

$$V_{GS} > V_{DS} + V_{tp}$$

$$-3.5 > -4.5 + (-1.0)$$

so the transistor is in saturated bias state and the solutions are correct.

EXAMPLE 3-7

Calculate I_D and V_{DS} for $V_{tp} = -0.6$ V, $K_p = 80 \, \mu A/V^2$, and $W/L = 10$.

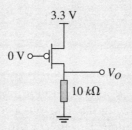

Assume a saturated bias state

$$I_D = \frac{\mu \varepsilon_{ox}}{2T_{ox}} \frac{W}{L} (V_{GS} - V_{tp})^2 = 80 \, \mu A(10)[0 - 3.3 - (-0.6)]^2$$

$$= 5.832 \text{ mA}$$

then

$$V_0 = I_D(10 \, k\Omega) = (5.832 \text{ mA})(10 \, k\Omega)$$

$$= 58.32 \text{ V}$$

This voltage is beyond the power supply value and is not possible. The saturated state assumption was wrong, so we must start again using the nonsaturated state equation

$$I_D = \frac{\mu \varepsilon_{ox}}{2T_{ox}} \frac{W}{L} \left[2(V_{GS} - V_{tp})V_{DS} - V_{DS}^2 \right] = 80 \, \mu A(10) \left[2(-3.3 + 0.6)V_{DS} - V_{DS}^2 \right]$$

Another equation is required so using the KVL (Ohm's law here)

$$I_D = \frac{V_D}{R_D} = \frac{V_{DD} + V_{DS}}{10\,k\Omega}$$

$$= \frac{3.3 + V_{DS}}{10\,k\Omega} = 80\,\mu A(10)\left[2(-3.3 + 0.6)V_{DS} - V_{DS}^2\right]$$

The two solutions are $V_{DS} = -75.70$ mV and -5.450 V. The valid solution is $V_{DS} = -75.70$ mV. Therefore

$$V_O = V_{DD} - V_{SD} = 3.3\text{ V} - 75.70\text{ mV} = 3.224\text{ V}$$

$$I_D = \frac{3.224\text{ V}}{10\,k\Omega} = 322.4\,\mu A$$

EXAMPLE 3-8

If $V_{tp} = -0.4$ V, $K_p = 65\,\mu A/V^2$, and $W/L = 6$, calculate V_G and I_D so that the transistor is on the sat-ohmic bias boundary.

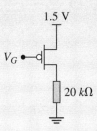

Use the saturated equation since it is simpler.

$$I_D = 65\,\mu A(6)\left[V_{GS} + 0.4\right]^2$$

At $V_{GS} = V_{DS} + V_{tp}$

$$I_D = 65\,\mu A(6)\left[V_{DS} + V_{tp} + 0.4\right]^2 = \frac{1.5 + V_{DS}}{20\,k\Omega}$$

Solve for $V_{DS} = -0.3791$ V and 0.5073 V

Since $V_{DS} = V_D + V_S = V_D - 1.5 = -0.3791$

and $V_D = 1.1209$ V

$$V_G = V_D + V_{tp} = 1.1209 - 1.5 = 0.7209\text{ V}$$

$$I_D = \frac{1.5 - 0.379}{20\,k\Omega} = 56.05\,\mu A$$

Self-Exercise 3-7

What value of R_D will raise V_O to half of the power supply voltage (i.e., $V_O = 0.5\ V_{DD}$), $V_{tp} = -0.7$ V, $K_p = 80\ \mu A/V^2$, and $W/L = 5$.

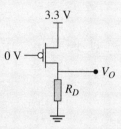

Answer: $R_D = 704.5\ \Omega$

Self-Exercise 3-8

Find I_D and V_O for $V_{tp} = -0.6$ V, $K_p = 20\ \mu A/V^2$, and $W/L = 3$.

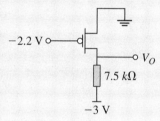

Answer: $I_D = 153.6\ \mu A$, $V_O = -1.848$ V

Self-Exercise 3-9

If $V_{tp} = -0.6$ V, $K_p = 75\ \mu A/V^2$, and $W/L = 2$:

(a) What value of R will place the transistor on the boundary between saturation and nonsaturated?

(b) If R doubles its value, what are I_D and V_O?

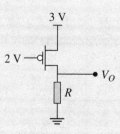

Answers: **(a)** $R = 108.3\ k\Omega$. **(b)** $V_O = 2.868$ V, $I_D = 13.2\ \mu A$

3.5. MOSFET with Source and Drain Resistors

We close the MOSFET analysis with circuits having source and drain resistors and either single or dual power supplies. This circuit is often called a phase splitter since the waveforms at the source and drain nodes are $180°$ out of phase. If the resistors are equal then the voltages changes are equal and opposite at the source and drain nodes as the gate voltage changes. The source voltage moves in phase with the gate signal, and the drain voltage moves $180°$ out of phase. A circuit of this type appears in memory sense amplifiers.

EXAMPLE 3-9

Calculate I_D and V_{DS}, and verify the assumed transistor bias state for $V_{tp} = -0.4$ V, $K_p = 60$ μA/V^2, and $W/L = 2$.

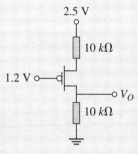

Assume a saturated bias state and

$$I_D = \frac{\mu \varepsilon_{ox}}{2 T_{ox}} \frac{W}{L} (V_{GS} - V_{tp})^2$$

Since V_{GS} is not known, we must search for another expression to supplement this equation.

$$I_D = 60 \, \mu\text{A}(2) \, [1.2 - V_S + 0.4]^2 = \frac{2.5 - V_S}{10 \, k\Omega}$$

$$V_S = 2.144 \text{ V}$$

$$I_D = \frac{2.5 - 2.144}{10 \, k\Omega} = 35.6 \, \mu\text{A}$$

V_{DS} is then

$$V_{DS} = I_D(10 \, k\Omega) - V_{DD}$$

$$= 35.56 \, \mu\text{A})(10 \, k\Omega) - 2.5 \text{ V}$$

$$= -1.789 \text{ V}$$

Verify the bias state

$$V_{GS} = V_G - V_S = 1.2 - 2.144 = -0.944 \text{ V}$$

Transistor is in saturation since

$$V_{GS} > V_{DS} + V_{tp}$$

$$-0.944 \text{ V} > -1.789 \text{ V} - 0.4 \text{ V}$$

Self-Exercise 3-10

The voltage drop across each of the two identical resistors and V_{DS} are equal. If $V_{tp} = -0.5$ V, $K_p = 100 \ \mu\text{A/V}^2$, and $W/L = 2$, find the value of the resistors. Note the negative -2 V power supply.

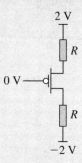

Answer: $R = 239.0 \ k\Omega$

These many examples and exercises with MOS transistors have a purpose. These problems when combined with a transistor family of curves plot should now allow you to think in terms of transistor reaction to its voltage environment. This is basic to transistor instruction. The techniques to solve these transistor–resistor problems should become reflexive.

3.6. Threshold Voltage in MOS Transistors

The bulk terminals are generally connected to either GND (*n*MOS) or V_{DD} (*p*MOS). However, when *n*MOS (*p*MOS) transistors are stacked in series, the source terminal of the upper (lower) transistors is not connected to the bulk. A major effect is on the threshold voltage, and it is called the *body effect*.

Figure 3-13 shows a cross section of two *n*MOS and one *p*MOS transistors connected in series. All devices are constructed on the same *p*-type silicon substrate. Since *p*MOS transistors are formed on an *n*-type substrate, there must be a region of the circuit that is oppositely doped to the initial *p*-doped substrate. This region is called a *well*. The *p*-type substrate for *n*MOS transistors is connected to zero volts (or ground, GND) while the *n*-type well is connected to V_{DD}, since the *n*-well forms the bulk of the *p*MOS transistors.

The source of the *n*MOS device N$_1$ is connected to ground, so that V_{tn} for the previous equations is valid for this device. Transistor N$_2$ source is connected to the drain of N$_1$ to make a series connection. The source of transistor N$_2$ is not grounded, and it can acquire

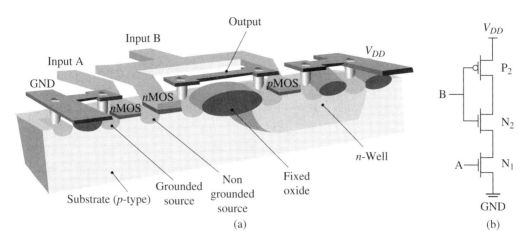

FIGURE 3-13.

(a) Structure for two series connected nMOS transistors and one pMOS transistor. (b) Circuit schematic.

voltages above ground while its substrate is connected to ground. Therefore, the condition $V_{SB} = 0$ will not hold in some bias cases for transistor N_2.

When the source and substrate voltages differ, the gate-source voltage is not fully related to the vertical electric field responsible for creating the channel. The effect of the higher source voltage above the substrate for an nMOS transistor is to lower the electric field induced from the gate to attract carriers to the channel. The result is an effective elevation of the transistor threshold voltage. The threshold voltage can be estimated as

$$V_t = V_{to} + \gamma \left(\sqrt{|2\,\phi_F| + V_{SB}} - \sqrt{|2\,\phi_F|} \right) \tag{3-14}$$

where

$$\gamma = \frac{\sqrt{2q \times N_A \times \varepsilon_{ox}}}{C_{ox}} \tag{3-15}$$

$$C_{ox} = \frac{\varepsilon_{ox}}{T_{ox}} = \frac{\varepsilon_r \varepsilon_0}{T_{ox}} = \frac{3.9(8.85 \times 10^{-14})}{T_{ox}} = \frac{3.45 \times 10^{-13}}{T_{ox}} \left(\frac{F}{cm^2} \right) \tag{3-16}$$

$$\Phi_F = -\left(\frac{kT}{q} \right) \ln \left[\frac{N_A}{n_i} \right] \tag{3-17}$$

V_{to} is the threshold voltage when the source and the substrate are at the same voltage, γ is the body-effect coefficient parameter dependent on the technology, ϕ_F is the Fermi potential, and V_{SB} is the source-bulk voltage. The positive sign in Eq. (3-14) is used for nMOS transistors and the negative sign for pMOS transistors. N_A is the p-substrate doping concentration. The Fermi potential is a semiconductor variable relating electron energy state distribution and temperature. A nominal calculation at room temperature (300 K) for

$N_A = 10^{16}$ cm^{-3} is $\phi_F \approx -0.35$ V. ϕ_F is negative in nMOS transistors and positive in pMOS transistors. V_{SB} is positive in nMOS transistors and negative in pMOS transistors. We will examine only the nMOS transistor body effect.

When the source and substrate are tied together, $V_{SB} = 0$ and the threshold voltage is constant. The significance of the threshold body effect lies with certain circuit configurations whose transistor thresholds will be altered leading to changes in transistor delay time.

EXAMPLE 3-10

(a) Calculate γ if substrate concentration is 3×10^{17} cm^{-3} and C_{ox} is 3.5×10^{-7} F/cm^2.
(b) Calculate V_t for this γ if V_{t0} is 0.55 V, the Fermi voltage is -0.45 V, and $V_{SB} = 0.5$ V.

$$\text{(a)} \quad \gamma = \frac{\sqrt{2 \times (1.6 \times 10^{-19})(3 \times 10^{17})(3.45 \times 10^{-13})}}{(3.5 \times 10^{-7})} = 0.520 \text{ V}^{1/2}$$

$$\text{(b)} \quad V_t = 0.55 + 0.520\left(\sqrt{|2 \times (-0.45)| + 0.5} - \sqrt{|2 \times (-0.45)|}\right) = 0.627 \text{ V}$$

The threshold voltage elevated 122 mV due to the body effect.

EXAMPLE 3-11

If $N_A = 10^{17}$cm^{-3}, $T = 360$ K, $C_{ox} = 3.5 \times 10^{-7}$, $\varepsilon_{ox} = 3.45 \times 10^{-13}$, $V_{t0} = 0.45$ V, and $V_t = 0.62$ V, calculate the reverse substrate bias V_{SB}.

$$V_t = V_{to} + \gamma\left(\sqrt{|2\phi_F| + V_{SB}} - \sqrt{|2\phi_F|}\right)$$

$$\gamma = \frac{\sqrt{2 \times 1.6 \times 10^{-19} \times 10^{17} \times 3.45 \times 10^{-13}}}{3.5 \times 10^{-7}} = 0.3002 \text{ V}^{1/2}$$

$$\phi_F = -86.17 \times 10^{-6}(360)\ln\left(\frac{10^{17}}{n_i}\right)$$

$$n_i = 5.23 \times 10^{15}(360)^{3/2}\exp\left[\frac{[-1.12]}{2 \times 86.17 \times 10^{-6} \times 360}\right] = 5.164 \times 10^{11} \text{ cm}^{-3}$$

Then $\phi_F = -0.3776$ V.

$$0.62 = 0.45 + 0.3002\left[\sqrt{(|2(-0.3776)| + V_{SB})} - \sqrt{|2(-0.3776)|}\right]$$

Solve for $V_{SB} = 1.305$ V.

Self-Exercise 3-11

If $V_{t0} = 0.5$ V, $V_t = 0.59$ V, $\Phi_F = -0.35$ V, and $V_{SB} = 0.8$ V, what is γ?

Answer: $\gamma = 0.232$ V$^{1/2}$

Reference

[1] D. Neamen, *Semiconductor Physics and Devices: Basic Principles*, 4th Edition, McGraw-Hill 2011.

3.7. Summary

The transistor–resistor circuit analysis should become reflexive through the numerous problems in the chapter. Engineers intuitively think in these concepts when designing, debugging, and testing CMOS ICs. The body effect is an important alteration and must be understood. Chapter 5 will combine the *n*MOS and *p*MOS transistors to form the most fundamental logic gate called the inverter.

Exercises

nMOSFET Biasing and Current-Voltage Analysis

3-1. For the three circuits, **(a)** Give the transistor bias state, **(b)** Write the appropriate model equation, **(c)** Calculate I_D, where $W/L = 2$, $V_{tn} = 0.4$ V, and $K_n = 200$ μA/V^2.

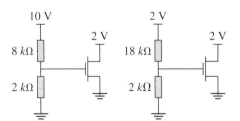

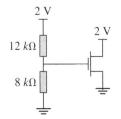

3-2. Given $K_n = 200$ μA/V^2, $W/L = 4$, $V_{tn} = 0.5$ V, and $V_D = 0.8$ V. Find
(a) Drain current I_D
(b) Find the value of R_D to satisfy these constraints

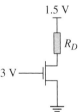

3-3. The transistor parameters in the following circuit are $K_n = 75$ μA/V^2 and $W/L = 4$. If $V_O = 1.2$ V, what is V_{IN}?

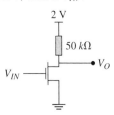

3-4. If $V_{tn} = 0.8$ V, $K_n = 100 \, \mu$A/V^2, and $W/L = 4$, calculate V_G so that $I_D = 200 \, \mu$A.

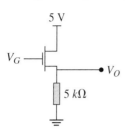

3-5. Given that $W/L = 3$, $V_{tn} = 0.6$ V, and $K_n = 75 \, \mu$A/V^2, calculate V_O and I_D.

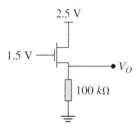

3-6. Given $V_{tn} = 0.8$ V, $K_n = 200 \, \mu$A/V^2, and $W/L = 4$, calculate V_O.

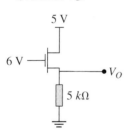

3-7. Given that $W/L = 2$, $V_{tn} = 0.4$ V, and $K_n = 80 \, \mu$A/V^2, what value of V_G sets $I_D = 50 \, \mu$A?

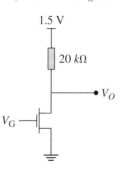

3-8. Given that $W/L = 5$, $V_{tn} = 0.25$ V, and $K_n = 110 \, \mu$A/V^2, find V_O and I_D.

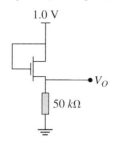

3-9. Given that $W/L = 20$, $V_{tn} = 0.5$ V, and $K_n = 120 \, \mu$A/V^2, find V_O and I_D.

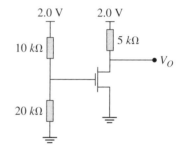

3-10. Given that $K_n = 250 \, \mu$A, $V_{tn} = 0.5$ V, and $W/L = 3$, find R_1 so that the transistor is on the saturated/nonsaturated bias boundary.

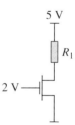

3-11. Given that $K_n = 250 \, \mu$A, $V_{tn} = 0.5$ V, and $W/L = 3$, what V_G makes transistor biased at the saturated/nonsaturated boundary?

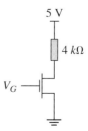

3-12. Calculate R_O so that $V_O = 2.5$ V, given $K_n = 300\ \mu A/V^2$, $V_{tn} = 0.7$ V, and $W/L = 2$.

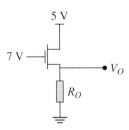

3-13. Adjust R_1 so that M1 is on the saturated/nonsaturated border where $V_{tn} = 0.5$ V.

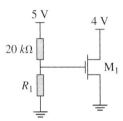

3-14. Transistors emit light from the drain depletion region when they are in the saturated bias state:
  (a) Show whether this useful failure analysis technique will work for the circuit, given that $V_{tn} = 0.6$ V, $K_n = 75\ \mu A/V^2$, and $W/L = 2$. R_{def} is a resistive defect.
  (b) Find I_D, V_{GS}, and V_{DS}.

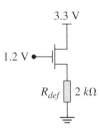

pMOSFET Biasing and Current-Voltage Analysis

3-15. For the three circuits: **(a)** Give the transistor bias state, **(b)** Write the appropriate model equation, **(c)** Calculate I_D, where $V_{tp} = -0.4$ V, $W/L = 4$, and $K_p = 100\ \mu A/V^2$. The *p*MOS source is higher voltage than drain.

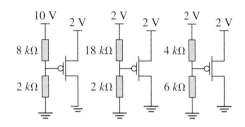

3-16. Calculate I_D and V_O for circuit where $V_{tp} = -0.8$ V, $K_p = 30\ \mu A/V^2$, and $W/L = 2$.

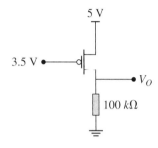

3-17. Repeat Problem 3.16 for $V_G = 1.5$ V.

3-18. Given that $W/L = 5$, $V_{tp} = -0.4$ V, and $K_p = 50\ \mu A/V^2$, calculate V_O and I_D.

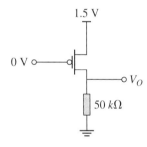

3-19. What value of R_D will place $V_D = 1.5$ V, given $K_p = 25\mu A/V^2$, $V_{tp} = -0.8$ V, and $W/L = 3$.

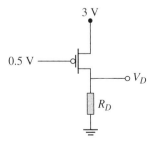

3-20. Given that $W/L = 20$, $V_{tp} = -0.6$ V, and $K_p = 30$ μA/V^2, calculate V_O and I_D.

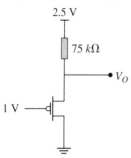

3-21. Given that $W/L = 6$, $V_{tp} = -0.3$ V, and $K_p = 40$ μA/V^2, calculate V_O and I_D.

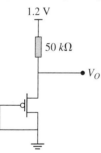

3-22. Given that $V_{tp} = -0.4$ V, $W/L = 4$, and $K_p = 100$ μA/V^2, **(a)** Give the transistor bias state, **(b)** Calculate I_D.

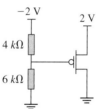

3-23. Given $V_{tp} = -0.8$ V and $K_p = 75$ μA/V^2, what is the required W/L ratio and what is R_D if M1 is to pass 0.25 A and keep $V_{SD} < 0.2$ V.

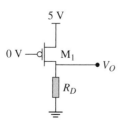

3-24. Given
$$K_p = 40 \ \mu\text{A/V}^2$$
$$V_{tp} = -0.4 \text{ V}$$
$$W/L = 5$$

(a) The capacitor is initially uncharged at $t = 0$. At $t = 0^+$ the gate voltage has changed state from 1.8 V to 0 V. What is the initial surge of current at $t = 0^+$.
(b) At $t = \infty$ what is the bias state on the transistor?
(c) How much energy is dissipated in the charge movement, and where does the heat loss occur?

3-25. Given $V_{tp} = -0.4$, $K_p = 50$ μA/V^2, and $W/L = 8$, the transistor is biased at the saturated/nonsaturated boundary. The power in the resistor is 10 μW. What is the value of the resistor R?

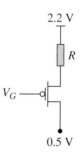

Two Resistor MOSFET Circuits

3-26. Given $V_{tp} = -0.6$ V and $K_p = 75$ μA/V^2, and $W/L = 5$, **(a)** Solve for source voltage V_S, **(b)** Solve for drain voltage.

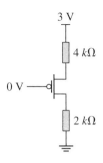

3-27. Given that $W/L = 4$, $V_{tn} = 0.4$ V, and $K_n = 95 \, \mu\text{A/V}^2$, calculate V_{O1}, V_{O2}, and I_D.

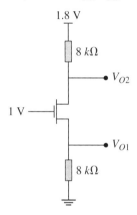

3-28. Given that $W/L = 8$, $V_{tp} = -0.5$ V, and $K_p = 43 \, \mu\text{A/V}^2$, calculate V_O and I_D.

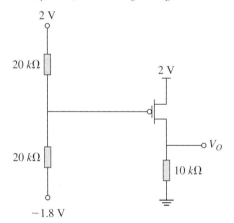

3-29. Given that $W/L = 4$, $V_{tn} = 0.4$ V, and $K_n = 95 \, \mu\text{A/V}^2$, calculate V_{O1}, V_{O2}, and I_D for
   (a) $V_G = +0.5$ V
   (b) $V_G = -0.3$ V

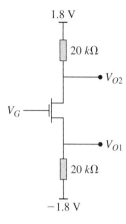

3-30. Given that $R_1 = R_2$, $W/L = 3$, $V_{tn} = 0.6$ V, and $K_n = 200 \, \mu\text{A/V}^2$, determine the resistance values so that $V_D = 1$ V and $V_S = -1$ V.

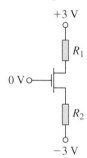

3-31. $V_{tp} = -0.5$ V, $W/L = 5$, and $K_p = 50 \, \mu\text{A/V}^2$, design a value for V_G such that $V_O = 0.3$ V.

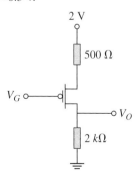

3-32. Given that $V_{tp} = -0.6$ V, $K_p = 75 \, \mu\text{A/V}^2$, and $W/L = 4$,
   (a) What gate voltage will put the transistor at the saturated/nonsaturated bias state boundary? Calculate V_O.

(b) Calculate the drain current.

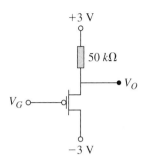

+3 V

50 $k\Omega$

V_O

V_G

−3 V

3-33. Given $V_{tp} = -0.6$ V, $K_p = 50\,\mu\text{A/V}^2$, $W/L = 3$, and $V_D = 0.8$ V, if $V_O = 1.2$ V, what are R and V_G?

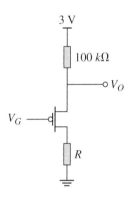

3 V

100 $k\Omega$

V_O

V_G

R

3-34. Given $V_{tp} = -0.6$ V, $K_p = 50\,\mu\text{A/V}^2$, $W/L = 3$, and $V_D = 0.8$ V, what is R?

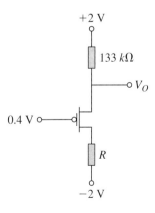

+2 V

133 $k\Omega$

V_O

0.4 V

R

−2 V

3-35. If $K_n = 90\,\mu\text{A/V}^2$, $V_{tn} = 0.5$ V, and $W/L = 10$, calculate I_D and V_O.

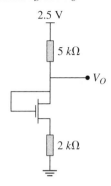

2.5 V

5 $k\Omega$

V_O

2 $k\Omega$

Body Effect and Threshold Voltage

3-36. The nMOSFET has: $V_{tn0} = 0.5$ V, $K_n = 200$ $\mu\text{A/V}^2$, $\phi_F = 0.35$ V, $W/L = 3$, and the body effect constant $\gamma = 0.1$ V$^{1/2}$. The bulk voltage is at -0.3 V with respect to the source. Calculate I_D.

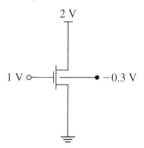

2 V

1 V   −0.3 V

3-37. An nMOSFET threshold voltage is measured as 0.62 V when it should be 0.60 V. A parasitic source to substrate voltage is suspected of raising V_{tn}. If $\gamma = 0.4$ V$^{1/2}$ and $\phi_F = -0.35$ V, what would be the V_{BS} of this suspected mechanism?

3-38. $V_{t0} = 0.6$ V, $\gamma = 0.25$ V$^{1/2}$, and $\phi_F = -0.35$ V. Calculate V_O.

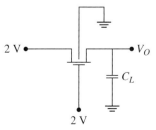

2 V V_O

C_L

2 V

CHAPTER 4

Metal Interconnection Properties

Computers in the future may weigh no more than 1.5 tons.

Popular Mechanics (1949)

The metal lines connecting logic gates and their transistor nodes require careful study. Previous simple interconnect technologies are now complicated by small nanometer (nm) sizes, tiny nm spacing between wires, faster rise and fall times of the logic pulses, operating hot spots over 150°C, and the total length of wires on some integrated circuits (ICs) approaching 10 miles. Integrated circuits stack several layers of metal connection above the silicon substrate with the transistors. Current pulses in power rails can be several amperes making wire resistance a concern. The fast rise and fall times make parasitic capacitance and inductance critical in the design. Clearly, we must learn more about metal properties.

Figure 4-1 shows an oblique magnification of the two lower metal structures of a microprocessor IC. The broad metal lines are the power and ground (GND) lines. The fine metal pattern between the power rails is signal lines connecting to transistor structures beneath. The *RLC* properties of these signal and power supply lines are the subject of this chapter.

Ideal interconnects are conductive wires without parasitic *RLC* electronic properties that guarantee two (or more) connected circuit nodes are always at the same voltage. But real interconnects have *R*, *C*, and *L* properties that deviate from ideal behavior. In practice, two connected metal nodes of an electronic system may not always be at the same voltage, either during the static or the dynamic operation periods. Interconnects may transmit noise by capacitive and inductive coupling.

Two other conducting materials in an IC are polysilicon that transports electrons in the gate, and the diffusion regions that make up the drain and source. These two regions are important, but it is the metal that most impacts the quality of signal transmission.

For years, transistor theory dominated electronics instruction, and the interconnect wires were treated as almost ideal components. Load capacitances were always taken into account but with relatively simple models. We emphasized transistor physics and their

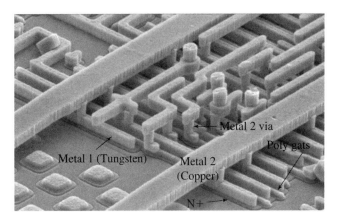

FIGURE 4-1.

Lower metal levels of a microprocessor showing ridge-shaped contacts to gate, drain, and source, and the cylindrical vias. (Reproduced by permission of Chipworks Corp.)

current-voltage models, but now we must include the fundamental nature of metal lines. We begin with the metal physical properties and then present the circuit-level models most commonly used to describe these interconnect effects.

4.1. Metal Interconnect Resistance

The dominant integrated circuit metals are copper and aluminum, but there is an important secondary group of metals that support copper (Cu) and aluminum (Al). Cu technologies use tungsten (W) for *contacts* and also for the first level metal. A contact is the vertical metal structure that connects the transistor semiconductor gate, drain, and source to the first level metal. Aluminum systems use W for the upper-level *vias* that connect the stacked horizontal metal layers, while Cu technologies use Cu itself for the upper-level vias. Al and Cu use barrier metals to enclose portions of the metals. These barrier metals—tantalum (Ta), tantalum nitride (TaN), and cobalt tungsten phosphide (CoWP)—have a protection purpose described later.

Metal resistance is critical in ICs with fast transition current spikes. Instantaneous currents in some power and GND lines can be over 100 amperes. The ohmic voltage drop in the V_{DD} line and the ohmic voltage rise in the GND line can markedly weaken the available voltage to drive transistors. Figure 4-2a shows an interconnect system of wires with no parasitic resistance to alter voltages. Ideally the V_{DD} nodes all have equal voltages, as do the GND nodes.

Figure 4-2b shows a network with parasitic resistances that alter node voltages. Current leaving the power supply has a higher voltage than at node-1 or node-3. The power supply voltage at its logic gate entry will be even lower due to local line resistance. GND current will elevate the IC ground node voltage. Nodes-2 and -4 will be higher than 0 V ground potential due to the parasitic resistance in the GND current paths. The V_{DD} and

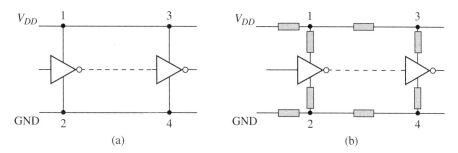

FIGURE 4-2.

Voltage changes sites on power lines alter expected normal V_{DD} and GND potential. (a) Perfect interconnect system. (b) Real system with parasitic resistance in lines.

GND potential difference delivered to the logic gates are shrunk. ICs could then suffer instantaneous loss of speed capability in what is sometimes called brownout.

Resistance is a measure of the difficulty that electrons have when traveling through a material. Resistance depends on the cross-sectional area of the material, its length, temperature, and a physical property called *resistivity*. Electrical resistivity (ρ) is defined by Eq. 4-1.

$$R = \rho \frac{l}{w \times t} \tag{4-1}$$

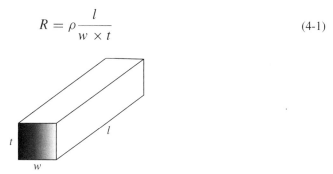

where R is the metal resistance in ohms (Ω), l is the length, w is the width, and t is the thickness of the material. ρ has units of ohm $\cdot$ m, but it is common to use $\mu\Omega \cdot$ cm for metals and cm for l, w, and t. Table 4-1 gives the resistivity (ρ) and thermal coefficient of resistivity (α (TCR)) of several metals. These values were taken for pure metals from the *CRC Press Handbook of Chemistry and Physics, 91 Ed.* But Internet search engines locate physical constants faster.

Al was dominant in ICs for many years using W as the via material that connected one metal layer to another. Cu was introduced about the year 2000, and its lower resistivity and more economic fabrication led a surge in converting from Al/W to Cu interconnection systems. But both metals are used in the IC industry.

Cu technology uses tungsten as the first level metal and for contacts. The W contacts and W first-level metal isolate upper-level copper from the immediate environment of the transistors. Cu diffuses readily in dielectrics and in silicon. If Cu diffuses into the channel

TABLE 4-1 Metal Resistivity at 20°C and Thermal
Coefficient of Resistivity (TCR)

Element	Resistivity ρ ($\mu\Omega \cdot$ cm)	α (TCR) (°C^{-1})
W	5.60	4.50×10^{-3}
Al	2.82	3.90×10^{-3}
Au	2.44	3.40×10^{-3}
Cu	1.72	3.93×10^{-3}
Ag	1.59	3.80×10^{-3}

and drain-source regions, it will degrade if not functionally kill the transistor. W has a relatively high resistivity, but signal delay is managed by keeping the W signal lines short in metal-1.

Silver (Ag) has the lowest resistivity and might be the IC interconnect metal of choice except it is too reactive in CMOS processes and is more expensive. Gold (Au) is used in certain printed circuit board interconnections (called traces) because it is virtually inert to corrosive chemical reaction in adverse environments.

EXAMPLE 4-1

A metal line has the geometry shown in the figure. Compare the resistance if the line is made of **(a)** aluminum, **(b)** copper, **(c)** tungsten.

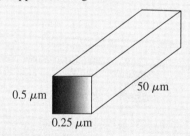

$0.5\ \mu m$ $50\ \mu m$

$0.25\ \mu m$

$$\textbf{(a)}\quad R_{Al} = 2.82\ \mu\Omega \cdot cm \times \frac{50\ \mu m}{0.25\ \mu m \times 0.5\ \mu m} \times \frac{10^4 \mu m}{cm} = 11.28\ \Omega$$

$$\textbf{(b)}\quad R_{Cu} = 1.72\ \mu\Omega \cdot cm \times \frac{50\ \mu m}{0.25\ \mu m \times 0.5\ \mu m} \times \frac{10^4 \mu m}{cm} = 6.88\ \Omega$$

$$\textbf{(c)}\quad R_{W} = 5.60\ \mu\Omega \cdot cm \times \frac{50\ \mu m}{0.25\ \mu m \times 0.5\ \mu m} \times \frac{10^4 \mu m}{cm} = 22.40\ \Omega$$

The resistivities in Table 4-1 were measured from relatively large, pure bulk samples of the metals. But resistivity increases in the small, very thin film metal lines of an IC. Al typically elevates to about 3.3 $\mu\Omega \cdot$ cm when a small percentage of copper is alloyed. Cu may elevate to 1.9 $\mu\Omega \cdot$ cm or higher for the very small metal geometries and also from the impurities added to the Cu electroplate solution that creates the Cu line. We will use the bulk material resistivities given in Table 4-1 unless otherwise stated.

4.1.1. Resistance and Thermal Effects

Table 4-1 gives the thermal coefficient of resistivity (α) for some metals. Modern ICs can operate well above 100°C with chip hot spots near 150°C, and these hot spots can impact timing accuracy and reliability when line resistance increases. Designers must anticipate the resultant different thermal values of interconnect resistance, since the line RC time constant changes. The resistance change with temperature is

$$R(T) = R_o(1 + \alpha(T - T_o)) \tag{4-2}$$

where R_o is a value at a reference temperature T_o that is typically 20°C.

EXAMPLE 4-2

Calculate the resistance at 20°C and at 125°C for the Cu line shown in the figure.

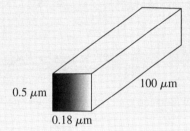

0.5 μm   100 μm

0.18 μm

At 20°C

$$R = 1.72 \ \mu\Omega \cdot \text{cm} \times \frac{100 \ \mu\text{m}}{0.18 \ \mu\text{m} \times 0.5 \ \mu\text{m}} \times \frac{10^4 \ \mu\text{m}}{\text{cm}} = 19.11 \ \Omega$$

At 125°C $R(125°C) = 19.11 [1 + (3.93 \times 10^{-3})(125 - 20)] = 27.0 \ \Omega$
   This is a 37% increase in line resistance.

Self-Exercise 4-1
A Cu metal line is shown. At what temperature will the resistance be 20 Ω?

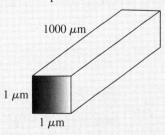

1000 μm

1 μm

1 μm

Answer: T = 61.42°C

Self-Exercise 4-2
Assume equal-sized Cu and Al metal lines. At what temperature will the lines have equal resistances?

Answer: $T = -239°C$

Self-Exercise 4-3
A metal line measures 10 Ω at 20°C. If the temperature sensitivity of the metal line is 39.3 mΩ/°C, identify the metal element.

Answer: It is copper.

4.1.2. Sheet Resistance

Engineers often prefer a shorthand method to estimate interconnect line resistance. Each metal layer has a constant thickness, t. The thickness may vary from layer to layer, but the constant height for a given layer allows a rapid way to estimate line resistance. We arrange Eq. (4-1) as

$$R = \rho \frac{l}{w \times t} = \left(\frac{\rho}{t}\right) \frac{l}{w} = R_{sheet} \frac{l}{w} \qquad (4\text{-}3)$$

where $\left(\dfrac{\rho}{t}\right)$ is a constant for a given metal layer and defines the sheet resistance, R_{sheet}:

$$R_{sheet} = \frac{\rho}{t} \qquad (4\text{-}4)$$

and l/w is the number of squares in a line. Since sheet resistance is a constant for each metal layer, it is easy for a computer program to read the dimensions of a metal line and convert that to a resistance. Sheet resistance also allows you to do rapid estimates in your head. Example 4.3 illustrates a sheet resistance calculation.

EXAMPLE 4-3

If the line is made of Cu, calculate the line resistance using sheet resistance at $T = 20°C$.

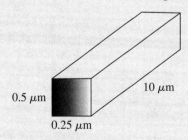

0.5 μm

0.25 μm

10 μm

The sheet resistance is

$$R_{square} = \frac{\rho}{t} = \frac{1.72\ \mu\Omega \cdot cm}{0.5\ \mu m} \times \frac{10^4 \mu m}{cm} = \frac{34.4\ m\Omega}{square}$$

And the total number of squares is $l/w = 10\ \mu m/0.25\ \mu m = 40$ squares. The total resistance is then

$$R = 34.4\ m\Omega \times 40 = 1.376\ \Omega$$

You get the same result from Eq. (4-1), but deal with one more constant in the sheet resistance calculation.

EXAMPLE 4-4

A via connects two levels of aluminum metal at $T = 20°C$. Ignore the via resistance, and calculate the resistance from node A to B (R_{AB}). The resistivity of the Al lines is 3.3 $\mu m \cdot cm$. Use the sheet resistance method. The figure is not drawn to scale.

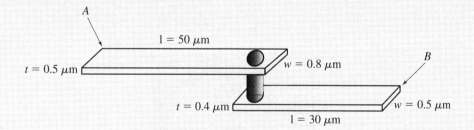

The sheet resistances are

$$R_{A\ square} = \frac{3.3 \times 10^{-6}\ \mu\Omega \cdot cm}{0.5\ \mu m \times 10^{-4} cm/\mu m} = 66\ \frac{m\Omega}{square}$$

$$R_{B\ square} = \frac{3.3 \times 10^{-6}\ \mu\Omega \cdot cm}{0.4\ \mu m \times 10^{-4} cm/\mu m} = 82.5\ \frac{m\Omega}{square}$$

$$R_{AB} = R_A + R_B$$

$$R_{AB} = 66\ m\Omega \times \frac{50\ \mu m}{0.8\ \mu m} + 82.5\ m\Omega \times \frac{30\ \mu m}{0.5\ \mu m}$$

$$R_{AB} = 4.125\ \Omega + 4.95\ \Omega = 9.075\ \Omega$$

EXAMPLE 4-5

If $R_{sheet} = 60$ mΩ/square, estimate the line resistance in your head.

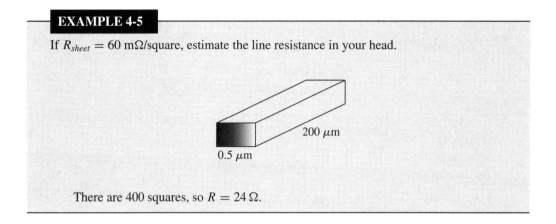

There are 400 squares, so $R = 24\ \Omega$.

4.1.3. Via Resistance

Typical ICs may have 4 to 10 horizontal metal layers depending upon the number of logic gates in the IC. The reason for several layers is that large numbers of transistors require many interconnections in a small area. There is not enough horizontal space on a single level of metal to accommodate all of the interconnections.

Vias are the small cylindrical vertical metals that connect one metal interconnect layer to another (Figure 4-3). There are billions of vias in large ICs, and the via resistance typically lies in series with the logic signal conduction path. Therefore, via resistance directly impacts the RC time constant. We will calculate some good via resistances to

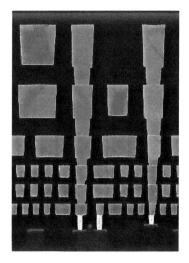

FIGURE 4-3.

Cross section of IC structure showing five vertically aligned vias with a contact (white) at bottom that connects to the transistors. The rectangular structures are metal interconnect cross section. The via sizes decrease closer to the transistors at the bottom. (Reproduced by permission of LSI Corp.)

get a feel for normal resistance. An important and common reliability failure mechanism occurs when the via has a defective structure that elevates its resistance. This is an acute reliability problem considering the number of vias, their small size, and the complex construction of the metal system. Cu resistivity can differ from the bulk Cu value in Table 4-1 since the Cu electroplate solution has impurities added and small dimensions introduce other resistive mechanisms. The examples that follow will illustrate how via resistance changes with geometry.

EXAMPLE 4-6

A tungsten via has a diameter of 0.4 μm, a height of 1 μm, and a resistivity $\rho_W = 5.6\,\mu\Omega \cdot$cm. Calculate the via resistance.

$$R_{via} = \rho \frac{l}{area} = \rho_W \frac{h}{\pi r^2} = 5.6\,\mu\Omega \cdot \text{cm} \times \frac{1\,\mu\text{m}}{\pi \times 0.2^2\,\mu\text{m}^2} \times \frac{10^4\,\mu\text{m}}{\text{cm}} = 445.6\text{ m}\Omega$$

0.4 μm

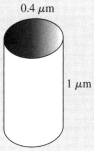

1 μm

Self-Exercise 4-4
Repeat Example 4-6 and calculate the via resistance for a copper technology where the measured Cu resistivity of Cu is $\rho_{Cu} = 1.9\ \mu\Omega \cdot$cm.

Answer: $R_{via} = 151.2$ mΩ. Notice the resistive advantage of Cu vias.

Self-Exercise 4-5
Calculate the via resistance for a copper technology where the measured Cu resistivity is $\rho_{Cu} = 2.4\ \mu\Omega \cdot$cm. The diameter $D = 100$ nm and height $h = 0.8\ \mu$m.

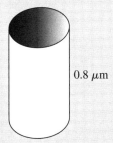

0.8 μm

Answer: $R_{via} = 2.44\ \Omega$

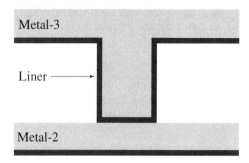

FIGURE 4-4.

Interconnect cross section showing two levels of Cu, a Cu via, and its protective barrier metal liner (dark line). The horizontal Cu lines are laid in a trench surrounded by an intermetal dielectric.

Copper lines have a thin barrier metal lining their bottom and sides (dark lines in Figure 4-4). These liners are typically tantalum nitride (TaN) or titanium nitride (TiN). The barrier is less than 200 Å thick, and it prevents Cu from diffusing through the oxide dielectric and poisoning the transistors. Also if an open circuit develops in the Cu portion of the interconnect line, then the barrier metal acts as a protective shunt path. This can reduce the risk of an open circuit in the Cu, but the downside is that the resistivity of the barrier metal may be 100 times or more than that of the copper. Recent technologies are covering the top of the Cu structures with CoWP. Materials complexity is growing, and these structures affect the signal quality. Example 4-7 illustrates the barrier metal effect on overall resistance of a Cu line.

EXAMPLE 4-7

A Cu metal line *trench* is 10 μm long, 0.18 μm wide, and 0.5 μm thick, and it is filled with a barrier metal and electroplated copper. The Cu resistivity is 1.9 $\mu\Omega \cdot$ cm, and the barrier metal is TaN with a resistivity of $\rho_{TaN} = 200$ $\mu\Omega$ cm. The barrier metal is 200 Å thick (20 nm $= 0.02$ μm). Calculate the line resistance including the barrier metal. Remember the barrier metal covers the two sides and bottom of the Cu.

Calculate the Cu resistance and then the Ta material in parallel.

$$R_{Cu} = 1.9 \ \mu\Omega \cdot \text{cm} \left[\frac{10 \ \mu\text{m}}{[0.18 - (2)(0.02)] \ (0.5 - 0.02) \ \mu\text{m}^2} \right] \frac{10^4 \ \mu\text{m}}{\text{cm}} = 2.827 \ \Omega$$

TaN has two lateral sides whose individual resistances are

$$R_{TaN}(side) = 200 \ \mu\Omega \cdot \text{cm} \left[\frac{10 \ \mu\text{m}}{(0.02)(0.5) \ \mu\text{m}^2} \right] \frac{10^4 \ \mu\text{m}}{\text{cm}} = 2 \ k\Omega$$

The bottom TaN side is

$$R_{TaN}(bottom) = 200 \,\mu\Omega \cdot cm \left[\frac{10 \,\mu m}{0.02 \,(0.18 \,\mu m - 2(0.02 \,\mu m))} \right] \frac{10^4 \,\mu m}{cm} = 7.143 \,k\Omega$$

Ta alters the line resistance as a parallel path to the Cu. The equivalent TaN resistance is

$$R_{TaN} = 2 \,k\Omega // 2 \,k\Omega // 7.143 \,k\Omega = 877.2 \,\Omega$$

The overall resistance is $R_{Cu} // R_{TaN} = 2.827 \,\Omega // 877.2 \,\Omega = 2.818 \,\Omega$.

EXAMPLE 4-8

Calculate the overall cylindrical via resistance for the Cu/TaN structure. The Cu resistivity is $1.72 \,\mu\Omega \cdot cm$, and the barrier metal is TaN with a resistivity of $\rho_{TaN} = 200 \,\mu\Omega \cdot cm$. The barrier metal thickness is 100 Å thick ($0.01 \,\mu m$).

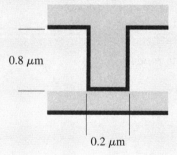

$$R_{TaN}(bottom) = 200 \,\mu\Omega \cdot cm \left[\frac{0.01 \,\mu m}{\pi \left(\frac{0.20 \,\mu m}{2} \right)^2} \right] \frac{10^4 \,\mu m}{cm} = 0.6366 \,\Omega$$

$$R_{TaN}(sleeve) = 200 \,\mu\Omega \cdot cm \left[\frac{0.8 \,\mu m - 0.01 \,\mu m}{\pi \left[\left(\frac{0.2}{2} \right)^2 - \left(\frac{0.18}{2} \right)^2 \right] \mu m^2} \right] \frac{10^4 \,\mu m}{cm} = 264.7 \,\Omega$$

$$R_{Cu} = 1.72 \,\mu\Omega \cdot cm \left[\frac{0.8 \,\mu m - 0.01 \,\mu m}{\pi \left(\frac{0.18}{2} \right)^2 \mu m^2} \right] \frac{10^4 \,\mu m}{cm} = 0.534 \,\Omega$$

$$R_{via} = R_{bottom} + R_{sleeve} // R_{Cu}$$
$$R_{via} = 0.6366 + 264.7 // 0.534 = 1.17 \,\Omega$$

Self-Exercise 4-6

The etched hole for a copper via has a diameter of 0.4 μm, height $= 1$ μm, and the barrier metal is TaN with a resistivity of 200 $\mu\Omega \cdot$ cm. The barrier metal surrounds the sides and bottom of the via. If the Ta wall thickness is 10% of the via diameter, what is the via resistance?

Answer: 841 mΩ

Self-Exercise 4-7

Chips experience "missing vias" from an error in fabrication. If a Cu via core is missing and all of the 1 mA current is carried by the TaN barrier in Self-Exercise 4-6:

(a) What is power dissipated in the structure? Calculate the area of the TaN sleeve.
(b) Calculate the power density of the sleeve in watts per cm^2. Neglect the bottom TaN plate.

Answers: **(a)** 43.08 μW. **(b)** 1.905 kW/cm^2. This is large, and it can damge an IC when it happens.

4.2. Capacitance

4.2.1. Parallel Plate Model

The capacitance parallel plate model is

$$C = \varepsilon_r \varepsilon_o \left(\frac{A}{d} \right) \tag{4-5}$$

where ε_r is the relative permittivity of the insulator, ε_o is the permittivity of free space ($\varepsilon_o = 8.854 \times 10^{-12}$ F/m), A is the area of the plates used to construct the capacitor, and d is the dielectric distance separating the plates. Permittivity relates the ability of material molecules to polarize in response to the field and thus reduce the field inside the material.

Table 4-2 gives the relative permittivity of several insulators used in CMOS fabrication. The relative permittivity allows easy comparison of dielectrics. SiO$_2$ remains the most abundant IC dielectric material despite having a higher $\varepsilon_r = 3.9$. We desire a low dielectric

TABLE 4-2 Relative Permittivity ε_r of Some Interconnect Insulators Used in CMOS Technology

Material	ε_r
SiO$_2$	3.9
SiOF	3.6
SiOC	2.9
P-MSQ	2.3
Nanoporous dielectrics (Xerogels)	1.2–2.2
Air vacuum	1.0

constant to reduce the interconnect RC time constant and the cross-talk capacitive coupling between metal lines. Dielectric constants below 3.0 are called low-k. Low-k dielectrics are challenging to manufacture, since these porous materials can have weaker mechanical strength and thermal cycling properties. Some low-k dielectrics are derivatives of SiO_2 with a different chemical group hung on one of the Si bonds. Recent experiments have used vacuum air gaps between metals but manufacturing is a challenge.

EXAMPLE 4-9

What is the total capacitance to ground of a metal line with a line 100 μm long, 1.0 μm wide, and an effective SiO_2 dielectric spacing of 0.25 μm?

$$C = \varepsilon_r \varepsilon_o \frac{\text{Area}}{\text{d}} = 3.9 \left(8.854 \times 10^{-12} \frac{\text{F}}{\text{m}} \right) \left(\frac{100\,\mu\text{m} \times 1.0\,\mu\text{m}}{0.25\,\mu\text{m}} \right) \left(\frac{10^{-6}\,\text{m}}{\mu\text{m}} \right) = 13.81\,\text{fF}$$

Capacitance appears as a parasitic element between transistor nodes or between the interconnection metals. Transistor and line capacitance contribute to the time for a transistor to switch between On- and Off-states and to the propagation delay between logic gates. Capacitance also allows a type of noise to appear in high-speed circuits called *cross-talk*, in which the voltage at one interconnection line is capacitively coupled to another close interconnection line. When one line switches, a current $i = Cdv/dt$ moves charge in or out of a neighbor line. This can slow or speed up the waveform depending on signal transition polarity.

EXAMPLE 4-10

Consider the static situation in the figure where line-1 is at 1.5 V and line-2 is floating. What is the induced voltage on line-2?

$$V_2 = \frac{10}{10 + 5} 1.5\,\text{V} = 1.0\,\text{V}$$

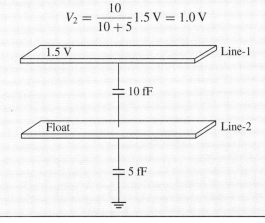

Self-Exercise 4-8
A copper interconnect line has $R_{sheet} = 19$ $\mu\Omega$ per square is 1.0 μm thick, 200 μm long, 0.13 μm wide, and has an effective dielectric spacing of 0.15 μm.

(a) What relative dielectric constant will limit the line capacitance to 3.3 fF?
(b) What is the line RC constant?

Answers: (a) $\varepsilon_r = 2.15$. (b) $RC = 96.5$ as (the unit a is atto $= 10^{-18}$).

Self-Exercise 4-9
What is the induced voltage on line-2?

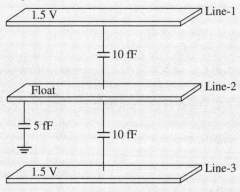

Answer: $V_2 = 1.2$ V

Self-Exercise 4-10
A SiOF dielectric isolates the metal interconnections. A long clock line has $l = 2$ mm, $w = 2$ μm, and $t = 1$ μm. If its effective distance to the V_{DD} and GND lines is 1.2 μm, what is the total line capacitance?

Answer: $C_{total} = 212.5$ fF

4.2.2. Capacitive Power

Chapter 1 derived the power to charge and discharge a capacitance and the heat generated by the driving circuitry. This is significant in IC design, so two review self-exercises are given.

Self-Exercise 4-11
A NAND gate makes a single logic transition during one clock cycle. The load capacitance of the NAND gate is 100 fF and $V_{DD} = 1.2$ V.

(a) What is the energy needed to charge the capacitance?
(b) If the logic gate transitions at a rate of 2 GHz, what is the power dissipated by the NAND gate when it charges or discharges this load?
(c) If the IC has 500 k logic gates also transitioning once on each clock cycle, what is the heat generated in the IC?

Answers: **(a)** $W = 72$ fJ. **(b)** $P = 144\,\mu$W. **(c)** $P = 72$ W

Self-Exercise 4-12
A low-power IC design must not dissipate more than 100 μW. The combinational logic effective load capacitance is 100 pF, and the clock logic circuitry has an effective load of 500 pF. If $V_{DD} = 1.5$ V, what is the maximum clock frequency?

Answer: $f_{clk} = 80.81$ MHz

4.3. Inductance

4.3.1. Inductive Voltage

An inductance occurs in a conducting structure whose current is linked to the energy needed to create a magnetic field around the inductor. The inductor can be a straight wire, or it can be one wire shaped into a coil. *Inductance* is a constant of proportionality between the magnetic flux generated by the inductor and the current change in the inductor. The inductance unit is the henry, and in ICs we often see pico henry (pH $= 10^{-12}$ henry) values. The parasitic straight wire inductance of the IC metal wires is an important parameter to control in high-speed circuits.

Figure 4-5 shows a time varying voltage source driving a resistor and inductor network. The inductor voltage to current relation (v_L, i_L) is derived from Faradays law and is

$$v_L(t) = L\frac{di}{dt} \tag{4-6}$$

and its integral form is

$$i_L(t) = \frac{1}{L}\int_0^t v_L(t')\,dt'$$

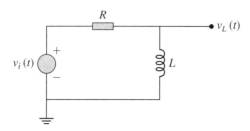

FIGURE 4-5.

Time varying voltage source driving a resistor and inductor network.

How does this relate to integrated circuits? The measured *di/dt* in modern integrated circuits can go higher than 10^{10} A/s. This is alarming since inductive voltages can be large for even small line inductances.

EXAMPLE 4-11

The inductance in a particular IC connecting metal is 200 pH. What is the inductive voltage generated during a current rise time of 10^9 A/s?

$$v_L = L\frac{di}{dt} = 200 \times 10^{-12} \times 10^9 = 200 \text{ mV}$$

Many ICs use power supply voltages on the order of 1.0 V and less. A 200 mV inductive bite is a severe temporary weakening of the normal voltage that drives logic circuitry.

EXAMPLE 4-12

A single 200 pH inductance has a voltage of 200 mV when the current rise time was 10^9 A/s. What is the inductive voltage generated if **(a)** two identical inductors are put in parallel for the same rise time, **(b)** 100 identical inductors are put in parallel?

(a) $L_{eq} = 100$ pH

so $v_L = L_{eq}\dfrac{di}{dt} = 100 \times 10^{-12} \times 10^9 = 100$ mV

(b) $L_{eq} = \dfrac{200 \text{ pH}}{100} = 2$ pH

and

$$v_L = L_{eq}\frac{di}{dt} = 2 \text{ pH} \times 10^9 = 2 \text{ mV}$$

Parallel V_{DD} and GND lines are common designs. Often each signal line is surrounded by one V_{DD} and one GND line to reduce inductance.

EXAMPLE 4-13

Given four inductors in parallel. What is the equivalent inductance?

$$L_{eq} = \frac{1}{\dfrac{1}{150} + \dfrac{1}{85} + \dfrac{1}{125} + \dfrac{1}{170}} pH = 31\ pH$$

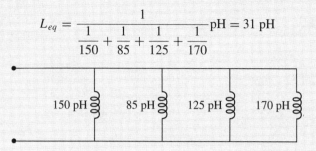

Self-Exercise 4-13
What low value of line inductance must be achieved to keep the inductive voltage to less than 100 mV if the current rate of change is 10^9 A/s?

Answer: $L = 100$ pH

Self-Exercise 4-14
A power supply current has a current rise time of 10^8 A/s, and the line inductance is 400 pH. How many parallel lines are needed to drop the inductive voltage to 10 mV?

Answer: Four lines.

4.3.2. Line Inductance

High-speed integrated circuits are sensitive to inductance on their mostly straight metal connection lines. An estimate for a high-frequency straight wire inductance is

$$L_{hi} = (2l) \left[\left(\ln \frac{2l}{r} \right) - 1.0 \right] \tag{4-7}$$

where l = the wire length (cm), r = the wire radius (cm), and L_{hi} is the high-frequency inductance in nH. Metal conductors have a rectangular cross section, so r is given as an effective radius [1].

EXAMPLE 4-14

If a wire is 50 μm in length and has an effective radius of 0.5 μm, what is the high-frequency inductance?

$$L_{hi} = (2 \times 50\ \mu m) \times 10^{-4} \frac{cm}{\mu m} \left[\left(\ln \frac{2 \times 50}{0.5} \right) - 1.0 \right] (10^{-9}\ H) = 43\ pH$$

Self-Exercise 4-15

A wire length is 10 μm and its effective radius $r = 100$ nm. What is the inductance?

Answer: 8.6 pH

Self-Exercise 4-16

A straight wire must have an inductance less than 200 pH, and it has an effective radius of 90 nm. What maximum length must the wire not exceed?

Answer: $\approx$ <141.7 μm

4.3.3. Inductive Power

The energy stored in the magnetic field surrounding the wire redelivers that energy back to the circuit when current decreases, or it can store that energy under direct current (DC) conditions. Heat is generated in the circuit energy source and not in the megnetic field where energy is stored. The derivation for inductive power p and energy w is

$$p = vi = L\left(\frac{di}{dt}\right)i \tag{4-8}$$

$$p\,dt = L\,i\,di \tag{4-9}$$

The integrals over the time period T and a constant current source I give the total energy W to energize the magnetic field

$$W = \int_0^T p\,dt = \int_0^I L\,i\,di \tag{4-10}$$

$$W = \frac{1}{2}LI^2 \tag{4-11}$$

If the current source is stopped, the magnetic field collapses its energy back into the circuit. That discharge energy is also $W = (1/2)LI^2$. This is the storage and release of potential energy. The associated heat is generated in the driving circuitry, not in the inductance.

EXAMPLE 4-15

A wire in an IC has an inductance of 20 pH. It energizes a magnetic field from a current of 1 A. What is the energy then stored in the magnetic field?

$$W = \frac{1}{2}LI^2 = \frac{1}{2}(20 \text{ pH})\,1\text{A} = 10 \text{ pW}$$

This inductive power analysis example shows that inductive power is generally negligible in contrast to the capacitive charge and discharge power.

4.4. Interconnect *RC* Models

IC interconnect models depend on the characteristics of the logic gates and the interconnect line length and parasitic parameters. Line length affects the model since capacitance and resistance increase with length. Accurate modeling requires R, C, and L elements for high-speed circuits with long interconnects. We will restrict this discussion to *RC* line models.

4.4.1. *C*-model for Short Lines

The short-line capacitance model is a first-order model of logic gate to logic gate delay. The very high resistance of the MOSFET transistor gate oxide allows us to model the load effect of a CMOS logic gate as a capacitor. One inverter driving another inverter (Figure 4-6) can be modeled with the first inverter driving the equivalent input impedance of the second one (i.e., a lumped capacitor). The line resistance and line capacitance can be neglected for this short-line application.

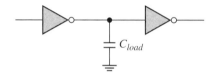

FIGURE 4-6.

Inverter driver with capacitive load.

The first inverter can improve its model with a voltage source and an internal resistance R driving a load gate with input capacitance C (Figure 4-7).

The Kirchhoff laws for current and voltage can be applied to circuits with capacitors as we did with resistors. Thus, once the switch is closed, the KVL applies at any time

$$v_{in} = v_R + v_C$$

Kirchhoff's current law applied to this circuit states that the current through the resistor must be equal to the current through the capacitor, or

$$\frac{v_R}{R} = C\frac{dv_C}{dt}$$

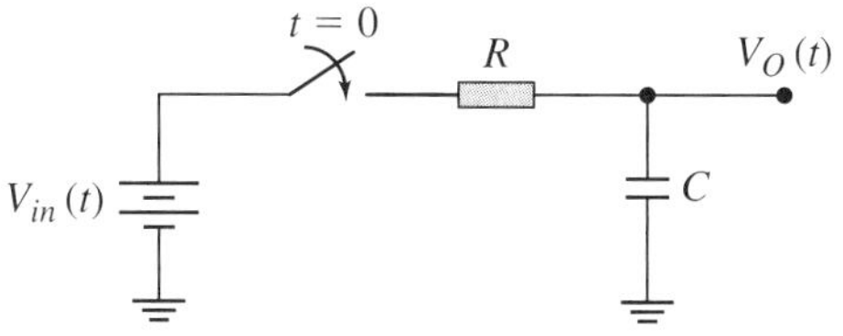

FIGURE 4-7.

Inverter driver load modeled as *RC* elements.

Using the KVL equation we can express the voltage across the resistor in terms of the voltage across the capacitor, obtaining

$$\frac{v_{in} - v_C}{R} = C\frac{dv_C}{dt}$$

This equation relates the input voltage to the voltage at the capacitor. The solution gives the time evolution of the voltage across the capacitor, that is,

$$v_C = v_{in}(1 - e^{-t/(RC)}) \qquad (4\text{-}12)$$

The current through the capacitor is

$$i_C = i_R = \frac{v_{in} - v_C}{R}$$

or

$$i_C = \frac{v_{in}}{R}e^{-t/(RC)} \qquad (4\text{-}13)$$

At $t = 0$, the capacitor voltage is zero and the current is v_{in}/R, while in DC conditions (for $t \to \infty$) the capacitor voltage is v_{in} and the current is zero. This example shows that the current evolution is exponential when discharging a capacitor through a resistor. The *time constant* is defined for $\tau = RC$, which is the time required to increase the capacitor voltage to $(1 - e^{-1})$ of its final value, or 63%. Figure 4-8 plots the voltage and current response. For $v_{in} = 5$ V.

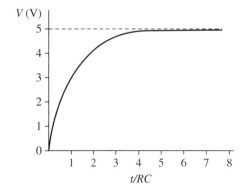
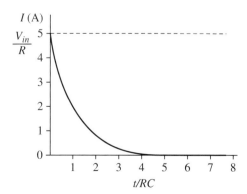

FIGURE 4-8.

Capacitive response to step voltage.

This derivation reviewed the time constant delay for a single stage *RC* connection. We will next analyze an interconnect containing a series of *RC* segments that provide a more accurate estimation.

4.4.2. *RC* Model for Long Lines

A long interconnect has considerable resistance that must be taken into account. ICs have many relatively long interconnects that can be as long as the chip dimensions. Big chips can be from 1 cm to over 2 cm on a side. The longer upper metal levels typically carry long clock, control, and data bus lines. An accurate estimate of the *RC* time constant for such a line requires careful analysis.

We can estimate the total resistance and capacitance of a long line by the sheet resistance and parallel plate methods studied so far. However, a simple multiplication of the lumped *RC* total value is inaccurate. A better model breaks the line into *RC* distributed segments. Figure 4-9a shows a lumped *RC* model for the long line. Figure 4-9b shows the same circuit but with the line segmented in a distributed cluster of equivalent *RC* subnetworks.

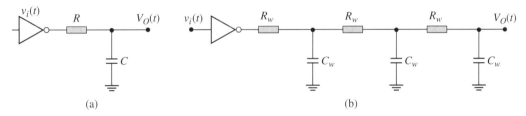

(a)   (b)

FIGURE 4-9.

(a) Lumped *RC* network. (b) Distributed *RC* network.

Elmore solved the delay estimate for the distributed model solution using what is called the first moment of the impulse function [2]. An example will illustrate this more accurate estimate of the time constant using the distributed *RC* model. It adds an *RC* term for the series resistance that each capacitance node sees to the voltage source driving the *RC* network. The formal statement of the Elmore time constant delay for a series delay is

$$\tau_i = \sum_i^k C_i \sum_{j=1}^i R_j \qquad (4\text{-}14)$$

where i is the node number, C_i is the capacitance at node-i, R_i is the sum of resistances seen from node-i to the driving source, k is the number of *RC* segments, and τ_i is the time constant to be estimated to node-i. An example will clarify Eq. (4-14). Elmore's analysis gets more complex when branches feed off the simple series example, but that topic is for the next course.

EXAMPLE 4-16

Use Elmore's analysis to estimate the output line time constant at node-3.

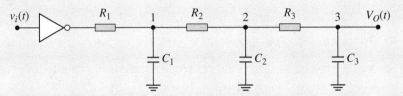

$$\tau_i = \sum_i^k C_i \sum_{j=1}^i R_j = R_1C_1 + (R_1 + R_2)C_2 + (R_1 + R_2 + R_3)C_3$$

Notice each capacitor has a time constant that is the product of that capacitance and the sum of the resistances seen to the input voltage source.

Let us now look at an overall procedure for analyzing a long line using numbers. The process first estimates the R and C of a long line wire with equations we used for wire resistance and capacitance. We divide the wire RC values by the wire segment length in μm to get Ω/μm and aF/μm (a is atto $= 10^{-18}$). We then divide the wire into a chosen number of segments (i.e., 5, 10, 20). The accuracy increases as more segments are used, but so does computation time. The R_w and C_w segments are found by multiplying segment length by the per μm values of R and C.

EXAMPLE 4-17

A Cu wire is 5 mm long, 1 μm wide, and 1 μm thick. The dielectric is SiO_2. Assume the capacitance is determined by the bottom surface to ground with an effective capacitance separation of 0.5 μm. Calculate the per micron resistance and capacitance of the wire. Then break the line into three segments and estimate the propagation delay using Elmore's model.
The total wire resistance and capacitance and their unit values are

$$R = \rho \frac{l}{wt} = 1.72 \times 10^{-6}\,\Omega \cdot cm \left[\frac{5 \times 10^3 \mu m}{1\,\mu m \times 1\,\mu m} \right] 10^4\,\frac{\mu m}{cm} = 86\,\Omega$$

$$R_w = \frac{86\,\Omega}{5 \times 10^3\,\mu m} = 17.2\,\frac{m\Omega}{\mu m}$$

and

$$C_{total} = \varepsilon_r \varepsilon_0 \frac{Area}{d} = 3.9\,(8.854 \times 10^{-12}) \left[\frac{(1\,\mu m)\,(5000\,\mu m)}{0.5\,\mu m} \right] 10^{-6}\,\frac{m}{\mu m} = 345.3\,fF$$

$$C_w = \frac{345.3\,fF}{5\,mm} = 69.1\,\frac{aF}{\mu m}$$

For ease of hand calculation we pick three segments although 10–20 would be more accurate. A segment is then $5000/3 = 1667 \, \mu$m long, so each segment resistor R_{seg} is

$$R_{seg} = 17.2 \frac{m\Omega}{\mu m} \left(\frac{5 \times 10^3 \, \mu m}{3} \right) = 28.67 \, \Omega$$

$$C_{seg} = 69.1 \frac{aF}{\mu m} \left(\frac{5 \times 10^3 \, \mu m}{3} \right) = 115.1 \, fF$$

The circuit is

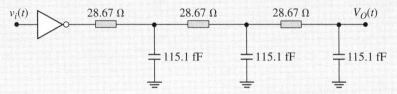

$$\tau_o = (28.67 \, \Omega)(115.1 \, fF) + (28.67 \, \Omega + 28.67 \, \Omega)(115.1 \, fF)$$

$$+ (28.67 \, \Omega + 28.67 \, \Omega + 28.67 \, \Omega)(115.1 \, fF) = 19.8 \, ps$$

The lumped model gives $\tau_o = 29.7$ ps, which is less accurate.

Self-Exercise 4-17
Let the line propagation delay be $\tau = 5$ ps. What is R?

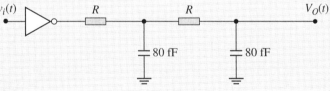

Answer: 20.83 Ω

Self-Exercise 4-18
An aluminum wire has $l = 10$ mm, $w = 1 \, \mu$m, and $t = 1 \, \mu$m. It has a SiO$_2$ dielectric and an effective capacitive plate distance of 0.5 μm with the dielectric. Compare the propagation delay times if you use **(a)** two segments in Elmore's model calculation, and **(b)** four segments.

Answers: **(a)** $\tau = 146.0$ ps. **(b)** $\tau = 121.7$ ps

---

An important conclusion is that long lines get exponentially worse as the line length increases. One solution inserts buffer inverters along the line to amplify the signal. This approach reduces overall propagation delay, but at the price of relatively large inverter buffers ($W/L = 20$–50) that produce considerable heat in clock lines. The technique works despite the contradiction that adding an inverter with its own propagation delay provides an overall reduction in metal line propagation delay. Long metal lines are typically found in the top metal layers, and the buffer inverters are down in the silicon. This increases the routing that reduces the available real estate intended for other wires. However, line buffers are a common solution for reducing delay on long interconnect lines.

4.5. Summary

Interconnect metal resistance, capacitance, and inductance electrical properties demand knowledge of physical properties. Resistivity, sheet resistance, thermal coefficient of resistance, and dielectric constant are optimized to minimize signal delay. Noise is another variable caused by R, L, and C properties aggravated in modern ICs by the high rates of change of current and voltage on the signal lines. The importance of the distributed RC model for interconnect lines was shown.

References

[1] E.B. Rosa, "The Self and Mutual Inductances of Linear Conductors", *Bulletin of the Bureau of Standards,* Vol. 4, No. 2, 1908, Page 301.

[2] W.C. Elmore, *J. Applied Physics,* **19**, pp. 55–63, 1948.

Exercises

Use values in Table 4-1 and Table 4-2 unless otherwise stated.

Metal Resistance

4-1. Tungsten is often used as a metal-1 material despite its high resistivity. If the M1 lines are kept short, then the RC time constant is manageable and will work. Tungsten offers a better protection to transistors from Cu poisoning. If a tungsten line effective capacitance is 10 pF and the line geometry is given in the figure, what is the RC time constant?

4-2. A 1000 μm Cu line is 0.5 μm wide and 0.8 μm tall.

(a) If its effective capacitance is 20 fF, what is the RC constant?

(b) The designer decides to reduce the line resistance by introducing an identical parallel line. What does this do to the overall *time* constant?

4-3. A Cu line is 200 nm wide and 500 nm high. If the line resistance must not exceed 12 Ω, what length must the line be limited to?

4-4. Given $\rho_{Cu} = 1.72\ \mu\Omega \cdot$ cm, $R_{via} = 2\ \Omega$, metal layers with 1 μm thickness, and a via resistance of 2 Ω, what is the total resistance of the interconnect structure?

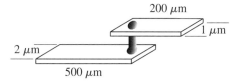

Resistance and Thermal Effects

4-5. Given a rectangular Cu interconnect height $t = 0.5\,\mu$m, width $w = 1.5\,\mu$m, $\alpha = 3.93 \times 10^{-3}$, $T_0 = 20°$C, $\rho_{CU} = 1.72 \times 10^{-6}\,\Omega \cdot$cm, and $R(T = 120°C) = 10\,\Omega$, what is the interconnect length?

4-6. Two Cu lines of equal dimensions are widely separated on a chip: $l = 200\,\mu$m, $w = 0.2\,\mu$m, and $t = 0.5\,\mu$m. One metal line is located in a thermal hot spot of 150°C, while the other is relatively cool at $T = 50°$C. The reference temperature is 20°C. What are the resistances of these identical metal structures?

4-7. A tungsten line is 25 μm long, 200 nm wide, and 400 nm high. Two such identical lines are also subjected to a temperature difference. One line is at 50°C and the other at 150°C. What are the resistances of these two lines?

4-8. Assume equal Al and Cu metal line geometries. At what temperature will R_{Cu} be 60% of R_{Al}?

Sheet Resistance

4-9. A tungsten line is 10 μm long, 200 nm wide, and 500 nm tall.
(a) What is the sheet resistance?
(b) What is the line resistance?

4-10. A metal-2 layer is made of Cu with a metal height of 750 nm. A metal-1 Cu line is 400 nm high.
(a) What is the sheet resistance of the M1 line?
(b) What is the sheet resistance of the M2 line?

4-11. An M1 layer uses tungsten with a height of 400 nm and width of 300 nm. M2 uses copper with a 600 nm height and a 400 nm width. What length of M2 line will equal the resistance of a 10 μm M1 line? Use sheet resistance in your calculation.

4-12. A Cu line is 180 μm long, 0.5 μm wide, and its resistance at 100°C is 11.5 Ω. What is the height of the Cu line?

4-13. Consider the copper line in the figure that is 100 μm long, 1 μm wide, and 1 μm tall. A void is shown that is 90% of the line width and 1 μm long. Use sheet resistance to
(a) Calculate the line resistance if there is no void.
(b) Calculate the resistance of the voided region and the line resistance when the void is present.

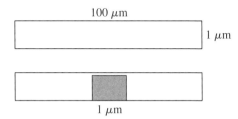

Via Resistance

4-14. An Al technology has an M1 to M2 via with a 500 nm diameter and is 1 μm tall. A later technology uses Cu vias from M1 to M2 with vias 100 nm in diameter and 500 nm tall. Compare the via resistances.

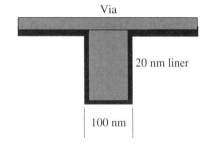

4-15. An etched Cu via hole is 600 nm high and 100 nm in diameter. It is filled with a 200 Å TaN liner with $\rho_{TaN} = 200\,\mu\Omega \cdot$cm resistivity. What is the via resistance?

Capacitance

4-16. A parallel plate capacitor has plates that are 100 μm long, 2.0 μm high, and a dielectric spacing of 0.25 μm. Compare the capacitance for SiO_2, SiOC, and air dielectrics.

4-17. An IC interconnect uses SiOC as a dielectric. The effective spacing is 200 nm, the line height is 1 μm, and the capacitance cannot exceed 10 fF. What is the maximum length allowed for this line?

RC Model of Interconnect

4-18. Solve for the transient response of V_O when the switch is closed at $t = 0$.

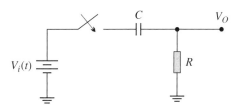

4-19. Use Elmore's relation to estimate the propagation delay from $V_i(t)$ to $V_O(t)$ if the inverter propagation delay is zero.

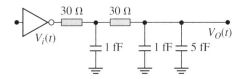

4-20. A Cu line is 15 mm long and has a lumped capacitance of 4.7 pF. It has a thickness of 0.4 μm and a width of 1.5 μm.
  (a) What is the lumped RC constant of this line?
  (b) Compare this value with that using Elmore's circuit model with three distributed line segments.

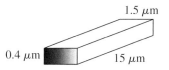

Inductance

4-21. A straight wire must have an inductance less than 30 pH. Its length is 100 μm. What is minimum radius of wire?

4-22. A straight interconnect line is 2 mm long and approximately 200 nm in diameter. How slow must voltage transitions be to hold the inductive voltage to less than 100 mV?

4-23. The V_{DD} line inductance looking into the chip for a single power line is 1 nH. What is the equivalent inductance of 100 local power lines run in parallel in the IC?

4-24. If a single ground line to an IC has an inductance of 125 nH, how many parallel ground lines must be run to get the GND inductance below 1 nH?

4-25. If an inductive current has a transition of 10^8(A/s), and the inductive voltage is to be kept to less than 100 mV, what must the inductance be?

CHAPTER 5

The CMOS Inverter

Hydrogen, photons, and electronic inverters are indivisible units of the elements, light, and computers.

Anonymous

Logic gate electronics begins with the inverter whose simple two-transistor appearance hides its complexity. The inverter has about a dozen important properties that are shared by multi-input gates such as NAND and NOR gates. We will become proficient in understanding these electronic properties.

CMOS refers to a particular method or technology for designing and building integrated circuits. The word complementary means that nMOS and pMOS transistor pairs are linked to make logic gates. Originally, CMOS integrated circuits (ICs) used metal for the transistor gate material. Then polysilicon replaced metal for many years, but new technology has returned to metal gates. CMOS is one of several technologies with which we can build digital circuits. It was first manufactured by the RCA Corp. in 1964. Its popularity grew slowly, but it has been the dominant digital technology since the early 1980s.

5.1. The CMOS Inverter

The inverter is the most abundant logic gate in digital ICs. An inverter converts a logic high input voltage, such as $V_{DD} = 1.5$ V to a low logic voltage of 0 V, and converts a logic low input voltage to a logic high voltage (i.e., 0 V to 1.5 V). V_{DD} is the standard symbol for CMOS power supply voltage. The inverter electronic symbol, truth table, and schematic are shown in Figure 5-1. The logic statement is $V_O = \overline{V}_{in}$. When V_{in} is a high voltage, the nMOS transistor turns on and the pMOS transistor turns off driving the output node to ground. A low input voltage turns the pMOS transistor *on* and the nMOS *off* driving the output node to a high logic voltage of V_{DD}. This *on–off* tandem operation guarantees there is no current from V_{DD} to ground (GND) when the logic values are settled. This is a significant low power feature of CMOS.

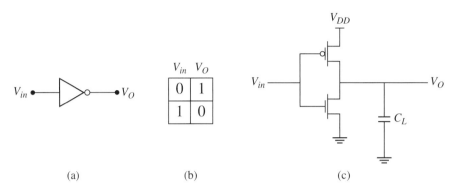

FIGURE 5-1.

Inverter. (a) Symbol. (b) Truth table. (c) Schematic.

Boolean values are detected in the inverter quiescent state when all signal nodes settle to their steady state. Only one transistor is *on* in the steady state connecting the output terminal V_O to one of the power rails. There is no current in the circuit since the other transistor is *off*, eliminating a direct current (DC) path between the rails. The capacitive load C_L shown in Figure 5-1 is unavoidable in any circuit. The capacitance is a lumped value including transistor internal nodes, load wiring, and downstream logic gates. C_L does not affect static properties, but hinders the speed of logic transitions. We will analyze the dynamic operation later.

V_{DD} varies with the application and advances in circuit design. V_{DD} may be lower than 1 V for certain leading-edge circuits or range from 1.2 V to 3.3 V for older still viable technologies. Battery operated digital circuits use V_{DD} values that are typically on the order of 1.5 V. This chapter uses a range of circuit power supply voltages in its examples to adjust our thinking to the diversity expected in real environments.

The logic gate output voltage responds to a small range of input voltages but functionally maps into just one of the two logic states. For example, a 2.5 V power supply technology has nominal logic levels of 2.5 V (high) and 0 V (low). When the input high ranges from 2 V to 2.5 V, the inverter output retains a stable logic low of about 0 V. An input logic-0 ranging from about 0 V to 0.5 V will deliver a stable 2.5 V output. Mapping an input voltage range to a stable output logic state implies noise immunity in the digital circuits. Logic circuit immunity to electronic noise is a design specification.

A third range of digital voltage levels is not mapped to any logic state and that is the voltage transition range that occurs during a logic change of state. Nodes voltages in the transition have no logic meaning. They are the voltages between a logic-1 and logic-0. But the transition region is extremely important since it determines the time to change logic states.

The transistor cross section for an inverter is shown in Figure 5-2. The *n*-channel transistors have a *p*-bulk to attract minority carriers. The *p*-doped bulk is called a *p*-well. Notice the *n*MOS source is tied to the grounded *p*-well and the *p*MOS source is tied to

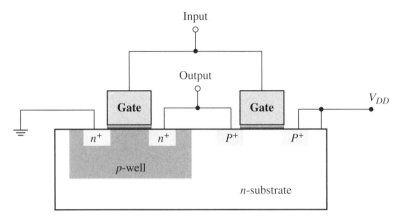

FIGURE 5-2.

CMOS inverter cross section.

the substrate at V_{DD}. The diagram sketches a p-well, but n-wells are also used with the p-channel transistors lying in an n-well. The p-channel transistors have an n-substrate to attract minority carriers.

5.2. Voltage Transfer Curve

The *voltage transfer curve* (VTC) measures the output voltage over the range of input voltages. The VTC in Figure 5-3a shows that $V_O = V_{DD} = 1.2$ V when $V_{in} = 0$ V, and that $V_O = 0$ V when $V_{in} = V_{DD} = 1.2$ V. The general inverted S-shape of the VTC does not change much as V_{DD} changes. The V_{DD} on IC products may range from 0.8 V to 5 V. Modern V_{DD} values lie at the lower end. $V_{DD} = 1.2$ V was chosen as a typical value in this VTC example.

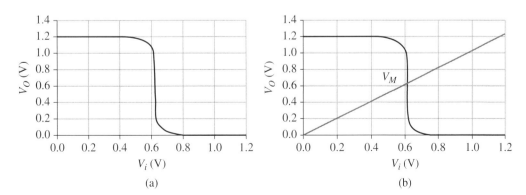

FIGURE 5-3.

(a) Voltage transfer curve (VTC) of CMOS inverter. (b) VTC showing threshold logic point V_M.

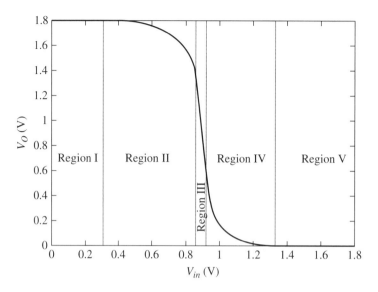

FIGURE 5-4.

V_O versus V_{in} voltage transfer curve (VTC) with five bias states. The input voltage V_{in} is swept from $0 - V_{DD}$ while V_O is measured.

Inverter designs often seek a symmetric static voltage transfer characteristic (Figure 5-3a), so $V_O = V_{in}$ at about $V_{DD}/2$. To be more exact, a plot of $V_{in} = V_O$ forms a 45° straight line from the origin in Figure 5-3b. The voltage at which $V_{in} = V_O$ is the logic threshold voltage V_M. The intersection of the measured VTC and the 45° line locates V_M, which is $V_M \approx 0.62$ V for this inverter.

The intermediate points in the VTC convey much about the dynamics of the inverter (Figure 5-3). Transistors experience the *off*, saturated, and nonsaturated bias states during the sweep. At $V_{in} = 0$ V, the *n*MOS transistor is *off* and the *p*MOS transistor is driven fully into the ohmic state with $V_{GSp} = -1.2$ V. As V_{in} increases, the *n*MOS transistor remains off until V_{tn} is reached. At that point, the two transistors compete with each other for control of the output voltage. As V_{in} increases further, the *p*MOS drive weakens and the *n*MOS drive strengthens. The *n*MOS transistor now pulls the output voltage lower and eventually to 0 V as V_{in} goes to 1.2 V. At this point the *p*MOS transistor is *off* since $V_{GSp} = 0$ V.

Figure 5-4 shows an inverter static voltage transfer curve partitioned into five regions corresponding to five distinct transistor bias states. $V_{DD} = 1.8$ V in this example.

These five bias regions in Figure 5-4 are as follows.

> **Region I.** *nMOS off, pMOS ohmic*: This voltage range exists for $V_{in} < V_{tn}$. The *n*MOS transistor is off, and the *p*MOS transistor is driven into nonsaturation since $V_{GSp} \approx -V_{DD} < V_{DSp} + V_{tp}$. The *p*MOS drain node at V_O is pulled up to a logic high V_{DD} through the low impedance of the *p*MOS channel.

Region II. *nMOS saturated, pMOS ohmic*: When V_{in} goes just above the nMOS threshold voltage ($V_{in} > V_{tn}$), the nMOS transistor barely turns on and is in saturation ($V_{in} < V_O + V_{tn}$). Current now passes through both transistors and V_O drops as V_{in} increases. The pMOS transistor remains in the ohmic state, but with decreasing gate drive.

Region III. *nMOS saturated, pMOS saturated*: When $V_O < V_{in} - V_{tp}$ and $V_O > V_{in} - V_{tn}$, the nMOS and pMOS transistors are both in saturation and the region has a straight line. Since V_O and V_{in} are linearly related, analog amplification occurs here. Small changes in the input waveform are amplified by a value equal to the slope of the straight line. MOS analog circuit designs use this property. It is also good for digital circuits that demand rapid V_O change during logic transitions of V_{in}. A digital goal is to get through the transition region as quickly as possible, and what better way than to have the circuit behave as an amplifier?

Region IV. *nMOS ohmic, pMOS saturated*: As V_{in} increases, it approaches a value such that the difference between V_{in} and V_{DD} is close to the pMOS transistor threshold voltage. This is similar to Region II, but the transistor roles are reversed. The pMOS transistor is in saturation and the nMOS enters nonsaturation.

Region V. *nMOS ohmic, pMOS off*: When V_{in} goes to a logic high voltage, then $V_{in} \gg V_O + V_{tn}$. The nMOS transistor is in its ohmic state pulling the drain voltage to ground, and the pMOS transistor is off.

Voltages slightly less than the logic high or slightly more than logic low voltages are called *weak* logic voltages. Weak logic states are read correctly, but noise margins and gate driving voltage strengths are compromised. The inverter switching threshold voltage (V_M) at which $V_{in} = V_O$ is a unique condition since the theoretical logic state changes at a point as V_O moves through V_M. Inverter voltages in the linear region are not logically defined.

5.3. Noise Margins

The rectangles in Figure 5-5 represent the voltage levels and ranges for the logic high and low as logic gate-A output drives the inputs of logic gate-B. The shaded areas represent a range of logic high and low voltages that logic gates must recognize as valid.

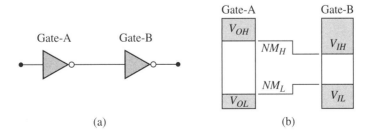

(a)   (b)

FIGURE 5-5.

(a) Series inverters. (b) Voltage ranges mapped to logic Boolean values.

The following terms are defined:

V_{OL} *Output low voltage*: maximum voltage at a gate output for a logic low to be read correctly by a load gate.

V_{OH} *Output high voltage*: minimum voltage at a gate output for a logic high to be read correctly by a load gate.

V_{IL} *Input low voltage*: maximum input voltage recognized as a logic low.

V_{IH} *Input high voltage*: minimum input voltage recognized as a logic high.

These logic voltages define the noise margin or immunity needed when connecting logic gates. The high and low *noise margins* (NM_H, NM_L) are obtained from these four voltage parameters and are typically specified by the manufacturer. NM_H and NM_L for the high and low logic values are

$$NM_H = V_{OH} - V_{IH}$$
$$NM_L = V_{IL} - V_{OL}$$

(5-1)

Noise margins must be positive for proper logic operation, and the larger these values, the better the circuit noise immunity. If an input voltage rises and makes NM_H negative, then there is risk of a noise-induced change of logic state. These parameters are an essential measurement during production testing of ICs. Board designers must know that ICs connected to each other are within specification and interface properly.

Noise margins reflect design and manufacturing capability. It is what a manufacturer can guarantee to a customer. Another NM definition picks the two points in the VTC where the slope is –1. This is arbitrary but does observe that when the absolute slope is >1, then amplification of V_{in} takes place, and that is not good from a noise standpoint. It is a high-risk region when a logic signal is moved to a point where the slope is greater than one. It is helpful to sketch the NM rectangles in Figure 5-5 and label values before solving the problems that follow.

EXAMPLE 5-1

An IC with $V_{DD} = 1.5$ V shows $V_{OH} = 1.35$ V, $V_{OL} = 0.2$ V, $V_{IH} = 1.2$ V, and $V_{IL} = 0.3$ V. Calculate the NM_L and NM_H for this IC.

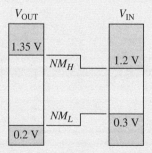

Sketch the NM rectangles
$NM_H = 1.35 - 1.2 = 150$ mV
$NM_L = 0.3 - 0.2 = 100$ mV

Self-Exercise 5-1

IC_A and IC_B have $V_{DD} = 2$ V. IC_A is sending data to IC_B and their input–output (I/O) specifications are

$$IC_A: \quad V_{IH} = 1.6 \text{ V}, V_{IL} = 0.4 \text{ V}, V_{OH} = 1.5 \text{ V}, \text{ and } V_{OL} = 0.3 \text{ V}.$$

$$IC_B: \quad V_{IH} = 1.6 \text{ V}, V_{IL} = 0.5 \text{ V}, V_{OH} = 1.5 \text{ V}, \text{ and } V_{OL} = 0.3 \text{ V}.$$

(a) Show calculations as to whether there is a problem.
(b) If there is a problem what should be done?

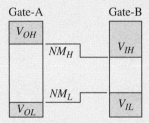

Answer: **(a)** The problem lies in the high voltage levels. **(b)** Answer intentionally not given.

Self-Exercise 5-2

(a) If $V_{DD} = 1.8$ V, $V_{OH} = 1.75$ V, and $NM_H = 130$ mV, calculate V_{IH}.
(b) If $V_{IL} = 0.35$ V and $NM_L = 180$ mV, calculate V_{OL}.

Answers: **(a)** $V_{IH} = 1.62$ V. **(b)** $V_{OL} = 170$ mV

5.4. Symmetrical Voltage Transfer Curve (VTC)

The aspect ratio W_p/W_n for a symmetric transfer characteristic is found by equating the saturation current for nMOS and pMOS transistors and setting the input voltage equal to $V_M = V_{DD}/2$. Eq. (5-2) gives the W_p/W_n ratio for achieving a symmetrical inverter where $V_M = V_{DD}/2$.

$$\frac{W_p}{W_n} = \frac{\mu_n}{\mu_p} \left[\frac{1 - \dfrac{2V_{tn}}{V_{DD}}}{1 - \dfrac{2|V_{tp}|}{V_{DD}}} \right]^2 \tag{5-2}$$

$V_M = 0.5\, V_{DD}$ in Eq. (5-2) when the pull-up and pull-down transistor current drive strengths are equal, but designers may need faster pull-up or pull-down in a given design situation. If $V_M < 0.5\, V_{DD}$ then the nMOS pull-down transistor is stronger than the pull-up, and the VTC is skewed to the left. If $V_M > 0.5\,V_{DD}$ then the pMOS pull-up is stronger, and the VTC is skewed to the right. In reality, it is difficult to make the VTC exactly symmetrical. Mobility's are uncertain, and W_p is typically not made a small fractional dimension relative to W_n. In our examples, we will pretend that exact symmetry can be attained.

EXAMPLE 5-2

Compute the ratio of nMOS and pMOS transistor width to obtain a symmetric inverter for a 0.18 μm technology in which $\mu_n = 360$ cm²/V·s, $\mu_p = 109$ cm²/V·s, $V_{tn} = 0.35$ V, $V_{tp} = -0.4$ V, and $V_{DD} = 1.8$ V.

$$\frac{W_p}{W_n} = \frac{360\text{ cm}^2/\text{V}}{109\text{ cm}^2/\text{V}} \left[\frac{1 - \dfrac{2(0.35)}{1.8}}{1 - \dfrac{2|-0.4|}{1.8}} \right]^2 = 4.0$$

Self-Exercise 5-3
Derive Eq. (5-2).

Self-Exercise 5-4
If $\mu_n = 360$ cm²/V·s, $\mu_p = 109$ cm²/V·s, $V_{tn} = 0.35$ V, $V_{tp} = -0.4$ V, and $V_{DD} = 1.5$ V, what W_p/W_n ratio provides the inverter with a symmetrical VTC?

Answer: $\dfrac{W_p}{W_n} = 4.31$

5.5. Current Transfer Curve

The DC power supply *current transfer curve* (ITC) is equally important. Figure 5-6 shows the I_{DD} versus V_{in} characteristic. At $V_{in} < V_{tn}$ in bias Region I, the nMOS transistor is off and no current passes through the circuit. When $V_{in} = V_{DD}$ in Region V, the pMOS transistor is off, and again no current passes from the power supply to ground. Typical inverter current at these long channel quiescent logic levels is in the low pA's and is mostly drain-substrate reverse bias saturation current. Virtually no power is dissipated in the quiescent logic states when the Off-current is this low.

The peak current near $V_{DD}/2$ depends upon transistor strength (the size of the width to length ratio, threshold voltage, and carrier mobility). Both transistors are in the saturated state at the current peak. When an inverter changes logic state, this transient current is wasted power. Peak total power supply currents in large microprocessor designs are many amperes since this is the sum of the transient currents of millions of switching logic gates.

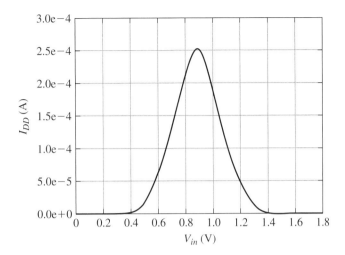

FIGURE 5-6.

Inverter power supply current transfer curve.

Peak current is easy to calculate. The nMOS and pMOS transistors are both in saturation at $V_{DD}/2$ for a symmetrically designed inverter. Either saturation equation for the nMOS and pMOS transistors will solve for the peak current. Examples for symmetrical inverters will illustrate.

EXAMPLE 5-3

If $V_{DD} = 2$ V, $V_{tn} = 0.5$ V, $V_{tp} = -0.5$ V, $K_n = 75$ μA/V^2, $K_p = 50$ μA/V^2, $(W/L)_n = 2$, and $(W/L)_p = 3$, calculate I_{peak}.

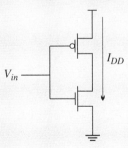

The nMOS saturated state equation gives

$$I_{peak} = I_{Dn} = 75 \left(\frac{\mu A}{V^2}\right) \left(\frac{2}{1}\right) [1 - 0.5]^2 = 37.5 \ \mu A$$

The pMOS saturated state equation also gives

$$I_{peak} = I_{Dp} = 50 \left(\frac{\mu A}{V^2}\right) \left(\frac{3}{1}\right) [1 - 2 + 0.5]^2 = 37.5 \ \mu A$$

Self-Exercise 5-5
If $V_{DD} = 1.5$ V, $V_{tn} = 0.4$ V, $V_{tp} = -0.4$ V, $K_n = 100$ μA/V^2, and $K_p = 50$ μA/V^2:

(a) The peak current of 35 μA occurs at 0.6 V. What is $(W/L)_n$?
(b) What is the $(W/L)_p$?

Answers: **(a)** $\left(\dfrac{W}{L}\right)_n = 8.75$, **(b)** $\left(\dfrac{W}{L}\right)_p = 2.8$

Self-Exercise 5-6
Given an inverter with $V_{tn} = 0.6$ V, $V_{tp} = -0.7$ V, $\mu_n = 1350$ V^2/(V $\cdot$ s), $\mu_p = 350$ V^2/(V $\cdot$ s), and $W_p/W_n = 9$, calculate V_{DD} for a symmetrical CMOS inverter.

Answer: $V_{DD} = 1.78$ V

5.6. Graphical Analysis of VTC

5.6.1. Static Transfer Curves

Figure 5-7 relates the bias state regions with the voltage transfer function and two 45° lines. Eq. (5-3) is the bias boundary condition for the nMOS transistor and is a straight line (a) when plotted on Figure 5-7. The lower unity-slope line-a defines the nMOS saturation and nonsaturation regions. All points on the transfer curve lying above line-a represent the nMOS transistor in either saturation or the off-state. All points below line-a represent

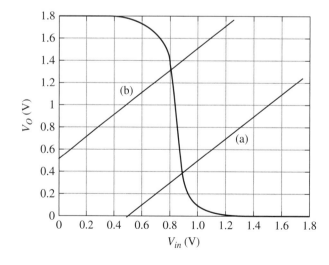

FIGURE 5-7.

Inverter VTC and transistor state.

the nMOS transistor in the ohmic state. V_{tn} is located on the x-intercept.

$$V_{GS} = V_{DS} + V_{tn} \tag{5-3}$$

or

$$V_{in} = V_O + V_{tn} \tag{5-4}$$

and

$$V_{tn} = V_{in} \text{ at } V_O = 0$$

A similar derivation leads to the pMOS transistor bias boundary line labeled (b) in Figure 5-7 and given in Eq. (5-4). The y-axis intercept is $V_O = -V_{tp}$. The pMOS transistor is either saturated or off for all points on the curve below line-b and is in the ohmic state above line-b. Both transistors are in saturation in the important region between the two lines.

$$V_{in} = V_O + V_{tp}$$

$$V_{tp} = -V_{in} \text{ at } V_O = 0 \tag{5-5}$$

EXAMPLE 5-4

Given a simulated inverter transfer curve. Estimate V_{tn} and V_{tp} using bias line concepts.

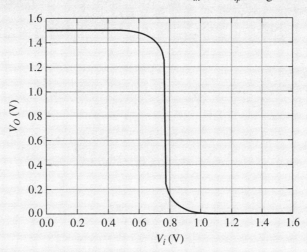

CMOS inverter voltage transfer curve.

Put a small mark on the estimated ends of the linear region and draw 45° lines. The threshold values are the intercepts approximately. $V_{tn} \approx 0.52$ V and $V_{tp} \approx -0.43$ V.

Graphical analysis allows visualization of transistor states during logic transitions. The maximum gain region occurs when both transistors are saturated as seen between the two bias lines (a, b) in Figure 5-7. An example emphasizes this thinking.

EXAMPLE 5-5

When V_O switches in an inverter from V_{DD} to 0 V, estimate the fraction of V_{DD} that the nMOS transistor is in saturation. Let $V_{tn} = 0.2\,V_{DD}$, $V_{tp} = -0.2\,V_{DD}$, and $K'_n = K'_p$, where $K'_n = K_n(W/L)_n$, and $K'_p = K_p(W/L)_p$.

We know $I_{Dn} = |I_{Dp}|$ for all the points in the static curve including the point on line-a where the nMOS transistor leaves saturation. At this point both transistors can be treated in the saturation state. This is the transition between Regions III and IV, and we will calculate V_O at that point.

$$K'_n\,(V_{GS} - V_{tn})^2 = K'_p(V_{GS} - V_{tp})^2$$

Substitute for $V_{GS} = V_{in}$, $K'_n = K'_p$, $V_{in} = 0.2 \times V_{DD}$, $V_{in} = V_O + V_{tn}$ and get

$$K'_n\,(V_{in} - V_{tn})^2 = K'_p(V_{in} - V_{DD} - V_{tp})^2$$

$$V_O^2 = V_O^2 - 1.2\,V_{DD}(V_O) + 0.36\,V_{DD}^2$$

$$V_O = 0.3\,V_{DD}$$

The fraction is

$$\frac{V_{DD} - V_O}{V_{DD}} = 0.7 = 70\%$$

The nMOS is in saturation for about 70% of the transition. This saturation bias is significant when we model the inverter logic transition speed.

EXAMPLE 5-6

If $K_p = 55\mu A$, $V_{in} = 1.0$ V, and $I_{DD} = 15$ μA, and the inverter is symmetrical, what is $(W/L)_p$?

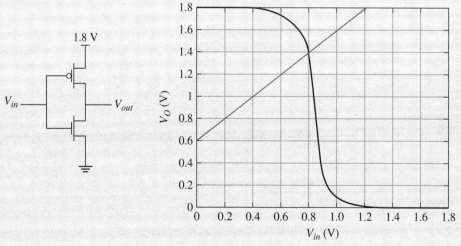

Use the VTC to estimate $V_{tp} \approx 0.6$ V

$$15\,\mu\text{A} = 55\,\mu\text{A}(W/L)_p \,(1.0 - 1.8 + 0.6)^2$$
$$(W/L)_p = 6.8$$

Self-Exercise 5-7
A CMOS inverter has transistor parameters: $K_n\,(W/L)_n = 265\,\mu\text{A/V}^2$, $K_p\,(W/L)_p = 200\,\mu\text{A/V}^2$, $V_{tn} = 0.55$ V, $V_{tp} = -0.63$ V, and $V_{DD} = 2.5$ V. What fraction percentage of the total output voltage swing will the nMOS transistor be in saturation?

Answer: 75.4%

Self-Exercise 5-8
The ITC and VTC are shown for an inverter. If $(W/L)_n = 50$, estimate V_t and solve for $K_n = \dfrac{\mu\varepsilon}{2T_{ox}}$ from the data.

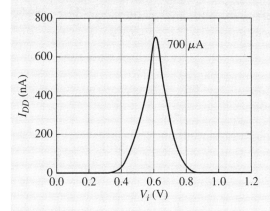 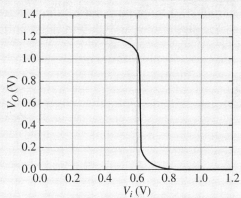

Answer: $K_n \approx 350\,\mu\text{A/V}^2$

Self-Exercise 5-9
A CMOS inverter has transistor parameters $K_n\,(W/L)_n = 265\,\mu\text{A/V}^2$, $K_p\,(W/L)_p = 200\,\mu\text{A/V}^2$, $V_{tn} = 0.55$ V, $V_{tp} = -0.63$ V, and $V_{DD} = 2.5$ V. What fraction of the total output voltage swing will the pMOS transistor be in saturation?

Answer: 71.8%

5.6.2. Dynamic Transfer Curves

The voltage transfer curve of Figure 5-4 gave essential inverter information but was swept slowly (static curve). It did not represent the circuit behavior during its rapid transition. The input signal switching time for modern inverters can be tens of picoseconds, and parasitic capacitance of the transistors and the external wiring load alter the phase relation between V_i and V_O. An inverter and its transfer curve phase relation are shown in Figure 5-8 for different input speed transitions, showing that for rapid transitions the drain voltage lags the input gate voltage changes.

The circuit model for the inverter dynamic analysis in Figure 5-8a shows two parasitic capacitances that are important during the transition. C_L represents the combined effects of the drain diffusion region capacitance, interconnect wiring, and a load gate input capacitance. The second capacitance is the input–output capacitance (called the coupling capacitance C_{coup}). It is strongly bias dependent and is the parallel overlapping gate-drain capacitance from the nMOS and pMOS devices.

Figure 5-8b shows simulated transfer curves for static, (1) medium speed, (2) higher speed, and (3) very high-speed input voltage transitions. These curves show hysteresis.

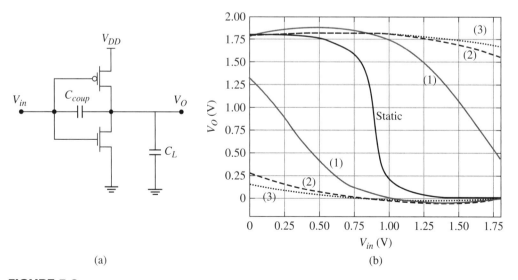

(a)   (b)

FIGURE 5-8.

(a) Dynamic CMOS inverter circuit model. (b) Transfer curves for output high to low and output low to high for different input ramp speeds.

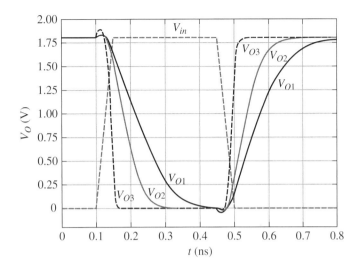

FIGURE 5-9.

Timing responses for different cases for a dynamic inverter transition.

The forward VTCs of V_O are shown in the upper portion of Figure 5-8b, and the reverse transition is shown in the lower portion. Curves (2) and (3) in Figure 5-8b show that the output drain node remains at a relatively high voltage when the gate input has almost completed its low to high transition. The same phenomenon holds when the gate input switches rapidly from high to low. The drain remains in a low voltage state until the input has almost completed its transition. The circuit parasitic capacitance causes this phase relation. A slow transition of the static curve allows time for the drain nodes to exactly follow the input gate voltage in time. Curve (1) is an intermediate case where the output is larger than $V_{DD}/2$ when the input reaches its final value, although it is far from the static transfer curve. These curves are dominantly a function of C_L and the switching speed.

Figure 5-9 plots the output voltage timing responses for different values of C_L when the input transition times are set to a constant rate of change. Curve V_{O3} corresponds to a small output capacitance, and the output voltage is almost zero when the input reaches V_{DD}. This case is similar to the static transfer curve. As the load capacitance increases for curve V_{O2} the output transition time increases. Further increase in load capacitance slows the transition time even more (V_{O1}). In all cases the output voltage initially goes beyond V_{DD} due to overshoot caused by the charge injected from the input through the coupling capacitance C_{coup}. During this period there is a small current from the output node through the pMOS transistor back to the supply terminal.

For high-speed transitions the coupling capacitance C_{coup} tries to maintain its initial voltage difference between the input and the output ($-V_{DD}$ for a low-high input transition and $+V_{DD}$ for a high-low one). But charge is delivered through C_{coup} to node V_O, and C_L cannot respond other than to accept new charge. The charge injection is more rapid than

C_L can move charge through the transistor paths to the rails. The change in output node voltage is $dv = dq/C$. This temporarily drives the output voltage beyond V_{DD} (overshoot) for an input rising transition and below ground for a falling edge (undershoot) (Figure 5-9). Circuit simulators are the best tool to compute the timing waveforms.

5.7. Inverter Transition Speed Model

An exact calculation of the propagation delay of an inverter requires complex differential equations. We will derive a simple model assuming that the transistor is an ideal current source, i.e. the transistor is always in saturation during the transition. The current source I_o in Figure 5-10a represents the pMOS transistor in saturation since that bias state dominates the transition rise time and C_L is the load capacitance. The nMOS transistor is assumed off for this simple model of rise time. The saturated state model for a pMOS pull-up is

$$I_O = I_{Dp} = \frac{\mu_n C_{ox}}{2} \frac{W}{L}(V_{GS} - V_{tp})^2 \tag{5-6}$$

where $C_{ox} = \dfrac{\mu \varepsilon}{T_{ox}}$

The capacitance expression for current, voltage, and time is

$$i(t) = C_L \frac{dV(t)}{dt}$$

If $i(t) = I_O$ (a constant current) and we approximate $dv(t)/dt = \Delta v(t)/\Delta t$, then

$$\Delta V(t) = \frac{I_o}{C_L} \Delta t \tag{5-7}$$

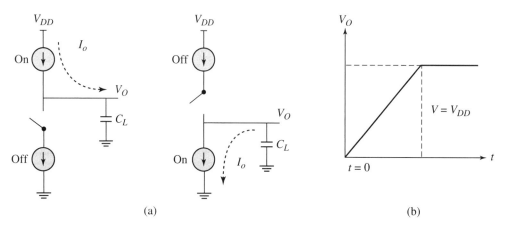

(a)   (b)

FIGURE 5-10.

(a) Circuit model to estimate rise and fall delays in a CMOS inverter. (b) Capacitance voltage response to constant current.

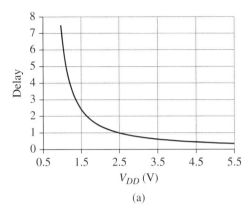
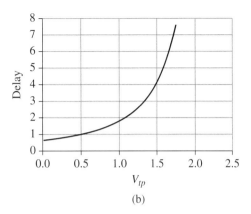

(a)   (b)

FIGURE 5-11.

(a) Normalized delay versus supply voltage for a constant V_t for the model in Eq. (5-9).
(b) Normalized delay versus threshold voltage for a constant V_{DD} for the model in Eq. (5-9).

If Δt is the delay or rise time τ_r in Figure 5-10b for the signal to rise to $\Delta V(t) = V_{DD}$ then Eq. (5-7) is rewritten as

$$\tau_r = \frac{C_L V_{DD}}{I_o} \tag{5-8}$$

Substituting Eq. (5-6) into Eq. (5-8) with $V_{GS} = V_{DD}$ gives

$$\tau_r = C_L V_{DD} \left(\frac{2L}{W \mu_p C_{ox}} \right) \frac{1}{(-V_{DD} - V_{tp})^2} \tag{5-9}$$

Eq. (5-9) shows that time delay is asymptotically related to the difference in V_{DD} and the threshold voltage. Figure 5-11a plots the time delay versus V_{DD} for $C_L = 20$ fF, $K_p = 75$ μA/V^2, $V_{tp} = -0.5$ V, and $W/L = 2$. Time delay asymptotically approaches infinity as V_{DD} approaches V_{tp}. The plot also shows a significant property when $V_{DD} \gg |V_{tp}|$. The change in transition delay time is small for small variations in the power supply level. If V_{DD} drops from 5 V to 4.5 V, the change in delay is small. This is an important noise protection since power supply and GND levels are noisy from clock to clock period.

The result is similar if V_t varies for a fixed V_{DD}. Figure 5-11b is similar to Figure 5-10a, but with time delay plotted against V_t *for* $V_{DD} = 2.5$ V. The plots show that circuit delay is very sensitive to the difference in V_{DD} and V_t in certain regions of the curves.

A similar derivation for the pull-down delay or fall time τ_f using the nMOS transistor gives

$$\tau_f = C_L V_{DD} \left(\frac{2L}{W \mu_n C_{ox}} \right) \frac{1}{(V_{DD} - V_{tn})^2} \tag{5-10}$$

EXAMPLE 5-7

(a) An inverter has $C_L = 100$ fF and a current drive $I_O = 400\,\mu$A. What value of V_{DD} is required to hold the signal rise time to 300 ps?

Use Eq. (4.7)

$$V_{DD} = \frac{\tau_r \times I_O}{C_L} = \frac{(300 \text{ ps})(400\,\mu\text{A})}{100 \text{ fF}} = 1.2 \text{ V}$$

(b) If $V_{DD} = 1.5$ V, what is τ_r?

$$\tau = \frac{(100 \text{ fF})(1.5 \text{ V})}{400\,\mu\text{A}} = 375 \text{ ps}$$

In this case, τ_r increased to 375 ps because the transition output voltage amplitude increased with V_{DD} increase. The example assumed that the drive current was constant.

EXAMPLE 5-8

Given that $C_L = 10$ fF, $\mu\varepsilon/2T_{ox} = 59\,\mu$A/V^2, $W/L = 6$, and $V_{DD} = 2.3$ V, initially $V_{tn} = 0.6$ V. If V_{tn} is reduced to 0.2 V, what is the ratio decrease in fall time?

You can substitute the values into Eq. (5-9) and take the ratio or divide Eq. (5-9) by itself substituting $V_t = 0.6$ V and $V_t = 0.2$ V. You get

$$\frac{\tau_f\,(V_t = 0.6 \text{ V})}{\tau_f\,(V_t = 0.2 \text{ V})} = \frac{(2.3 - 0.2)^2}{(2.3 - 0.6)^2} = 1.526$$

Self-Exercise 5-11

A pMOS transistor has $V_{tp} = -0.5$ V, $K_p = 70\,\mu$A/V^2, $(W/L)_p = 4$, and $V_{DD} = 2$ V. Estimate the rise time delay for an inverter with a load of 100 fF.

Answer: $\tau_r = 317.5$ ps

Self-Exercise 5-12

An nMOS transistor has $V_{tn} = 0.35$ V, $K_n = 120\,\mu$A/V^2, $(W/L)_n = 3$, and $V_{DD} = 1$ V. Estimate the 90% fall time delay (0.9 V_{DD}) for an inverter with a load of 10 fF.

Answer: $\tau_f = 59.2$ ps

Self-Exercise 5-13

Transistors in a CMOS inverter have $C_L = 50$ fF, $V_{tn} = 0.4$ V, $V_{tp} = -0.4$ V, $K_n = 100$ μA/V^2, $K_p = 50$ μA/V^2, and $V_{DD} = 1.5$ V. What should the W/L ratios be for both transistors if the circuit is to have equal rise and fall times of 200 ps?

Answer: $\left(\dfrac{W}{L}\right)_n = 3.1$

$\left(\dfrac{W}{L}\right)_p = 6.2$

The time delay τ is actually slower than calculated by the simple model of a single transistor pulling the output node up or down. One reason is that the pMOS and nMOS transistors as a pair siphon charge from each other that is intended for C_L. While the pMOS is trying to charge C_L, the nMOS bleeds some of that charge to ground thus lengthening the change of state. However, the simple model does identify the major parameters that control logic transitions.

5.8. CMOS Inverter Power

An inverter has dynamic and static energy components. Dynamic dissipation has two major components: (1) the charge–discharge of the logic gate load capacitance (transient component); and (2) the short-circuit current from the supply to ground created during the transition (Figure 5-6). Static dissipation for long channel transistors is due mainly to reverse bias drain-substrate (well) pn junction leakage current from transistors in the off-state. The dynamic power calculation requires computation of transient and short-circuit components. This is typically small for a single logic gate, but it can be significant when switching power includes millions of gates. Off-state leakage is a major concern for modern, very short channel transistors, but that is a topic for an advanced course.

5.8.1. Transient Power

The dynamic power (P_d) to charge and discharge a capacitance C_L for a clock period T_{clk} is

$$P_d = \frac{1}{T_{clk}} \int_0^{T_{clk}} i_{load}(t) v_o(t) dt \tag{5-11}$$

In one period, the output voltage changes from 0 V to V_{DD} and then from V_{DD} to 0 V. Eq. (5-11) is rewritten using

$$i_{load}(t) = \frac{dq}{dt} = C\frac{dv_{load}}{dt} = C\frac{dv_o}{dt}$$

and

$$P_d = \frac{1}{T_{clk}} \left[\int_0^{V_{DD}} C_L v_o dv_o + \int_{V_{DD}}^0 C_L \left(V_{DD} - v_o \right) dv_o \right] \qquad (5\text{-}12)$$

giving

$$P_d = \frac{C_L V_{DD}^2}{T_{clk}} = C_L V_{DD}^2 f_{clk} \qquad (5\text{-}13)$$

Eq. (5-13) shows that reducing the output capacitance, the supply voltage, or the operating frequency will lower transient power. Since the power dependence on the supply voltage is quadratic, lowering V_{DD} is more efficient for reducing power dissipation than the other two parameters. Notice Eq. (5-13) assumes an up and down transition in one clock period. If a single transition occurred during one clock period, then P_d is reduced by half. A single transition is typical for combinational logic circuits. A double transition per clock period is typical of logic gates in the clock timing networks. Eq. (5-13) is often used as an estimate of gross IC power despite inaccuracies as to how many gates are double transitioning, or what fraction of the gates actually switch at all in a given clock period.

When a clock pulse drives the registers in an integrated circuit, a single switching occurs in the combinational logic load gates. However, not every logic gate in the IC switches. Typically, about 5–30% of the total combinational logic gates in an IC switch and draw dynamic power in a single clock pulse. We define an activity coefficient α as the fraction of logic gates in the IC that are expected to change state (switch) on a clock pulse. The power equation Eq. (5-13) for a single transition period becomes

$$P_d = \frac{1}{2} \alpha C_L V_{DD}^2 f_{clk} \qquad (5\text{-}14)$$

$\alpha = 1$ in a typical clock network, but some networks in the IC are intentionally gated off from the clock to reduce power, and their activity coefficient is zero.

EXAMPLE 5-9

Given that a gated IC clock network has $V_{DD} = 1.5$ V, $\alpha = 0.84$, a total load capacitance $C_L = 5$ nF, and $f_{clk} = 2.6$ GHz, what is the power dissipation in the clock network?

$$P_d = (1)(0.84)(5 \times 10^{-9})(1.5^2)(2.6 \times 10^9) = 24.57 \text{ W}$$

Self-Exercise 5-14

For a combinational logic gate, $\alpha = 0.2$, $C_L = 150$ fF, $V_{DD} = 2$ V, and $f_{clk} = 2$ GHz.

(a) Calculate the circuit power dissipation.
(b) Calculate the power dissipation if $V_{DD} = 1.5$ V.

Answer: (a) $P_d = 120 \ \mu$W. (b) $P_d = 67.5 \ \mu$W

Self-Exercise 5-15

An IC has six combinational logic blocks driven by a clock. Each block has a 750 pF load, $V_{DD} = 1.1$ V, and $f_{clk} = 3$ GHz. How many blocks must be turned off at any instant if the total power of the IC clock network is to be less than 5 W. Use the gross assumption that all the logic gates make either a single charge or discharge during a clock cycle.

Answer: Power per block is 1.36 W so only three blocks can be on at one time.

Self-Exercise 5-16

V_{DD} goes from 1.2 V to $\frac{2}{3}V_{DD}$ in a circuit with $\alpha = 1$.

(a) Calculate the power dissipated in a combinational logic gate for both V_{DD} values when $C_L = 10$ nF, and $f_{clk} = 1.5$ GHz.
(b) Calculate the power dissipated in a clock logic gate when $C_L = 10$ nF, $V_{DD} = 1.2$ V, and $f_{clk} = 1.5$ GHz.

Answer: **(a)** $P = 4.8$ W, $P = 10.8$ W. **(b)** $P = 21.6$ W

5.8.2. Short-Circuit Power

Both transistors conduct when the input is changing, and the transition voltage is between V_{tn} and $V_{DD} - |V_{tp}|$. This creates a circuit current path from V_{DD} to ground (Figure 5-6) that may generate from 5 to 30% of the total switching power. This short-circuit power component depends on device current strength, input transition time, and output capacitance. If the transition time is long, then CMOS short-circuit current is active for a longer time. The exact computation of short-circuit current is complex so we present an approximation.

Consider a symmetric inverter (i.e., $K_n' = K_p'$, and $V_{tn} = -V_{tp}$) with no output load and an input voltage transition having equal rise and fall times. The time interval when both transistors conduct is from t_1 to t_3 in Figure 5-12. During the interval $t_1 - t_2$ the short circuit current increases from zero to its maximum value I_{max} and the nMOS transistor is saturated. Its drain current is

$$I_D = K_n \frac{W}{L}(V_i - V_{tn})^2 \quad \text{for } 0 < I < I_{max}$$

Since the inverter was assumed symmetric with no load, the maximum current occurs at $V_{in} = V_{DD}/2$ and its shape is symmetric along the vertical axis at $t = t_2$. We compute a mean current by integrating from $t = 0$ to $t = T$ and dividing by the period T. There are

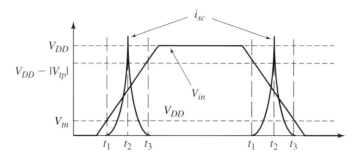

FIGURE 5-12.

A simplified view of the short-circuit current contribution over one clock period.

four equal area current segments to integrate in Figure 5-12 over the whole period T.

$$I_{mean} = \frac{1}{T}\int_0^T I(t)dt = \frac{4}{T}\int_{t_1}^{t_2} K_n\frac{W}{L}\left(V_{in}(t) - V_{tn}\right)^2 dt \qquad (5\text{-}15)$$

If the input voltage is a linear ramp of duration τ

$$V_{in} = \frac{V_{DD}}{\tau}t \qquad (5\text{-}16)$$

t_1 and t_2 are given by

$$t_1 = \frac{V_t}{V_{DD}}\tau \quad \text{and} \quad t_2 = \frac{\tau}{2} \qquad (5\text{-}17)$$

Substituting Eqs. (5-16 and 5-17) into Eq. (5-15)

$$I_{mean} = K_n\left(\frac{W}{L}\right)\frac{1}{T}\int_{\frac{V_T}{V_{DD}}\tau}^{\frac{\tau}{2}}\left(\frac{V_{DD}}{\tau}t - V_{tn}\right)^2 dt \qquad (5\text{-}18)$$

This integral is of the type $\int x^2 dx$ with $x = (V_{DD}/\tau)t - V_t$ so the result is

$$I_{mean} = \frac{1}{6}K_n\left(\frac{W}{L}\right)\frac{1}{V_{DD}}(V_{DD} - V_{tn})^3\frac{\tau}{T} \qquad (5\text{-}19)$$

Finally the power contribution is given by

$$P_{sc} = V_{DD}\,I_{mean} \qquad (5\text{-}20)$$

EXAMPLE 5-10

What is the short-circuit power in an inverter if $V_{DD} = 2$ V, $V_{tn} = -V_{tp} = 0.5$ V, $T = 500$ ps, $f_{clk} = 2$ GHz, $K_n = 140$ μA/V^2, $W/L = 2$, and the pulse rise and fall times are 100 ps?

$$I_{mean} = \left(\frac{1}{6}\right)(140 \ \mu A)\left(\frac{2}{1}\right)\left(\frac{1}{2V}\right)(2 - 0.5)^3\left(\frac{100 \text{ ps}}{500 \text{ ps}}\right) = 15.75 \ \mu A$$

so

$$P_{sc} = 2 \text{ V} \times 15.75 \ \mu A = 31.5 \ \mu W$$

Self-Exercise 5-17
The short-circuit current pulses in a CMOS inverter are modeled as rectangular pulses of 400 μA peak and 420 ps width. The clock frequency is 800 MHz and $V_{DD} = 1.2$ V. Calculate the short circuit power dissipation using Eq. (5-18).

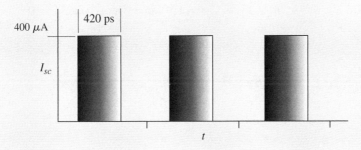

Answer: $P_{sc} = 161.3 \ \mu W$

5.8.3. Quiescent Leakage Power

The current transfer curve in Figure 5-6 showed that the quiescent current at the two logic states (0, V_{DD}) was quite low in the low nA range. Low leakage transistors lead to low power ICs. However, as the transistors become smaller, other leakage mechanisms kick in and the leakage can increase by orders of magnitude leading to ICs with DC leakages ranging from 20 to 40 amps in high-frequency ICs such as in Internet servers. It is called short channel leakage. Battery-operated ICs cannot stand this and use many tricks to bring the current leakage under control. This is a huge problem and is beyond this introductory course.

5.9. Power and Power Supply Scaling

The ratio of V_{tn} to V_{DD} impacts several inverter properties. The saturated current equation is

$$I_D = \frac{\mu_n \varepsilon}{2T_{ox}} \frac{W}{L}(V_{GS} - V_{tn})^2 \qquad (5\text{-}21)$$

When $V_{GS} = V_{DD}$ for logic circuits, Eq. (5-21) becomes

$$I_D = \frac{\mu_n \varepsilon}{2T_{ox}} \frac{W}{L} (V_{DD} - V_{tn})^2 \qquad (5\text{-}22)$$

When V_{DD} decreases, several trade-offs occur:

- The voltage difference in the parenthesis (the gate overdrive) is smaller so the current drive is less, and transistor speed is degraded. Load capacitance is charged and discharged slower.
- As V_{DD} drops, the the logic voltage traverses a smaller range giving a slight speed compensating feature.

When $V_{DD} < (V_t - V_{tp}) \approx |2V_t|$ the transition slows, but interestingly only one transistor is on at a time. There is essentially no transient current spike. Figure 5-13 shows a transfer curve measurement at $V_{DD} = 1$ V and $V_{DD} = 0.5$ V for transistors with thresholds on the order of 0.35 V. There is no current spike for the $V_{DD} = 0.5$ V measurement since $V_{DD} < V_{tn} + |V_{tp}|$. The power reduction for this condition is large.

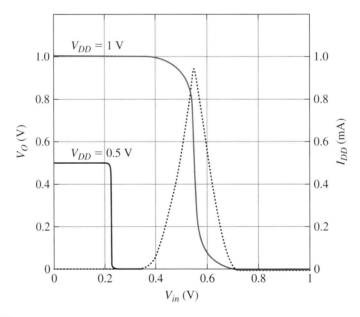

FIGURE 5-13.

Inverter transfer curves at two V_{DD} values. Notice the absence of a short-circuit current spike for $V_{DD} = 0.5$ V where V_{in} was swept from 0.5 V to 0 V. You can understand this low power operation if you sketch an inverter with $V_{DD} = 0.5$ V. Follow both transistor's on–off action as V_{in} goes from 0 V to 0.5 V.

Low-power, battery-operated products such as electronic watches and medical implants use this technique.

EXAMPLE 5-11

$V_{tn} = 0.4$ V and $K'_n = 400$ μA/V^2. What is the instantaneous peak power dissipation in a symmetrical inverter during a switching event for (a) $V_{DD} = 2$ V, and (b) $V_{DD} = 1.5$ V?

(a) $I_D = 400\,\mu\text{A} \left(\dfrac{2.0\,\text{V}}{2} - 0.4 \right)^2 = 144\,\mu\text{A}$

   $P_{sc} = (1\,\text{V})(144\,\mu\text{A}) = 144\,\mu\text{W}$

(b) $I_D = 400\,\mu\text{A} \left(\dfrac{1.5\,\text{V}}{2} - 0.4 \right)^2 = 49\,\mu\text{A}$

   $P_{sc} = (0.75\,\text{V})(49\,\mu\text{A}) = 36.75\,\mu\text{W}$

The power and peak current show a marked reduction with decrease in V_{DD}.

Self-Exercise 5-18
An inverter is driven by $V_{DD} = 1.0$ V pulses.

(a) What is the mean drain current limit to keep power dissipation to $<1\,\mu$W per pulse.
(b) If 10^5 inverters on a chip switch simultaneously for the values in part (a), what is the mean current of the chip over one clock period.

Answers: (a) $I_{mean} = 1\,\mu$A. (b) $I_{mean} = 100$ mA

Self-Exercise 5-19
The parameters for a symmetrical inverter are $V_{tn} = 0.6$ V, $V_{tp} = -0.6$ V, $K_n = 100\,\mu$A/V^2, $K_p = 50\,\mu$A/V^2, off-state leakage current is 1 pA, $(W/L)_n = 2$, and $(W/L)_p = 5$. Compare peak current and peak power at

(a) $V_{DD} = 2$ V
(b) $V_{DD} = 1.4$ V
(c) $V_{DD} = 1.0$ V

Answers: (a) $I_{peak} = 32\,\mu$A, $P_{peak} = 32\,\mu$W. (b) $I_{peak} = 2\,\mu$A, $P_{peak} = 1.4\,\mu$W. (c) $I_{peak} = 1$ pA, $P_{peak} = 0.5$ pW

Self-Exercise 5-20

Figure 5.13 showed an inverter voltage transfer curve for $V_{DD} = 0.5$ V. The curve actually is for a reverse sweep of V_i from 0.5 V to 0 V.

(a) Draw the transfer curve if the sweep for V_i goes from 0 V to 0.5 V. Approximate this forward sweep estimating V_{tp} from the 1.0 V VTC, and assume that $V_{tn} = -V_{tp}$.
(b) What is the width of the hysteresis zone?

Answer: **(b)** 0.26 V

5.10. Sizing Inverter Buffers to Drive Large Loads

A problem exists when a small logic gate must drive a large capacitive load C_L. A fast charge–discharge of a large load capacitance requires a large W/L of the driving transistors. However, a large W/L is defeating, since its larger transistor gate area increases its own logic input capacitance. Working backward, that would cause all preceding logic gates to have ever larger W/L ratios. A better solution exists.

Instead, one approach for driving large loads at high speed uses successively larger channel widths in a cascade of inverters to sufficiently increase the current drive of the last stage. A circuit driving a large load is commonly known as a buffer, and a circuit designed with successively larger inverters is known as a tapered buffer (Figure 5-14). When the area of each stage increases by the same factor the circuit is called a fixed tapered buffer.

The fixed tapered buffer structure was proposed by Linholm in 1975 [1]. He used a simple capacitance model making the W/L of a stage proportional to the size of the input capacitance of the next stage, while the area of each inverter was proportional to the channel width of the transistors. The overall buffer delay was optimized by minimizing the delay of each stage.

Let each succeeding stage in the buffer in Figure 5-14 have transistor widths larger than the previous one by a scale factor α. The first inverter is the smallest with an input

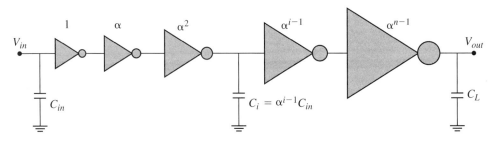

FIGURE 5-14.

Tapered buffer structure.

capacitance C_{in}, while the i-th stage has an input capacitance given by

$$C_i = \alpha^{i-1} C_{in} \quad i = 1, 2, \ldots, n$$

The number of stages n is computed from

$$C_L = \alpha^n C_n$$

then

$$\alpha^n = \frac{C_L}{C_{in}}$$

and

$$n = \frac{\ln\left(\dfrac{C_L}{C_{in}}\right)}{\ln \alpha} \tag{5-23}$$

α is computed by optimizing the delay. Assuming that the delay of the first stage driving an identical one is τ_0, the delay of the i-th stage is

$$t_{di} = \alpha \tau_0 \quad i = 1, 2, \ldots, n$$

The global delay of the n stages is

$$t_d = \sum_{i=1}^{n} t_{di} = n\alpha\tau_0$$

giving

$$t_d = \ln\left(\frac{C_L}{C_{in}}\right) \frac{\alpha}{\ln \alpha} \tau_0 \tag{5-24}$$

Differentiating (5-24) with respect to α and equating to zero, gives the optimum α_{opt} as

$$\alpha_{opt} = e \approx 2.7$$

while the optimum number of stages n_{opt} is

$$n_{opt} = \ln\left(\frac{C_L}{C_{in}}\right) \tag{5-25}$$

This section seeks to impress that care must be taken in a design when a logic gate drives a large capacitance. This occurs when an IC drives large capacitive loads such as an off-chip board capacitance or on-chip bus lines.

EXAMPLE 5-12

How many buffer stages are need to optimally drive a 1 pF load if the driving gate has an input capacitance of $C_{in} = 25$ fF?

$$n = \ln\left(\frac{C_L}{C_{in}}\right) = \ln\left(\frac{1\,\text{pf}}{25\,\text{fF}}\right) = 3.7$$

Therefore, a total of four buffer stages are needed.

Self-Exercise 5-21

An IC with a tapered buffer drives a load capacitance on a board that is 100 pF. The input capacitance of the logic gate originating the signal is 100 fF, and that gate has $W/L = 4$.

(a) How many buffer gates are required to optimally drive that load using the fixed tapered buffer model?

Answer: $n = 6.9$

Seven total stages are needed. We must insert five tapered stages between the original gate and the output driver.

(b) Write the equation that predicts the W/L ratio of the final buffer in terms of the scaling factor and the originating gate W/L

(c) What is the W/L ratio of the final output buffer driver to the board?

Answer: $W/L = 1,614$

Self-Exercise 5-22

A tapered buffer design has an input capacitance of 20 fF and a load capacitance of 5 pF.

(a) What is the required number of buffer stages to minimize the propagation delay?

(b) If the propagation delay of the first stage is 1.5 ns, what is the overall delay?

Answers: (a) $n_{opt} = 5.52 \Rightarrow 6$ stages. (b) $T_d = 24.5$ ns

5.11. Summary

This chapter examined detailed electronic properties of the inverter. The inverter properties align with NAND, NOR, and other multi-input logic gates studied in the next chapter. Static and dynamic transfer curves explain much of the speed and power behavior of integrated circuits. Tapered buffers are commonly used in design to match small logic gate drive to larger high input capacitance load gates. The next chapter expands these concepts to show how the inverter leads to multi-input logic gates.

References

[1] F. A. Linholm IEEE J. Solid State Circuits, SC-10, 2, pp. 106–109, April 1975.

Exercises

Inverter Static Voltage Characteristics

5-1. A CMOS inverter has $V_{DD} = 1.2$ V. $V_{OH} = 1.15$ V, $V_{OL} = 0.15$ V, $V_{IH} = 1.05$ V, and $V_{IL} = 0.2$ V. Calculate NM_H, NM_L, and draw the noise margin map with appropriate labels of numbers.

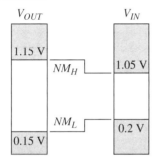

5-2. A logic gate noise margin parameters are $V_{IH} = 1.6$ V, $V_{IL} = 0.3$ V, $V_{OH} = 1.7$ V, and $V_{OL} = 0.2$ V.
(a) Calculate NM_H.
(b) Calculate NM_L.
(c) The input voltage is down to 1.7 V and a negative 50 mV noise spike appears. What happens to the circuit fidelity?
(d) The input voltage is down to 1.7 V and a negative 150 mV noise spike appears. What happens to the circuit fidelity?

5-3. Given the logic gate noise margins: $NM_H = 100$ mV, $NM_L = 75$ mV, and $V_{DD} = 2$ V.
(a) If $V_{IH} = 1.75$ V, what is V_{OH}?
(b) If $V_{IL} = 0.3$ V, what is V_{OL}?

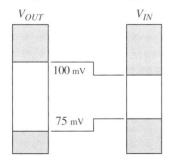

5-4. A CMOS inverter uses $V_{DD} = 0.9$ V. $V_{OH} = 0.8$ V, and $V_{OL} = 0.1$ V. If the noise margins must be 20% of V_{DD}, what are V_{IL} and V_{IH}? Draw the noise margin map and label.

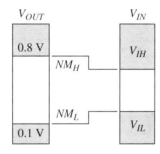

5-5. Graphically determine the change in logic threshold of the CMOS inverter transfer curve in the figure if the curve shifts 0.2 V to the right in the midregion.

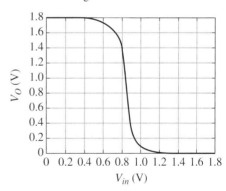

5-6. (a) Design the W_p/W_n ratios of a CMOS inverter for symmetrical static voltage transfer characteristic. $\mu_n = 1400$ cm^2/V $\cdot$ s, $\mu_p = 500$ cm^2/V $\cdot$ s, $V_{tn} = 0.35$ V, $V_{tp} = -0.35$ V, and $V_{DD} = 1.3$ V.

(b) Redesign if $V_{tp} = -0.45$ V.

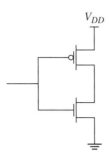

5-7. An inverter with a symmetrical voltage transfer curve has a restriction that $W_p/W_n = 4.6$. $V_{DD} = 1.2$ V, $\mu_n = 1530$ cm^2/V·s, $\mu_p = 540$ cm^2/V·s, and $V_{tp} = -0.4$. What must V_{tn} be set to satisfy this condition?

5-8. A CMOS inverter has transistor parameters: $K_n(W/L)_n = 100$ μA/V^2, $K_p(W/L)_p = 300$ μA/V^2, $V_{tn} = 0.7$ V, $V_{tp} = -0.75$ V, and $V_{DD} = 2.5$ V. What fraction of the total output voltage swing will the nMOS transistor be in saturation?

5-9. A CMOS inverter has its nMOS transistor in nonsaturation and its pMOS transistor in saturation. Given $K_n = 50$ μA/V^2, $K_p = 25$ μA/V^2, $V_{tn} = 0.5$ V, $V_{tp} = -0.6$ V, $(W/L)_n = 2$, $(W/L)_p = 4$, $I_{DD} = 11$ μA, and $V_{DD} = 2$ V, calculate the inverter output voltage V_O.

Inverter Static Current Characteristics

5-10. Given an inverter with $V_{DD} = 1.5$ V, $V_{tn} = 0.4$ V, and $V_{tp} = -0.4$ V, calculate the peak current during the transition if $(W/L)_n = 3$, $(W/L)_p = 7.5$, $K_p = 50\,\mu$A/V^2, and $K_n = 125\,\mu$A/V^2.

5-11. An inverter has $V_{DD} = 2$ V, $V_{tn} = 0.5$ V, $V_{tp} = -0.5$ V, $K_n = 300\,\mu$A/V^2, $K_p = 200\,\mu$A/V^2, $(W/L)_n = 2$, and $(W/L)_p = 3$.
(a) If $V_{IN} = 0.8$ V, what is I_{DD}?
(b) The I_{DD} solution in part (a) appears twice in the current transfer curve. Use the pMOS equations to calculate the other V_{IN} value to satisfy the current in part (a).

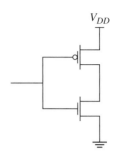

5-12. Given $V_{DD} = 1.5$ V, $K_p = 70$ μA. $K_n = 120$ μA, $(W/L)_p = 150$, and $(W/L)_n = 75$, use the ITC to calculate V_{tp}.

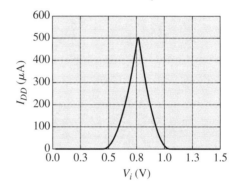

5-13. Given that $V_{DD} = 1.8$ V, $V_{tn} = 0.5$ V, and $K_n = 100$ μA/V^2, what is W/L of the nMOS transistor?

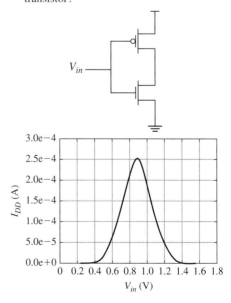

5-14. $I_{DD} = 40\ \mu\text{A}$, $(W/L)_n = 2$, $K_n = 100\ \mu\text{A}$, $V_{tn} = 0.5\ \text{V}$, $K_p = 50\ \mu\text{A}$, $V_{tp} = -0.5\ \text{V}$, and $V_i < 1.5\ \text{V}$. What is V_i?

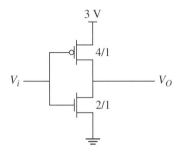

5-15. Given an inverter with $V_{tn} = 0.4\ \text{V}$, $V_{tp} = -0.35\ \text{V}$, $K_n = 200\ \mu\text{A/V}^2$, $K_p = 100\ \mu\text{A/V}^2$, $(W/L)_n = 2$, and $(W/L)_p = 3$, calculate the peak drain current I_{peak} during an inverter transition for **(a)** $V_{DD} = 1.5\ \text{V}$ and **(b)** $V_{DD} = 1.0\ \text{V}$.

Inverter Speed Property

5-16. Use the transition time delay model where $C_L = 30\ \text{fF}$, $V_{DD} = 1.5\ \text{V}$, $(W/L)_n = 2$, $K'_n = 100\ \mu\text{A/V}^2$, $(W/L)_p = 5$, $K'_p = 25\ \mu\text{A/V}^2$, $V_{tp} = -0.35\ \text{V}$, and $V_{tn} = 0.35\ \text{V}$. What is the difference between rise and fall time of the transition if defined between 0 V and 1.5 V?

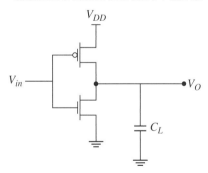

5-17. If a *p*MOS transistor in an inverter has $\mu\varepsilon/2T_{ox} = 28\ \mu\text{A/V}^2$, $V_{tp} = -0.6\ \text{V}$, and $W/L = 6$, what is the expected additional rise time delay if the gate power supply voltage is reduced from a normal $V_{DD} = 2.5\ \text{V}$ to $V_{DD} = 1.8\ \text{V}$ with $C_L = 25\ \text{fF}$.

5-18. A CMOS inverter has $W/L = 6$ for both transistors, $V_{tn} = 0.6\ \text{V}$, $V_{tp} = -0.6\ \text{V}$, and $V_{DD} = 2.3\ \text{V}$. If V_t is reduced to $|V_t| = 0.2\ \text{V}$ for both transistors, what is the percent decrease in speed of transition?

Inverter Power

5-19. Calculate the power dissipated by a cardiac pacemaker circuit if $f_{clk} = 32.6\ \text{kHz}$, $\alpha = 0.1$, $V_{DD} = 1.5\ \text{V}$, C_L (per gate) $= 300\ \text{fF}$, and the number of logic gates $= 10\ \text{k}$.

5-20. A clock network has $C_L = 10\ \text{nF}$, $\alpha = 1$, and $V_{DD} = 1.2\ \text{V}$. The maximum power dissipation allowed is 5 W. What is the maximum clock frequency?

5-21. Use Figure 5-12. $V_{DD} = 0.9\ \text{V}$, $V_{tn} = 0.2\ \text{V}$, $V_{tp} = -0.2\ \text{V}$, $f_{clk} = 3\ \text{GHz}$, $W/L = 3$, $K_n = 250\ \mu\text{A/V}^2$, and $t_r = t_f = 40\ \text{ps}$. Calculate the mean current during the logic transition and the average power dissipated in the chip.

Power Supply Scaling

5-22. Given an inverter with: $V_{tn} = 0.4\ \text{V}$, $V_{tp} = -0.4\ \text{V}$, $K_n = 200\ \mu\text{A/V}^2$, $K_p = 100\ \mu\text{A/V}^2$, $(W/L)_n = 2$, and $(W/L)_p = 3$. Calculate the peak drain current I_{peak} during an inverter transition for **(a)** $V_{DD} = 1.5\ \text{V}$ and **(b)** $V_{DD} = 1.0\ \text{V}$.

5-23. P_{sc} must be kept under 1 W. The chip has $V_{DD} = 1.5\ \text{V}$, one million transistors, and $\alpha = 0.1$. Assume that 10^6 transistors represent an equivalent 500 k inverters for analysis. What is the mean drain current per inverter?

Sizing and Inverter Buffers

5-24. An output buffer has an input capacitance of 95 fF and a load capacitance of 100 pF. How many inverters are required in a fixed tapered design to minimize the propagation delay?

5-25. A fixed tapered buffer has an input capacitance of 1 pF. If the output stage must drive a load of 54 pF, how many stages are needed?

5-26. The number of tapered buffers in a design must be kept at no more than five to accommodate chip area constraints.
  (a) If the input gate capacitance is 50 fF, what is the maximum load capacitance that can be driven?
  (b) What is the width ratio of the last inverter W_L to the first inverter in the chain W_{in}?

CHAPTER 6

CMOS NAND, NOR, and Transmission Gates

Give me just inverters and NAND gates, and I can build a computer, in fact just give me NAND gates.

Anonymous

Multi-input gates such as the NAND, NOR, and exclusive-OR (XOR) share most of the inverter properties, plus a small number of their own. We will continue the bottom-up transistor-level approach to understanding computing circuits. This chapter concludes with the two-transistor CMOS transmission gate that is essential for most flip-flop designs. We will start by looking at a distribution of the different types of logic gates in an integrated circuit (IC).

Figure 6-1 shows the distribution of logic gate types typically found in existing IC designs. The four most common logic gates, in order of popularity, are inverters, flip-flops, 2-NANDs, and 2-NORs. There are, in fact, more inverters used than shown since each flip-flop typically has four inverters. The gates studied in this chapter are heavily used in IC designs.

6.1. NAND Gates

CMOS circuits naturally implement negated functions such as the inverter. This means that logic gate output signals are inverted with respect to one or more inputs. Simple examples are the inverter, the NAND gate, and the NOR gate. AND and OR gates implement positive functions, where the output signal is in phase with the input signal. But CMOS AND and OR gates add an inverter to a NAND or NOR gate to perform a double-negated function.

A 2NAND gate symbol and truth table are shown in Figures 6-2a and 6-2b, and there are two properties to note. The first is that any logic-0 to the inputs of a NAND gate causes a logic-1 output. The other property is more subtle but vital to logic design, debug, and

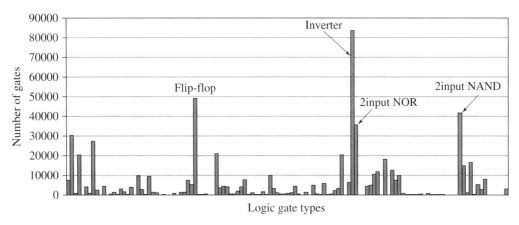

FIGURE 6-1.

Histogram of logic gate types in an ASIC design. (Reproduced by permission of LSI Corp.)

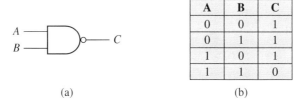

A	B	C
0	0	1
0	1	1
1	0	1
1	1	0

(a)   (b)

FIGURE 6-2.

(a) 2NAND gate symbol. (b) Truth table.

testing ICs. Certain input levels are called *noncontrolling states*. When $A = 1$ in rows 3 and 4, then the output-C is the negation or complement of B ($C = \overline{B}$). C depends only on the value of B if $A = 1$. Likewise, if $B = 1$, C is the complement of A. Control signals are typically set to the noncontrolling logic state prior to activating an event. This property is also essential for moving signals through a sequence of logic gates deep in the logic blocks and then observing the result at an observable circuit node such as an output pin. Designers and test engineers use this property when debugging their circuits or writing test programs to detect possible defective signal nodes.

6.1.1. Electronic Operation

Figure 6-3a shows the 2NAND gate transistor schematic. The electronic operation follows the truth table in Figure 6-3b. A logic-0 on any input line turns off an nMOS pull-down transistor and closes the path from the output C to ground. But a logic-0 input ensures that a pMOS is turned on passing V_{DD} to the output. Therefore, for any logic-0 on the inputs, the output is at logic-1, or $V_C = V_{DD}$. If both inputs are logic-1 ($V_A = V_B = V_{DD}$), then both nMOS transistors turn on, both pMOS transistors are off, and $V_C = 0$ V.

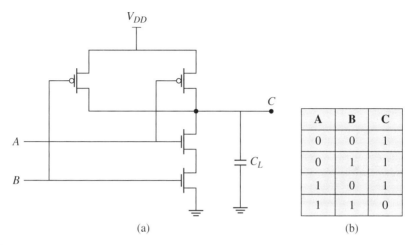

(a)   (b)

FIGURE 6-3.

(a) Transistor level structure of a static CMOS NAND gate. (b) Truth table.

6.1.2. NAND Noncontrolling Logic State

When the noncontrolling logic-1 state is set for a NAND input node, we could short the affected nMOSFET with a wire and remove the off-state pMOSFET in Figure 6-3a and get the same effect. If we set node-B in Figure 6-3 to its noncontrolling state $V_B = V_{DD}$ then a voltage sweep at node-A produces static and dynamic transistor curves similar to those measured for the inverter. A similar response is measured if node-A is shorted to node-B. The NAND gate has most of the inverter properties developed in Chapter 5.

AND gates can be made by adding an inverter to the output of a NAND gate, so the noncontrolling logic state for an AND gate is also logic-1. An example will illustrate the noncontrolling logic state concept with respect to selected signal transmission.

EXAMPLE 6-1

Let us observe an inner node state at the circuit output node O_1.

(a) What should node I_1 be to read the state of node-B?
(b) What should I_2 and I_3 be to read the state of node-A?
(c) What should I_1 and I_2 be to read the state of node-I_3 at output-O_1?

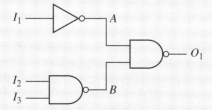

(a) To pass a signal from node-B to the output O_1 requires that node-A is set to the logic-1 noncontrolling state. Therefore, I_1 must be logic-0.

(b) To pass a signal from node-A to the output O_1 requires that node-B is set to the logic-1 noncontrolling state. Therefore, $I_2 I_3$ must be 00, 01, or 10.

(c) To pass a signal from node-I_3 to the output O_1 requires that node-I_2 be set at logic-1 and I_1 be set at logic-0.

EXAMPLE 6-2

Write the binary sequence for $AB\ CD\ E$ that will

(a) Make $F = f(D)$.
(b) Make $F = f(A)$.

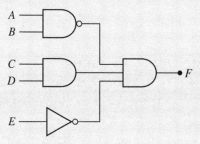

Answers: **(a)** E must set the noncontrolling logic state for the output 3AND gate so $E = 0$. The 2NAND gate must also set a noncontrolling logic state for the output 3AND gate so $AB = 00$. Node-C is 1. The x symbol indicates that either logic state can be used. The complete binary word to make $F = f(D)$ is $AB\ CD\ E = 00\ 1\ x\ 0$. Notice that $AB = 01$ or $AB = 10$ also satisfy $F = f(D)$. **(b)** The 2AND must deliver a noncontrolling logic state for the output 3AND gate so $CD = 11$, node-E is a logic-0, and node-B is logic-1. The complete binary word for $F = f(A)$ is $AB\ CD\ EF = x1\ 11\ 0$.

The noncontrolling logic state property is also useful when manually analyzing logic circuits. If a multiple input logic gate has noncontrolling logic states on one or more of its inputs, then those states can be ignored in the analysis. The following example will illustrate.

EXAMPLE 6-3

The circuit is a small example of what an engineer might be required to read. What is the logic signal at the output? Locate signals that are the noncontrolling logic state, and focus on those signals that are not.

Notice that three of the 4NAND inputs are logic-1, which is the noncontrolling logic state. Therefore, node-D is the only signal of influence on node-F, and $F = \overline{D} = 1$.

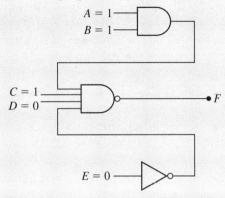

Self-Exercise 6-1

Write the binary values for $AB\ CD\ EF$ that make $G = f(F)$. Use the noncontrolling logic state concept.

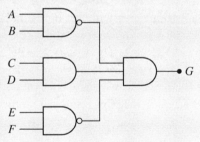

Answer: AB CD EF = 00 11 1×
AB can take one of three pairs: 00, 01, 10.

Self-Exercise 6-2

Use the noncontrolling state property to analyze the logic state of node-G.

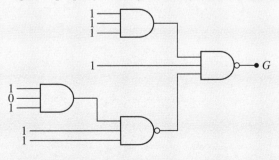

Answer: G = 0

6.2. NAND Gate Transistor Sizing

NAND gates have a major difference with the inverter, and that has to do with matching rise and fall times. The goal of matching inverter rise and fall times led to an inverter transistor symmetrical design of $K_n = K_p$ and $V_{tn} = -V_{tp}$. That led to making $(W/L)_p > (W/L)_n$ by a factor of about 2.0–3.0 to compensate for the lower pMOS transistor carrier mobility. Inverter pull-up and pull-down strengths could be made equal for both logic state transitions.

The NAND gate in Figure 6-3 is more complicated since it has more input signal possibilities to deal with if equal rise and fall times are desired. The pull-up current drive strength depends on the number of pMOS transistors that are activated. Two parallel pMOS transistors have twice the pull-up strength (PU) of a single pMOS. Also, when the pull-down path (PD) is activated, two series nMOS transistors have about half of the current drive strength of just a single nMOS.

How do we deal with this? Since symmetrical rise and fall times cannot be done for all logic states another approach is needed. A pragmatic design adjusts *worst-case rise time* and *worst-case fall time* transitions with respect to a balanced inverter. These criteria use the property that two identical parallel transistors can be treated as a single transistor with its W/L ratio multiplied by two. We approximate two identical series transistors by an equivalent single transistor with its W/L ratio divided by two.

Figure 6-4a shows a balanced inverter with the p-channel transistor W/L ratio sized at three times the n-channel transistor. Figure 6-4b shows the worst-case pull-up and pull-down configuration for the NAND gate. The worst-case PU is a single pMOS transistor, and the worst-case PD is the two series nMOS transistors. To maintain the same pull-down strength as the inverter, the n-channel transistor's W/L ratio must be doubled. The pull-up transistor can maintain its $W/L = 3$ since that retains the same PU strength as the reference inverter. Figure 6-4c shows the adjusted W/L ratios for the quasi-symmetrical 2NAND

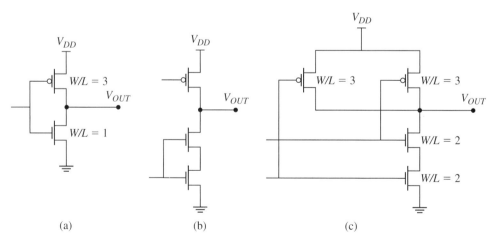

| (a) | (b) | (c) |

FIGURE 6-4.

(a) Balanced inverter. (b) Worst-case pull-up and pull-down configuration for a 2NAND gate. (c) 2NAND sized for equivalence to balanced inverter in (a).

gate. This design will have a faster best-case pull up when both inputs are equal to logic-0, but that imbalance is normally anticipated and acceptable.

A general equation to find the equivalent transistor width in a series stack of transistors is

$$W_{equiv} = \left[\frac{1}{\dfrac{1}{W_1} + \dfrac{1}{W_2} + \cdots \dfrac{1}{W_n}} \right] \qquad (6\text{-}1)$$

We use the same equation to solve for the individual NAND gate widths to match the desired inverter reference value. The nMOS transistor equivalent width in Figure 6-4 is

$$W_{equiv} = \left[\frac{1}{\dfrac{1}{W_1} + \dfrac{1}{W_2}} \right] = \left[\frac{1}{\dfrac{1}{2} + \dfrac{1}{2}} \right] = 1$$

The pull-up and pull-down strengths are equivalent in Figures 6-4a and 6-4c.

EXAMPLE 6-4

Size a 3NAND gate for worst-case pull-up and pull-down strength to be approximately that of an inverter whose p-channel transistor $W/L = 3$, and whose n-channel transistor $W/L = 1$.

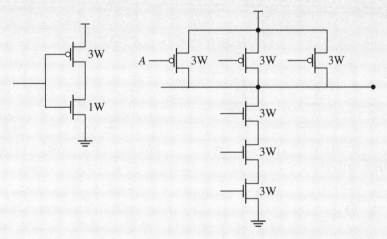

The worst-case pull up has a single p-channel transistor, and the worst-case pull down has three n-channel transistors in series. The 3NAND p-channel transistor W/L ratio will remain at $W/L = 3$ since that is the pull-up strength of the original inverter. The 3NAND n-channel transistor W/L ratio must be the pull-down strength of the original inverter. Calculate the enlarged nMOS transistor widths from Eq. (6.1) where $W/L = 3$.

$$W_{equiv} = \left[\frac{1}{\dfrac{1}{W_n} + \dfrac{1}{W_n} + \dfrac{1}{W_n}} \right] = \left[\frac{W_n}{3} \right] = 1 \qquad W_n = 3$$

Self-Exercise 6-3
Use Eq. (6-1) and show that the pull-down strengths are equal.

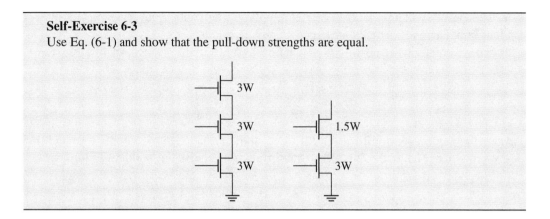

Self-Exercise 6-4
Use Eq. (6-1) and show that the pull-down strengths are equal.

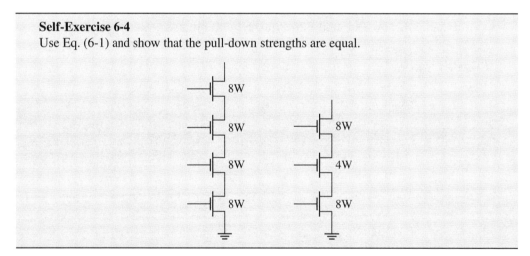

6.3. NOR Gates

A NOR gate symbol and truth table are shown in Figures 6-5a and 6-5b, and again there are two properties to note. The first is that any input logic-1 causes a logic-0 output. The noncontrolling logic states are different from the NAND and AND gates. When $A = 0$ in rows 1 and 2, then the output C is the complement of B ($C = \overline{B}$). A similar property is seen in rows 1 and 3 where the output $C = \overline{A}$ when $B = 0$. The noncontrolling logic states for a NOR and an OR gate are logic-0.

6.3.1. Electronic Operation

Figure 6-6a shows the 2NOR gate transistor schematic. The electronic operation follows the truth table in Figure 6-6b. A logic-1 on any input line turns off a pMOS pull-up transistor and closes the path from the output C to V_{DD}. But a logic-1 input ensures that an nMOS is turned on passing the output to ground. Therefore, for any logic-1 on the inputs,

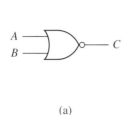

A	B	C
0	0	1
0	1	0
1	0	0
1	1	0

(a)   (b)

FIGURE 6-5.

(a) 2NOR gate symbol. (b) Truth table.

the output is at logic-0, or $V_C = 0$ V. If the inputs are logic-0 ($V_A = V_B = 0$ V), then both nMOS transistors turn off, both pMOS transistors are on, and $V_C = V_{DD}$.

6.3.2. NOR Noncontrolling Logic State

Figure 6-6a shows that when an input to a NOR gate is set to logic-0, then the effects of that signal node are removed with respect to signals on the other input lines. A logic-0 turns on a pMOS transistor but turns off its complementary nMOS transistor. In effect, a shorted wire can replace the pMOS, and the nMOS removed without changing the electrical result. OR gates can be made by adding an inverter to the output of a NOR gate, so the noncontrolling logic state of an OR gate is also a logic-0.

The NOR gate has static and dynamic inverter properties. These are seen by setting one of the inputs to its noncontrolling logic state and measuring a transfer curve at the other terminal or by shorting both inputs.

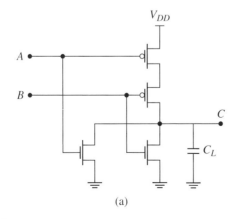

A	B	C
0	0	1
0	1	0
1	0	0
1	1	0

(a)   (b)

FIGURE 6-6.

(a) 2NOR transistor–level schematic. (b) Truth table.

EXAMPLE 6-5

Write the binary sequence for *AB CD EF* that will

(a) Make $G = f(D)$.
(b) Make $G = f(F)$.

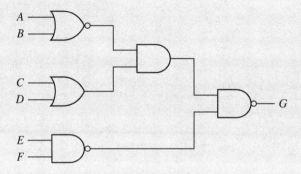

(a) *E* and *F* must set the noncontrolling logic state for the output 2NAND gate so $EF = 00$, 10, or 01. The 2NOR gate must set the noncontrolling logic state for the 2AND gate so $AB = 00$, node-*C* is 0. *x* denotes a variable state (logic-0 or logic-1 for node-*D*). A complete binary word is

$$AB\ CD\ EF = 00\ 0 \times 00$$

(b) The upper OR, NOR, and AND gates must deliver a noncontrolling logic state for the output 2AND gate so $AB = 00$, 10, or 01, $CD = 11$, 10, or 01. Node-*E* is a logic-1. One binary word solution is

$$AB\ CD\ EF = 00\ 11\ 1 \times$$

Self-Exercise 6-5

Use the noncontrolling logic state concept.

(a) Specify the input signals that allow node-*C* contents to be measured at node-O_1.
(b) Repeat for reading node-*B*.

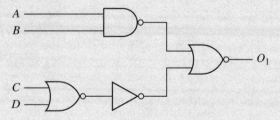

Answers: (a) $ABCD = 11 \times 0$. (b) $ABCD = 1 \times 00$

Self-Exercise 6-6
(a) Specify the input signals that allow node-C contents to be measured at F.

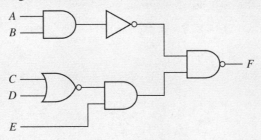

(b) Repeat for reading node-B.
(c) Repeat for reading node-A.

Answers: **(a)** AB CDE = 0000×. **(b)** AB CDE = 00 × 01. **(c)** AB CDE = ×1001

Self-Exercise 6-7
Write the binary values for *AB CD EF* that make $G = f(F)$. Use the noncontrolling logic state concept.

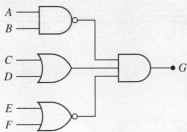

Answer: AB CD EF = 00 11 0×
AB can take one of three pairs: 00, 01, 10. *CD* can take one of three pairs: 11, 01, 10.

Self-Exercise 6-8
Use the noncontrolling state property to analyze and give the logic state of node-G.

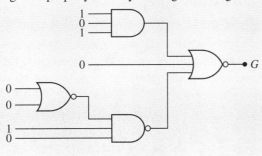

Answer: G = 0

Self-Exercise 6-9

What is the logic signal at the output? Locate signals that are in the noncontrolling logic state, and focus on those signals that are not.

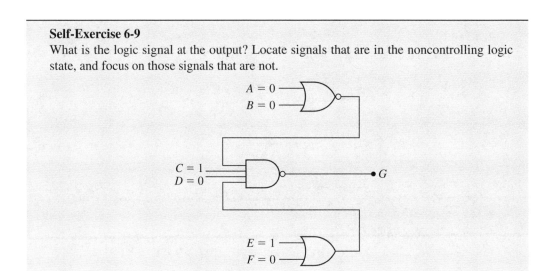

Answer: $G = 1$

Self-Exercise 6-10

Use the noncontrolling state property to analyze the logic state of node-G.

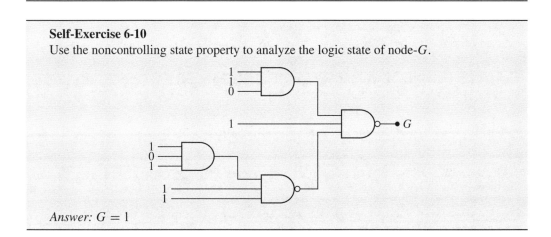

Answer: $G = 1$

6.4. NOR Gate Transistor Sizing

The NOR gate schematic in Figure 6-6 showed that any logic-1 input turns on an nMOS transistor forcing the output node-C to 0 V. If both inputs are driven high with V_{DD} then the pull-down strength is double since two nMOS transistors are in parallel. Also, two pMOS transistors in series are potentially very slow, so the W/L adjustments for equal rise and fall times require larger pMOS transistors.

The NOR gate general W/L sizing is similar to the NAND gate. The NOR transistors are sized to maintain the same strength as a balanced inverter in the worst-case configuration for rise and fall time. Figure 6-7a shows a balanced inverter with the p-channel transistor W/L ratio sized at three times the n-channel transistor. Figure 6-7b shows the worst-case

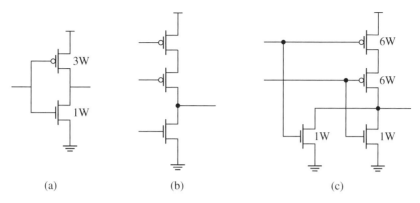

FIGURE 6-7.

(a) Balanced inverter. (b) Worst-case pull-up and pull-down configuration for a 2NOR gate. (c) 2NOR sized for equivalence to balanced inverter in (a).

pull-up and pull-down configuration for the NOR gate. A single pull-down transistor is matched against two pull-up transistors in series. To maintain the same pull-up strength as the inverter, the p-channel transistor's W/L ratio must be doubled to $W/L = 6$. The pull-down transistor can maintain its $W/L = 1$ since that retains the same strength as the inverter. Figure 6-7c shows the adjusted W/L ratios for the 2NAND. The NOR gate suffers a larger gate width in the pull-up transistors. A circuit with a need for a fast pull-down and a noncritical pull-up can relax the pMOS size.

EXAMPLE 6-6

Draw the worst-case pull-up and pull-down strengths and size a 3NOR gate for it to be that of an inverter whose p-channel transistor $W/L = 3$ and whose n-channel transistor $W/L = 1$.

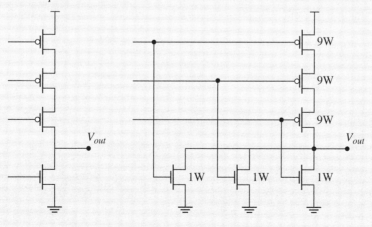

The worst-case pull-up has three p-channel transistors in series and the worst-case pull-down is a single n-channel transistor. The 3NOR p-channel transistors W/L ratio must collectively be that of $W/L = 3$ since that is the pull-up strength of the original inverter. Therefore, the p-channel transistors in series must have $W/L = 9$.

$$W_{equiv} = \left[\frac{1}{\dfrac{1}{W_p} + \dfrac{1}{W_p} + \dfrac{1}{W_p}} \right] = \left[\frac{W_p}{3} \right] = 3 \qquad W_p = 9$$

The 3NOR n-channel transistor W/L ratio must be the pull-down strength of the original inverter, so $W/L = 1$. Notice again that the 3NOR sizing requires quite large transistors increasing the gate capacitance and chip area. The total area of a 3NOR is considerably larger than a 2NAND.

Self-Exercise 6-11

If a symmetrical inverter has a $3:1$ width ratio of its pMOS and nMOS transistors, what should the logic gate transistor sizes be for the multi-input logic circuit?

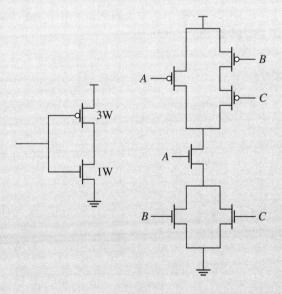

Answers: nMOS: $W_A = W_B = W_C = 2$. pMOS: $W_A = 3$, $W_B = W_C = 6$

Self-Exercise 6-12

Given the symmetrical inverter sizing of its pMOS and nMOS transistors in the inverter, what should the transistor sizes be of this logic gate?

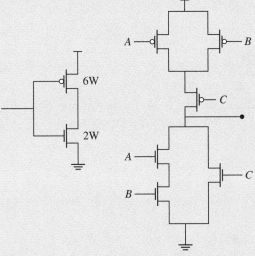

Answers: nMOS: $W_A = W_B = 4$, $W_C = 2$. pMOS: $W_A = W_B = W_C = 12$

Self-Exercise 6-13

Given the symmetrical sizing of its pMOS and nMOS transistors in the inverter, what should the transistor sizes be of this logic gate?

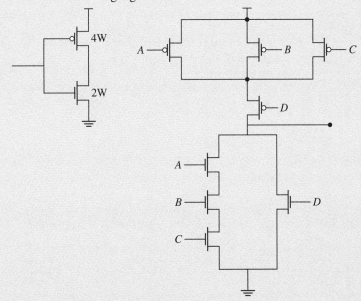

Answers: nMOS: $W_A = W_B = W_C = 6$, $W_D = 2$. pMOS: $W_A = W_B = W_C = W_D = 8$

The transistor sizing examples stressed a design that references PU and PD strength to a balanced inverter. However, certain local design situations may have only a PU or a PD that is speed sensitive. In such a case the designer may use smaller W/L sizes for the nonsensitive transition saving chip area.

EXAMPLE 6-7

(a) A balanced inverter gate is shown on the left. $K_n = 90\ \mu A/V^2$, $V_{tn} = 0.4$ V, $V_{tp} = -0.4$ V. What is the current drive constant K_p?

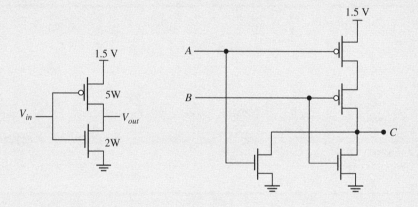

Model the p- and n-channel transistors in their saturated bias state and choose a bias voltage. Set the input signal at $V_{DD}/2$.

$$90\ \mu A(2)(0.75 - 0.4)^2 = K_p(5)(0.75 - 1.5 + 0.4)^2$$

$$K_p = 36\frac{\mu A}{V^2}$$

(b) If each transistor type in the 2NOR has the same W/L as the inverter, what is the impact on the speed of the 2NOR?

- The 2NOR nMOS transistors have $W/L = 2W$ and the worst-case PD is unaffected. The 2NOR pMOS transistors have $W/L = 5W$ so that the PU speed is slower by a factor of two.

EXAMPLE 6-8

The current drive strengths of the transistors are equal. Analyze the logic function. Expand on any anomalies you find.

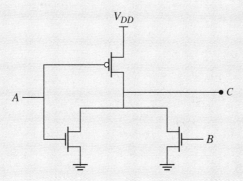

A	B	C
0	0	1
0	1	$0.5 \, V_{DD}$
1	0	0
1	1	0

The second logic state ($AB = 01$) puts A-pMOS in contention with B-nMOS. Since transistors have equal drive strengths, $C = 0.5 \times V_{DD}$ where both PU and PD transistors are in the saturated state. In this bias region the circuit acts as a linear amplifier, but it has no logic function. Also it has two signal inputs (A, B) that are both driven to the rails. Small signal changes on A or B would modulate I_{DD} and V_{out}.

6.5. Pass Gates and CMOS Transmission Gates

6.5.1. Pass Gates

Numerous situations require that a signal path be gated. A signal will pass to a next stage only on command from a control signal. Transmission gates (T-gates) perform this function. Single transistors acting as transmission gates are called pass transistors, but they have a weakness.

Assume that the n-channel transistor in Figure 6-8a drives C_L and node-$A = V_{DD}$. When the control input-G is high, the T-gate passes charge from the node-A drain to C_L. When the node-B source rises to a voltage one-threshold drop below V_G, then the n-channel transistor turns off, preventing the node-B source from rising above $V_G - V_{tn}$.

An n-channel transistor passes a weak logic high and a strong logic low. Similarly, a p-channel transistor in Figure 6-8b will have a strong logic-1 and a weak logic zero that is one threshold drop above ground.

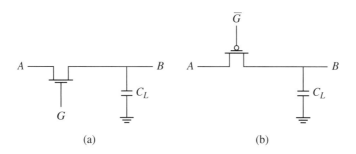

FIGURE 6-8.

Pass transistors: (a) nMOS. (b) pMOS.

EXAMPLE 6-9

Given that $V_{tn} = 0.5$ V and $V_{tp} = -0.5$ V:

(a) What is V_{out} ignoring the transistor body effect?

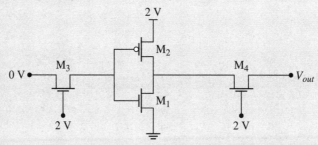

The input logic-0 passes unattenuated through M$_3$. That logic-0 becomes a strong logic-1 at the inverter output.

The strong logic-1 drops by one-threshold voltage as it passes through pass transistor M$_4$, so

$$V_{out} = 2 - 0.5 = 1.5 \text{ V}$$

(b) If the input voltage changes from 0 V to 2 V, what is V_{out}?

The 2 V input is attenuated to 1.5 V by pass transistor M$_3$. The inverter strengthens the weak 1.5 V input to a weak logic-0 near 0 V. The weak logic-0 passes M$_4$ unattenuated so $V_{out} \approx 0$ V.

Self-Exercise 6-14

Given the pass transistor circuit with $V_{tn} = 0.5$ V. Ignore the body effect. What is the output voltage?

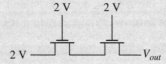

Answer: $V_{out} = 1.5$ V

Self-Exercise 6-15

Given the pass transistor circuit with $V_{tn} = 0.5$ V. Ignore the body effect. What is the output voltage?

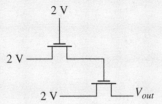

Answer: $V_{out} = 1.0$ V

Self-Exercise 6-16

Given that $K'_n = 200\ \mu A/V^2$, $K'_p = 400\ \mu A/V^2$, $V_{t0} = \pm 0.5$ V, $\gamma = 0.3$, $\phi_F = 0.35$ V, and

$$V_t = V_{t0} \pm \gamma \left(\sqrt{(2\phi_F) + V_{SB}} - \sqrt{2\phi_F} \right)$$

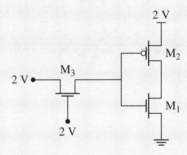

M_1 and M_3 have their source terminals tied to the substrate. M_3 has its bulk tied to ground and its source is tied to the inverter signal input.

(a) Calculate the voltage at the inverter input node.
(b) Calculate I_{DD}.

Answers: **(a)** $V_{SB} = 1.324$ V. **(b)** $I_D = 12.4\ \mu A$

6.5.2. CMOS Transmission Gates

A CMOS transmission gate (T-gate) solves the weak logic problem of a single pass transistor. Figure 6-9 shows the symbol, truth table, and schematic. The *n*MOS transistor passes a strong logic-0 and the *p*MOS transistor passes a strong logic-1. Putting the two transistors in parallel is a simple solution. Signal transmission is controlled by the gating or control signal G in Figure 6-9. When $G = 1$, both transistors turn on and the signal passes unattenuated to the output node B. When $G = 0$, no signal can pass since both transistors are off.

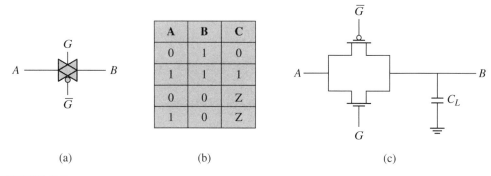

	A	B	C
	0	1	0
	1	1	1
	0	0	Z
	1	0	Z

(a)   (b)   (c)

FIGURE 6-9.

(a) Transmission gate symbol. (b) Truth table. (c) Transistor-level representation.

The T-gate shown in Figure 6-9c may cause circuit problems in combinational logic since it has a high impedance state (also called floating, hi-Z, or tristate) when the control signals are off. Floating nodes allow voltage drift on transistor load gates that can upset logic states. But CMOS T-gates appear abundantly in CMOS flip-flop designs. Typically there are four CMOS T-gates per flip-flop in an IC, and there are often millions of FFs. In these applications, the floating node is accounted for in the design and does not cause logic instability.

6.5.3. Tristate Logic Gates

The high-Z state is necessary when connecting many gate outputs to a single line, such as a data bus or address line. A potential conflict would exist if more than one gate output tried to simultaneously control the bus line. A controllable high impedance state circuit solves this problem. An alternative design may use multiplexers to select only one of many possible lines. Parallel high-Z output gates attached to a common node won't cause that node to float if one of the gates is driving the node to a strong logic state.

There are two ways to provide high impedance to CMOS gate outputs. One method provides a tristate output to a CMOS gate by connecting a transmission gate at its output (Figure 6-10). The control signal C sets the transmission gate conducting state that passes the non-tristated inverter output $\overline{out}$ to the tristated gate output out. When the transmission gate is off ($C = 0$), then its gate output is in the high impedance or floating state. When $C = 1$, the transmission gate is on and the output is driven by the inverter.

This transmission gate connected to the output provides tristate capability and also consumes unnecessary power. The design of Figure 6-10 contributes to dynamic power each time that the input and output ($\overline{Out}$) switch even when the gate is disabled in the tristate mode. Parasitic capacitors are charged and discharged. Since the logic activity at the input does not contribute to the logic result while the output is in tristate, the power consumption related to this switching is wasted.

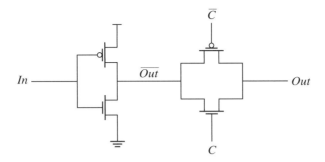

FIGURE 6-10.

Inverter with a transmission gate to provide tristate output.

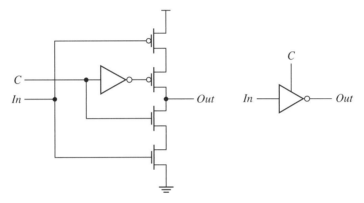

FIGURE 6-11.

Schematic and symbol where the transmission gate "inside" the inverter provides tristate output.

Power is reduced by putting a transmission gate "in series" with the inverter transistors (Figure 6-11). The pMOS and nMOS transistors of the transmission gate are in series within the conducting path between the power and ground rails and the inverter transistors. When the gate is in the tristate mode with $C = 0$, the inner transistor source nodes float. The activity at the inverter output signal node does not consume power as long as the gate is in the high-Z state $(C = 0)$. When $C = 1$, the outer inverter transistors act in a normal system mode.

6.6. Summary

This chapter examined detailed electronic properties of the NAND, NOR, and other multi-input logic gates. The noncontrolling logic state property was developed. The transistor sizing challenge was addressed with a design procedure. It is essential to understand the operation of NAND, NOR, and transmission gates at the transistor schematic level. The next chapter expands these concepts to show how design of higher functions is achieved.

Exercises

NAND Gates

6-1. $K'_n = K'_p = K$ for all transistors in the 2NAND gate, and input nodes A and B are tied together. $V_{DD} = 2.0$ V, $V_{tn} = 0.5$ V, and $V_{tp} = -0.5$ V. There is a point in the voltage transfer curve where the *n*MOS transistor is at the saturation/nonsaturation boundary and the *p*MOS transistor is in saturation. It is the boundary between bias Regions III and IV in Figure 5.3.

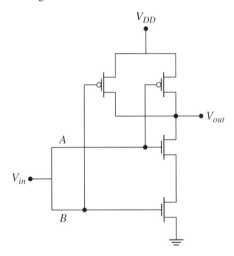

Calculate V_{in} and V_{out} at this bias point. *Hint*: The effective current drive strength of two transistors in parallel is double that of a single transistor, and the effective current drive strength of two transistors in series is half that of a single transistor.

6-2. If $K'_p = K'_n = 400$ μA/V^2, calculate the power supply current I_{DD} at the bias point in the previous example.

6-3. Let $K'_n = K'_p = K$ for all transistors in the 2NAND gate, and input nodes A and B are tied together. $V_{DD} = 2.0$ V, $V_{tn} = 0.5$ V and $V_{tp} = -0.5$ V. There is a point in the voltage transfer curve where the *p*MOS transistor is at the saturation/nonsaturation boundary and the *n*MOS transistor is in saturation. It is the boundary between bias Regions II and III in Figure 5.3.

Calculate V_{in} and V_{out} at this bias point.

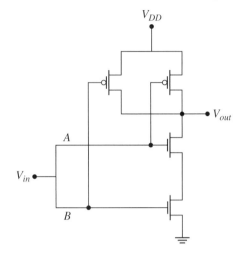

6-4. If $K'_p = K'_n = 400$ μA/V^2, calculate the power supply current I_{DD} at the bias point in Problem 6.3.

6-5. Why is the voltage transfer curve of the 2NAND in Exercise 6.1 skewed to the right of a symmetrical transfer curve?

6-6. Use your noncontrolling logic state skills to solve for logic states of node-A and node-B.

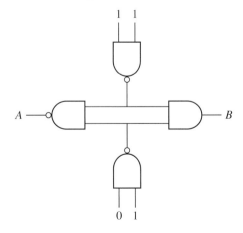

6-7. Use your noncontrolling logic state skills to solve for logic state of F.

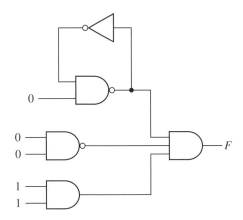

NOR Gates

6-8. *Let* $K_n' = K_p' = K$ *for all transistors in the 2NOR gate.* $V_{DD} = 2$ V, $V_{tn} = 0.5$ V, and $V_{tp} = -0.5$ V. *There is a point in the voltage transfer curve where the nMOS transistor is at the saturation/nonsaturation boundary and the pMOS transistor is in saturation. It is the boundary between bias Regions III and IV in Figure 5.3.*

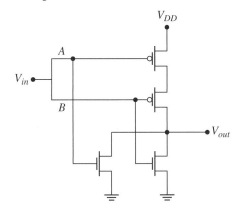

Calculate V_{in} and V_{out} at this bias point.

6-9. *If* $K_n' = K_p' = 400$ μA, *calculate the power supply current* I_{DD} *at the bias point in the previous problem.*

6-10. *Let* $K_n' = K_p' = K$ *for all transistors in the 2NOR gate.* $V_{DD} = 2$ V, $V_{tn} = 0.5$ V, and $V_{tp} = -0.5$ V. *Assume that A and B are tied together. There is a point in the voltage transfer curve where the pMOS transistor is at the satu-*

ration/nonsaturation boundary and the *n*MOS transistor is in saturation. It is the boundary between bias Regions II and III in Figure 5.3.

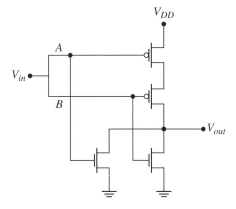

Calculate V_{in} and V_{out} at this bias point.

6-11. *If* $K_n' = K_p' = 400$ μA, *calculate the power supply current* I_{DD} *at the bias point in the previous problem.*

Noncontrolling Logic States

6-12. Given the logic circuit.

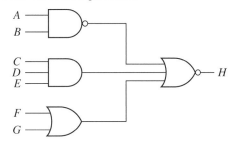

(a) What signal values must the inputs nodes be set in order to read the contents of node-*C* at the output-*H*?

(b) What signal values must the inputs nodes be set to in order to read the contents of node-*G* at the output-*H*?

6-13. (a) If node-*E* is a logic-1, write the logic values that allow that signal to be read at node-*H*. What is node-*H*?

(b) Write the logic input to read node-A at node-H.

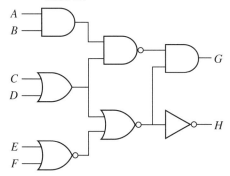

6-14. (a) An engineer must examine node-D, and read that logic signal at node-G. What should input vector be?

(b) What should input vector be to read contents of node-E at output node-G?

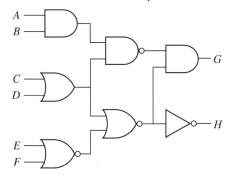

6-15. (a) Write the correct input signal values so that $F = f(E)$,

(b) If $E = 1$, what is F?

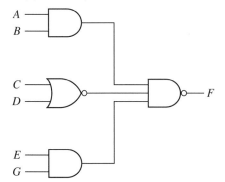

6-16. Use your noncontrolling logic state skills to solve for logic states of node-A and node-B.

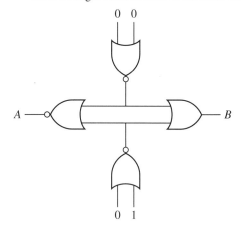

6-17. Use your noncontrolling logic state skills to solve for logic states of node-F using a truth table. What function does F perform?

A	B	$\overline{A} + B$	$A + \overline{B}$	F
0	0			
0	1			
1	0			
1	1			

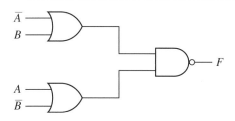

6-18. Create your own noncontrolling logic state problem as Problem 6.17 using a circuit with at least four multi-input logic gates.

6-19. Fill in the table for logic values at $ABCDEF$ such that

(a) $G = f(E)$

(b) $H = f(F)$

Use the logic symbols:
$1, 0, x, \overline{x}$

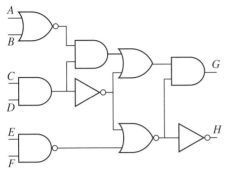

	A	B	C	D	E	F	G	H
$G = f(E)$								
$H = f(F)$								

Sizing Combinational Logic

6-20. Consider the circuit and
   (a) Provide the logic function that is implemented by the circuit.
   Hint: You can provide the truth table or analyze the operation of each transistor.
   (b) Size the transistors to have the same delay as a balanced CMOS inverter with $W_p = 3W_n$.

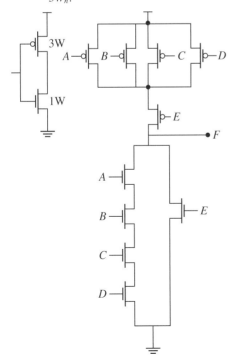

6-21. Given the function: $F = (\overline{C} + \overline{A})(B + AC) + AC$
   (a) Reduce and express this function to its minimum Boolean statement.
   (b) Draw the transistor schematic and size the transistors for worst-case pull-up and pull-down with respect to an inverter whose symmetrical voltage transfer curve has $W_p = 3W_n$.

6-22. Size the transistors in the circuit with respect to the symmetrical inverter.

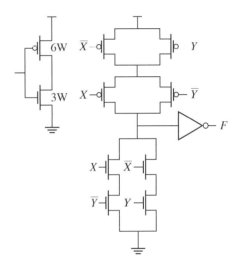

6-23. Size the transistors in the circuit with respect to a symmetrical inverter whose pMOS to nMOS width to length ratio is 3.

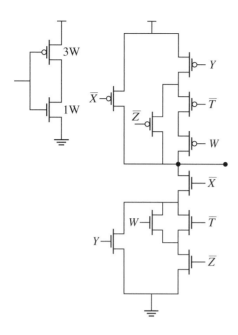

6-24. Given a symmetrical inverter on the left. Size the transistors in the circuit to the right for worst-case pull-up and pull-down that match the current drive of the inverter.

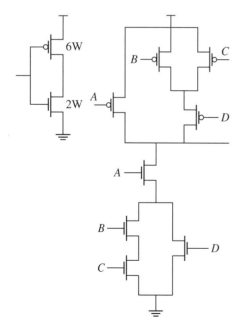

6-25. Create your own sizing problem for a CMOS combinational logic circuit that has five nMOS and five pMOS transistors.

Pass Gates

6-26. The figure shows a single transistor transmission gate. At $t = 0$ the gate voltage moves to $V_G = 2$ V. The load capacitance C_L initially is at zero volts. As the node capacitance (the source) charges V_{GS} becomes smaller until it is less than V_{tn}. The transistor cuts off and the node is less than the input drain voltage by $V_S = V_D - V_{tn}$. If the body coefficient is $\gamma = 0.2$ V$^{1/2}$, $\phi_F = -0.35$ V, and $V_{to} = 0.4$.

Calculate V_{GS} and V_S when the capacitor fully charges.

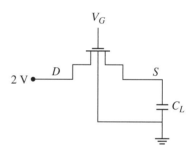

6-27. Given $V_{t0} = 0.25$ V, $\phi_F = 0.35$ V, $C_{ox} = 3.9 \times 10^{-7}$, and $N_A = 8 \times 10^{16}$ cm^{-3}, calculate V_O.

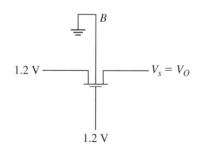

6-28. What are V_{tn} and V_{Ss1} for M1 during the time interval between 200 ns and 300 ns. Assume V_{S1} is in steady state for your analysis. Given: $V_{t0} = 0.5$ V, $\gamma = 0.2$ V$^{1/2}$, $\phi_F = 0.35$ V.

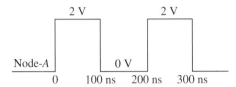

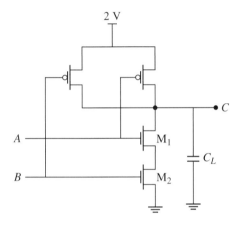

CHAPTER 7

CMOS Circuit Design Styles

Chapter 6 described basic CMOS logic gates emphasizing their electronic parameter characteristics. We now look at CMOS design styles that take different approaches to making an arbitrary logic function. Each design style has unique advantages in circuit speed, power, and physical size of the die. No design style is superior in all three characteristics.

CMOS combinational logic has several design style options. The complementary static design style is dominant in CMOS circuits (Chapter 6). It is a robust, lower power design, while high-performance circuits may use dynamic logic styles more suitable for high speed with less concern for power. Dynamic logic design may be used on smaller sections of the integrated circuit (IC) that requires repetitive high-speed operations, such as an arithmetic logic unit (ALU). A third logic design style uses pass-transistor or pass-gate elements as basic switches using fewer transistors to implement a function. We want to understand these combinational logic design styles and their trade-offs as options for the designer and for diagnostic knowledge necessary for test engineers, failure analysts, and reliability engineers.

7.1. Boolean Algebra to Transistor Schematic Transformation

Static, fully complementary CMOS designs can use inverter, NAND, NOR, and transmission gates as building blocks for more complex logic circuits. This chapter extends these simple designs to build "complex" combinational logic circuits that can be implemented for arbitrary Boolean functions. This complementary design style has good noise margins and low static power dissipation. CMOS static complementary gates have two transistor networks (nMOS and pMOS) whose topologies are related (Figure 7-1). Each nMOS transistor is paired with a pMOS transistor as are the simple gates of Chapter 6. The pMOS transistor net is connected between the power supply and the

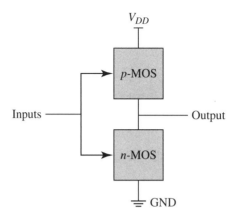

FIGURE 7-1.

Standard configuration of a CMOS complementary gate.

logic gate output, while the nMOS transistor topology is connected between the output and ground.

The nMOS transistor network is referred to as a *pull-down* network and the pMOS as a *pull-up* network (Figure 7-1). The output node pulls up to V_{DD} when the pMOS transistors are activated, while the pull-down net pulls the output node down to ground when the nMOS transistors are activated. We will learn an algorithmic technique that converts arbitrary Boolean algebra to a transistor CMOS schematic.

A Boolean function transforms to a transistor schematic with a straightforward design procedure:

1. Derive the nMOS transistor topology with the following rules:

    *   Product terms (AND operations) in the Boolean function are implemented with series-connected nMOS transistors.
    *   Sum terms (OR operations) are mapped to nMOS transistors connected in parallel.

2. The pMOS transistor network has a dual or complementary topology with respect to the nMOS net. This means that serial transistors in the nMOS net convert to parallel transistors in the pMOS net, and parallel connections within the nMOS block translate to serial connections in the pMOS block.

3. Add an inverter to the output if the nonnegated (positive) function is needed. An inverter added to a NAND or NOR function produces the AND and OR function. Some examples in this chapter require an inverter to fulfill the function.

This procedure is illustrated with three examples.

EXAMPLE 7-1

Design a complementary static CMOS 2NAND gate at the transistor level.

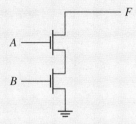

The previously listed rules define $F = \overline{AB}$ so the nMOS net has two series connected transistors, while the pMOS net has the complementary topology, that is, two transistors in parallel.

Step 1: Draw the nMOS net of two series transistors.

Step 2: Add the pMOS net of two parallel transistors as the complement.

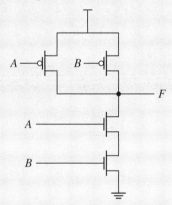

EXAMPLE 7-2

Design a complementary static CMOS XOR gate at the transistor level.
    The XOR gate Boolean expression F has four literals and is

$$F = X \oplus Y = X\overline{Y} + \overline{X}Y$$

F is the sum of two product terms. The design steps are as follows:

1. Derive the nMOS transistor topology with four transistors, one per literal in the Boolean expression. The transistors driven by X and $\overline{Y}$ are in series as well as the devices

driven by $\overline{X}$ and Y. These two transistor groups are in parallel, since they are additive in the Boolean function. The nMOS transistor net is

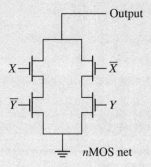

2. Implement the pMOS net as a dual topology to the nMOS net. The pMOS transistors driven by X and $\overline{Y}$ are in parallel, as well as the devices driven by $\overline{X}$ and Y. These transistor groups are then in series, since they are parallel in the nMOS net. The *out* node is the negated function $\overline{F}$.

3. Finally, add an inverter to obtain the function F, so that $F = \overline{out} = \overline{X}Y + X\overline{Y}$. The signal inputs have mutual connections that we omit for schematic clarity. Notice that to simplify the schematic we ignored the inverters needed for $\overline{x}$ and $\overline{y}$.

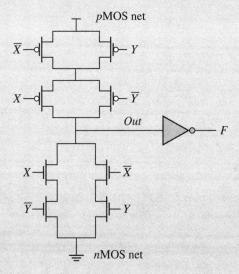

Steps 1–3 show that any Boolean function, regardless of its complexity, can be implemented with a CMOS complementary structure and an inverter. A more complicated example is developed.

EXAMPLE 7-3

Design the nMOS transistor net for the Boolean function $F = X + \{\overline{Y}[Z + T\overline{W}]\}$

The nMOS transistor network is connected between the output and ground terminals. The higher-level function F is a sum of two terms

$$F = X + \{operation\ A\}$$

where *operation A* stands for the logic within the braces of F. The transistor version of this sum is

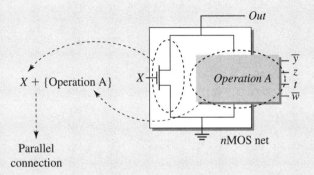

Parallel
connection

Now we design the transistor topology that implements the block *operation A* whose $\overline{Y}$ operation is an AND, that is,

$$operation\ A = \overline{Y} \cdot [operation\ B]$$

Hence the design topology is a transistor controlled by input $\overline{Y}$ in series with a third box that will implement *operation B*, that is,

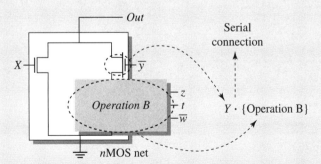

We then design the topology of *operation B*. This is a transistor controlled by input Z in parallel with two transistors in series: one controlled by input T, and the other by input $\overline{W}$. The complete nMOS network is shown.

The pMOS block is a dual topological structure and we then connect an inverter to its output to form the nonnegated function $F = X + \{\overline{Y}[Z + T\overline{W}]\}$.

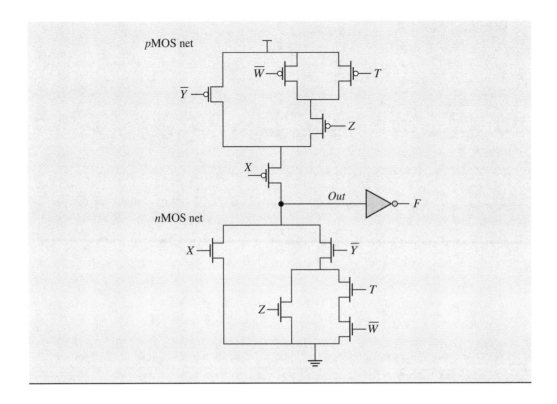

Self-Exercise 7-1

(a) Design the transistor schematic for $F = \overline{ABCD + E}$.
(b) Discuss the rise and fall time issues.

Self-Exercise 7-2
Design the transistor schematic for $F = \overline{(A + B)(C + D) + E}$.

Self-Exercise 7-3
Design the transistor schematic for $F = (X + Y)[Z + (WT)(\overline{Z} + X)]$.

7.2. Synthesis of DeMorgan Circuits

Thus far, the function F was created from the natural CMOS negated function $\overline{F}$ by adding an inverter at the output. There is another approach that uses DeMorgan's law to convert a NAND gate to an OR equivalent and a NOR to an AND gate. The technique is similar to that just described with examples, but eliminates the output inverter. The DeMorgan equivalent NAND gate to OR gate is

$$F = \overline{AB} = \overline{A} + \overline{B}$$

The DeMorgan equivalent on the right side of the equation turns our focus to the pMOS transistor network where $F = 1$ on any logic-0. Our thinking now asks what logic-0 signals will turn on the pMOS transistors and cause a logic-1 at the output.

Similarly, the NOR expression converts to the DeMorgan AND gate as

$$F = \overline{A + B} = \overline{A}\,\overline{B}$$

The right side of the equation activates the pMOS network F to a logic-1 when A and B are both logic-0. Our interest lies in a function that is not negated. The AND DeMorgan equivalent circuit is

$$F = AB = \overline{\overline{A} + \overline{B}}$$

The negated function on the right implies a NOR pull-down network with transistor inputs $\overline{A}$ and $\overline{B}$. The pull-up network has two series pMOS transistors also with inputs $\overline{A}$ and $\overline{B}$. The schematic is given in Figure 7-2. The circuit performs the function $F = AB$ without adding an output inverter to a NAND gate. The original AND gate is equivalent to a NOR gate if the signal polarities are reversed on the NOR gate.

At first it seems that we have created a need for two inverters to negate the inputs $\overline{A}$ and $\overline{B}$. This may be true, but signal polarities on an input node are not uniformly the positive literal value. They may be the literal complement. A flip-flop circuit inherently provides a logic signal and its complement. The assumption that the DeMorgan circuit requires additional inverters for each literal has no more meaning on an IC than if we

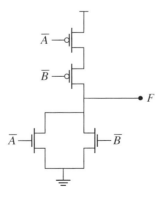

FIGURE 7-2.

An AND gate function as the equivalent DeMorgan NOR gate.

said that NAND and NOR gates needed signal polarity correction to get true values of the literals. Whether one implementation has fewer transistors will depend on the local signal situation. We are stressing flexible design to optimize local gate designs. These are tools at the designer's discretion.

EXAMPLE 7-4

Repeat the XOR using the DeMorgan conversion technique to implement the function without using an inverter. Verify the operation with a truth table.

$$F = X \oplus Y = X\overline{Y} + \overline{X}Y$$

The DeMorgan equivalent is

$$F = X\overline{Y} + \overline{X}Y = \overline{(\overline{X} + Y)(X + \overline{Y})}$$

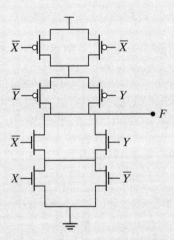

The schematic shows the pMOS net activating a logic-1 for $F = X\overline{Y} + \overline{X}Y$ when the signal polarities are inverted. The truth table is consistent with the transistor schematic.

X	Y	$\overline{X}$	$\overline{Y}$	$F = X\overline{Y} + \overline{X}Y$
0	0	1	1	0
0	1	1	0	1
1	0	0	1	1
1	1	0	0	0

Notice that the nMOS topology of Example 7.2 is the pMOS topology of the DeMorgan design given in this example. Likewise the pMOS topology is the same as the nMOS topology of the DeMorgan design. However, the signal polarities are reversed.

EXAMPLE 7-5

Redesign the function $F = X + \{\overline{Y}[Z + T\overline{W}]\}$ without using an inverter. Compare your solution with Example 7-3.

The DeMorgan equivalent is

$$F = \overline{\overline{X}[Y + \overline{Z}(\overline{T} + W)]}$$

whose circuit is on the right.

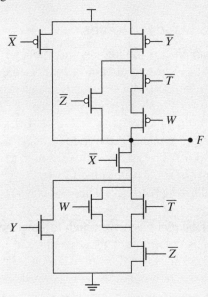

Self-Exercise 7-4

Derive the equivalent CMOS transistor circuit for $F = A + B$ without an output inverter.

Self-Exercise 7-5

Design the transistor level schematic for $F = (X + Y)[Z + (WT)(\overline{Z} + X)]$ without using an output inverter. Compare your solution with Self-Exercise 7-3.

Self-Exercise 7-6

Design an equivalent DeMorgan logic gate schematic for $F = \overline{A} + (B + C)\overline{D}E$.

Self-Exercise 7-7

Design an equivalent DeMorgan logic gate schematic for $F = W\overline{X}\overline{Y} + \overline{W}X\overline{Y} + \overline{W}\overline{X}Y$. What function is performed?

7.3. **Dynamic CMOS Logic**

Previous design techniques showed that conventional combinational CMOS circuits do not use a clock signal. If inputs are held stable, then the circuits retain their output state (all circuit nodes remain at their valid quiescent logic values) as long as power is maintained. Dynamic circuits require a clock signal, but they are faster, smaller, and are used for IC subcircuits that are data processing intensive. We will develop this design style exploring dynamic CMOS strengths and weaknesses.

7.3.1. Dynamic CMOS Logic Properties

A dynamic CMOS gate implements the logic with a block of transistors (usually nMOS for its higher carrier mobility). The output node is connected to ground through an nMOS transistor logic block and a single nMOS evaluation transistor (Figure 7-3). The output node is connected to the power supply through one precharge pMOS transistor. The logic gate has two phases: *precharge* and *evaluation*. A global clock drives the precharge and evaluation transistors. During precharge, the global clock goes low turning the pMOS transistor on and the nMOS off. The gate output goes high (it is being precharged) while the block of nMOS transistors are driven to the Off-state. All inputs to the nMOS logic block must have logic-0 during precharge. This disconnects the logic block transistors from the output precharge.

In the evaluation phase, the clock is driven high turning the pMOS device off and the nMOS on. The input signals determine if there is either a low or high impedance path from the output to ground. If the logic does not turn on transistors in the n-logic block giving a path from the *out* node to ground, then the *out* node floats at the precharged logic high voltage. If the nMOS logic transistors provide a path to GND, then the output node is logic-0.

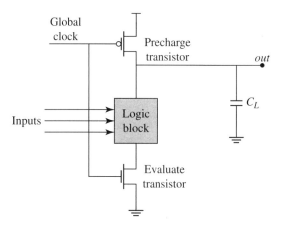

FIGURE 7-3.

Basic structure of a dynamic CMOS gate.

The precharge and evaluation transistors never conduct simultaneously, which eliminates through current with its power loss and speed degradation. The design also reduces the capacitance at the *out*-node, since there is only one *p*MOS transistor. For example, a static complementary NAND gate has more than one parallel *p*MOS transistor tied to the *out*-node and that extra drain capacitance loads the output node. Conveniently, if the logic state determined by the inputs is a logic-1 (V_{DD}) then the rise time is effectively zero. The lower capacitance, the near zero rise time for a logic-1 output, the reduced through current during a transition, and reduced area makes dynamic logic a favorite for high-speed applications.

Dynamic circuits with an *n*-input gate use only $n + 2$ transistors instead of the $2n$ devices required for the complementary CMOS static gates. Dynamic CMOS gates have a drawback. When the *out* in Figure 7-3 is evaluated high, then the output node is in the high-Z state, since there is no electrical path to V_{DD} or ground. This exposes the node to leakage induced drift, noise fluctuations, and possible charge sharing within the logic block thus degrading its voltage. The output load capacitor will slowly discharge due to transistor Off-state leakage currents, and may lose its logic value. This limits the low frequency operation of the circuit. The *n*MOS transistor gate inputs can only change only during evaluation since charge redistribution from the output capacitor to internal nodes of the *n*MOS logic block may drop the output voltage when it has a logic high. We will analyze charge sharing in the next section.

Dynamic CMOS logic families have the following properties:

Advantages

— They use fewer transistors and therefore less area than a complementary transistor design.
— Fewer transistors present smaller input capacitance to previous driving gates and therefore provide faster switching speed.
— Gates are designed and transistors sized for fast switching characteristics. High-performance circuits use these families.
— The logic transition voltage for output low-to-high is not a transition since the output node is precharged before the event. Therefore, the low-to-high transition is said to have zero rise time.
— There is no through or short-circuit current so power is reduced and speed is increased.

Disadvantages

— Each gate needs a clock signal that must be routed through the whole circuit. This requires precise timing control.
— Clock circuitry runs continuously, drawing significant power.
— The circuit loses its state if the clock stops.
— Dynamic circuits are more sensitive to noise.
— Clock and data must be carefully synchronized to avoid erroneous states.

We will next look at the dynamic circuit property of charge sharing, and then see how the domino version solves many of these problems.

7.3.2. Charge Sharing in Dynamic Circuits

The generic charge-sharing problem involves switches and charged capacitors and is a rearrangement of electrons when a switch is closed after a precharge. Switching exposes a fixed amount of charge to a different capacitive environment causing node voltage changes. This occurs in dynamic circuits. The circuit is first precharged high followed by an evaluation mode where the precharge can partially trickle into the n-logic block when certain nMOS transistors turn on but there is no direct path to ground. This reduces the voltage at the output node. Analysis uses the principle of charge conservation when switches isolate a circuit. Charge is $Q = CV$ and Q is constant in a closed system. Q has the unit of coulombs (C).

Figure 7-4 shows a dynamic circuit design of a 3NAND gate. The precharge phase with $Clk = 0$ charges capacitances connected to the output node. During precharge C_L accepts an initial charge $Q_{init} = C_L V_{DD}$. When the clock returns to the logic-high evaluation mode, then that initial charge will spread into the n-logic block if paths are open. Assume that only the top transistor is on during evaluation. Then, capacitance C_1 accepts charge from the output node. The total charge in the system in the same as the initial charge so the final voltage is calculated by

$$Q_{init} = C_L V_{DD} \tag{7-1}$$

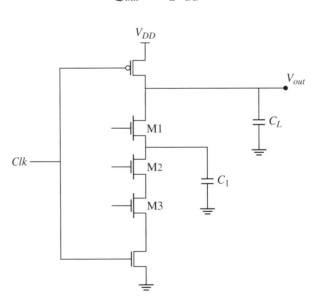

FIGURE 7-4.

Dynamic CMOS 3NAND gate.

When M1 turns on

$$Q_{init} = Q_{final} = C_L V_{DD} = C_L \, V_{final} + C_1 \, V_{final} \tag{7-2}$$

$$V_{final} = \frac{C_L}{C_L + C_1} V_{DD} \tag{7-3}$$

If the M1 source voltage attempts to rise higher than $V_{DD} - V_{tn}$, then transistor M1 turns off. V_{S1} cannot rise higher than $V_{G1} - V_{tn} = V_{DD} - V_{tn}$. When the circuit finishes its charge transfer there is no current in the transistor, so $V_D = V_S$. The following example will illustrate how to deal with this situation.

EXAMPLE 7-6

The 3NAND dynamic gate is precharged, and then the top transistor M1 turns on. Given $V_{DD} = 1.5$ V, $V_{tn} = 0.4$ V, $C_L = 150$ fF, $C_1 = 25$ fF, and $C_2 = 50$ fF, what is the final voltage?

$$Q_{init} = C_L V_{DD} = 150\,\text{fF} \times 1.5\,\text{V} = 225\,\text{fC}$$

$$Q_{final} = Q_{init} = 225\,\text{fC} = C_L \, V_{final} + C_1 \, V_{final}$$

$$225\,\text{fC} = (150\,\text{fF})(V_{final}) + (25\,\text{fF})(V_{final})$$

and

$$V_{final} = \frac{225\,\text{fC}}{175\,\text{fF}} = 1.286\,\text{V}$$

But as the source of M1 rises toward 1.286 V, it will stop at $V_G - V_{tn} = 1.5 - 0.4 = 1.1$ V. Therefore, $V_{final} = 1.286$ V is incorrect. When $V_{S1} = 1.1$ V, M1 turns off and $V_{D1} = V_{final}$. The charge on the drain is less than Q_{init} by the amount of charge leaked into C_1. That leaked charge is

$$Q_1 = 1.1 \text{ V}(25 \text{ fF}) = 27.5 \text{ fF}.$$

Then $V_{D1} = \dfrac{Q_{final}}{150 \text{ fF}} = \dfrac{225 \text{ fC} - 27.5 \text{ fC}}{150 \text{ fF}} = 1.317$ V

Self-Exercise 7-8

$V_{DD} = 2.0$ V and $V_{tn} = 0.5$ V. During evaluation, $\overline{Y}$ and T are activated. What is the final voltage when the evaluation mode settles? Write the Boolean expression for F.

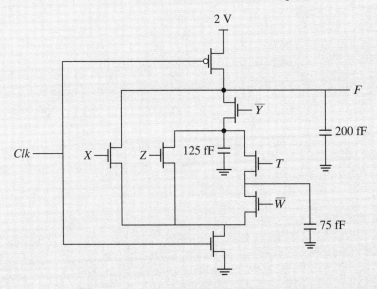

Answer: $V_{final} = 1$ V, which is an invalid logic state.

Self-Exercise 7-9

Calculate C_1 so that the final voltage is determined by M1 that is just at cut-off. $C_L = 175$ fF, $V_{DD} = 1.5$ V, and $V_{tn} = 0.4$ V.

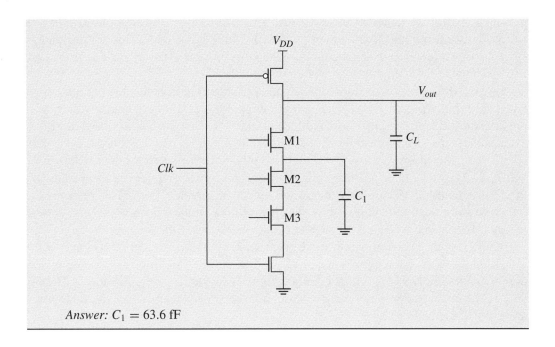

Answer: $C_1 = 63.6$ fF

7.4. Domino CMOS Logic

Domino CMOS eliminates the charge share problem and has other advantages. It was proposed in 1982 [1]. Figure 7-5 shows a domino design. The global clock, the precharge and evaluation transistors, and the *n*-logic block work as before, but an inverter and a

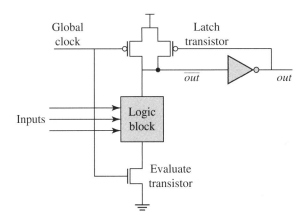

FIGURE 7-5.

Domino CMOS logic gate with a weak *p*MOS feedback transistor.

pull-up transistor are added to the $\overline{out}$-node. The inverter drives the domino output to a strong logic state forcing the circuit output to be 0 V during precharge. That is an advantage since the signal inputs at the next domino stage must be logic-0 during precharge so there is no path to GND. The inverter conveniently provides this.

A weak pMOS feedback transistor is added to latch the internal floating node high when $\overline{out}$ is high (Figure 7-5). The latch transistor holds $\overline{out}$ at V_{DD} so that its voltage will not drop from charge sharing or leakage currents through the logic block transistors during evaluation. If the nMOS logic block discharges the $\overline{out}$-node to GND during evaluation (Figure 7-5), then the inverter output signal out goes high, turning off the feedback pMOS. When $\overline{out}$ is evaluated or precharged high, then the inverter output goes low turning on the feedback pMOS device providing a low impedance path to V_{DD}. This prevents the $\overline{out}$-node from floating making it less sensitive to node voltage drift, noise, and current leakage. The latch transistor forms a positive feedback loop with the inverter.

The W/L of the latch transistor must be kept small with respect to the nMOS pull-down transistors to minimize the influence of contention during a high-to-low transition. The domino configuration is the most common form of dynamic gate achieving a 20–50% speed performance increase over static logic. It is a popular subcircuit for fast adders and arithmetic logic units ALUs.

Figure 7-6 shows domino 2OR and 2AND logic gates. The nMOS pull-down network is the same as for the complementary CMOS designs earlier in this chapter.

Domino CMOS with its output inverter allows logic gate cascading since all inputs are naturally set to zero during precharge avoiding erroneous evaluation from different delays (Figure 7-7). All logic blocks must have zero input signals during precharge to avoid any paths from $\overline{out}$ to ground.

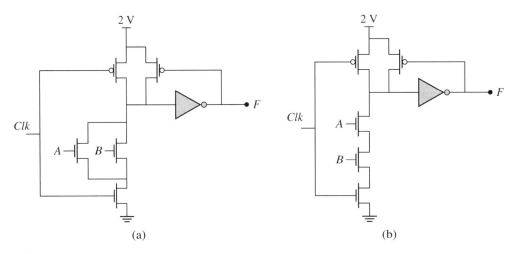

(a)   (b)

FIGURE 7-6.

Domino circuits for (a) 2OR and (b) 2AND.

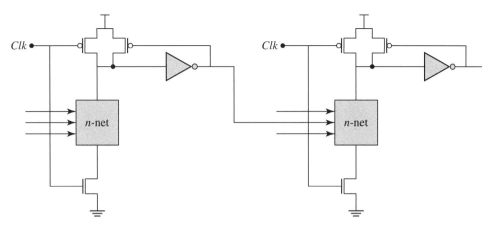

FIGURE 7-7.

Cascaded domino circuits.

Domino logic uses only noninverting gates making it an incomplete logic family. There are techniques to achieve an inverting domino circuit, but they will not be discussed in this book.

EXAMPLE 7-7

Draw the domino transistor schematic for $F = A + BC$.

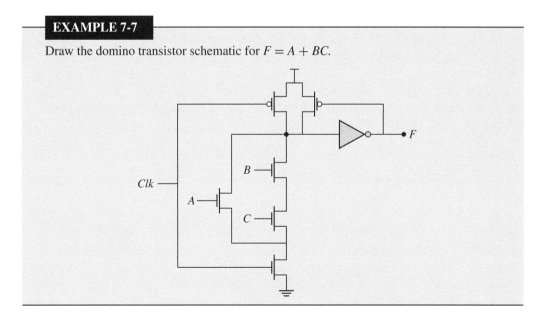

Self-Exercise 7-10

Draw the schematic for a domino circuit to perform $F = (A + B) C (ED + G)$.

Self-Exercise 7-11
A domino circuit performs $F = [AC + (BC)(AC)]B$.

(a) Draw the transistor circuit.
(b) Minimize the function and redraw the schematic.

Answer: $F = ABC$

Self-Exercise 7-12
A domino circuit performs $F = AB$. Draw and implement this function. Can the n-logic block be converted to a DeMorgan equivalent circuit and fulfill the AND function? Show your work using truth tables.

7.5. NORA CMOS Logic

A design modification to domino CMOS logic eliminates the output buffer without compromising race problems between clock and data that arise when cascading dynamic gates. NORA CMOS (No Race CMOS) avoids race problems by cascading alternate nMOS and pMOS blocks for logic evaluation. The cost is routing and controlling two complemented clock signals. The cascaded NORA gate structure is shown in Figure 7-8. The pMOS logic block takes its output from the evaluation transistor drain. When the global clock (GC) is low ($\overline{GC}$ is high), the nMOS logic block output nodes are precharged high, turning off the pMOS logic block transistors. The outputs of gates with pMOS logic blocks are precharged low at the nMOS precharge transistor drain. All logic blocks are in the precharge state until the clocks reverse their polarity. When the clocks change to the evaluate state, the first nMOS logic block is driven first, and it starts the logic operation cascade.

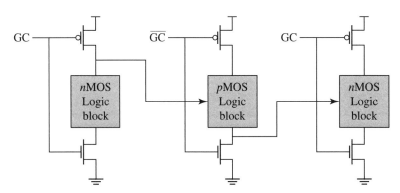

FIGURE 7-8.

NORA CMOS cascaded gates.

7.6. Pass Transistor Logic

Pass transistor logic uses either single transistors or transmission gates as switches to carry logic signals from node to node instead of connecting output nodes directly to V_{DD} or ground. Figure 7-9 shows a pass transistor logic circuit for $F = AB$. When $A = B = 1$, then $F = 1$. The pass transistor connecting node-F to ground keeps node-F from floating when $A = 0$ or $B = 0$.

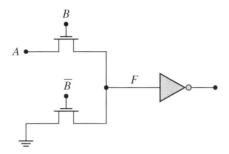

FIGURE 7-9.

Pass transistor logic AND gate.

A pass gate weakness is apparent when the nMOS transistor passes a logic-1 from node-A to node-F. Node-F suffers a drop in voltage $V_F = V_{DD} - V_{tn}$. An inverter placed at the end of the pass transistor restores strong voltages.

Figure 7-10 shows a 3AND whose first stage transistor connects to the gate terminal of a load transistor forming the function $F = ABD$. The voltage level at node-C is $V_{DD} - V_{tn}$. While the overall logic function is achieved the logic high voltage at the output node-F drops by another threshold level. The weak logic-1 voltage at the output is $F = V_{DD} - 2\,V_{tn}$. Designers must take care in how signals are connected.

AND gates can have many pass transistors in series with each input tied to one of the transistor gates. Figure 7-11 shows a 4AND logic gate design $F = ABCD$. The three parallel transistors pull node F to ground when $F = $ logic-0. Pass transistor logic appears simple, especially when compared to static complementary CMOS designs. But there

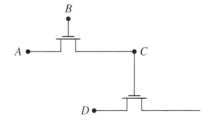

FIGURE 7-10.

Chaining pass transistor source output to gate input of load transistor.

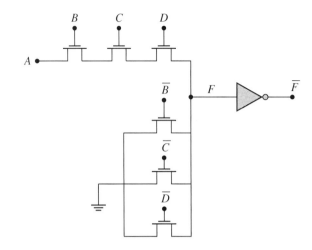

FIGURE 7-11.

Pass transistor logic gate. $F = ABCD$.

are two weaknesses and both relate to propagation delay. The chain puts significant load impedance on the signal path, degrading the high-frequency capability. The second weakness also relates to speed in that the driving signals may not be strong rail voltages, but a weakened logic voltage as in Figure 7-11. Despite these drawbacks, the node capacitances of smaller pass transistors can allow speed properties to sometimes be competitive with complementary CMOS logic. Pass transistor design is presented as a design option.

EXAMPLE 7-8

Design a 2OR gate using pass transistors where $C = A + B$. Include a pull-down network to account for the floating output logic-0 state.

Two parallel transistors with signals A and B are applied to the transistor gates. V_{DD} is applied to the drain inputs of both transistors. The pull-down network uses two pass transistors in series driven by $\overline{A}$ and $\overline{B}$.

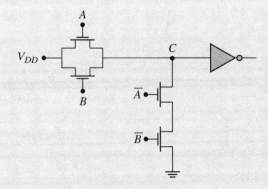

EXAMPLE 7-9

Draw a pass transistor schematic for $F = \overline{AB + CD}$. Include the circuitry that protects the circuit from a floating node and a weak logic state.

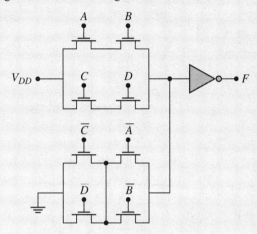

Self-Exercise 7-13
If $ABCD = 1.5$ V and $V_{tn} = 0.5$ V, what is the voltage at node-F in Figure 7-10?

Answer: $V_F = 0.5$. What does this mean?

Self-Exercise 7-14
Implement a pass transistor design for the XOR gate $F = \overline{X\overline{Y} + \overline{X}Y}$. Don't let nodes float.

Self-Exercise 7-15
Implement a pass transistor design for $F = \overline{AXYZ + W}$. Don't let nodes float.

7.7. CMOS Transmission Gate Logic Design

CMOS transmission gates avoid the weak logic voltages of single-pass transistors by using the two-transistor transmission gate. A typical CMOS transmission logic gate design is the gate-level multiplexer (MUX) shown in Figure 7-12a for a 2-to-1 MUX. A MUX selects one from a set of logic inputs to connect with the output. In Figure 7-12a, the logic signal C selects either A or B to activate the output (*out*). Figure 7-12b shows a MUX design with transmission gates. The complementary CMOS gates (Figure 7-12a) require 14 transistors (4 transistors for each NAND and 2 transistors for the inverter), while the transmission gate design requires only 6 transistors (4 transistors for T-gates and 2 for a C inverter).

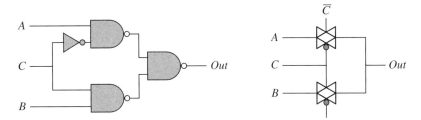

FIGURE 7-12.

(a) Standard 2-to-1 MUX design. (b) Transmission (pass) gate-based version.

Each transmission gate in Figure 7-12b has two transistors plus two more to invert the control signal. The control signal polarities can be generated by a master circuit that feeds these control signals to many subcircuits, thus lessening the transistor count.

Another pass gate design example is the XOR gate that produces a logic one output when only one of the inputs is logic high. If both inputs are logic one or logic zero, then the output is zero. Figure 7-13 shows an 8-transistor XOR gate having a tristate buffer and transmission gate with their outputs connected. Both gates are controlled by the same input through a complementary inverter (Input-A in this case). The XOR standard CMOS design built from primitive logic gates requires 12 transistors, whereas the design of Figure 7-13 requires only 8.

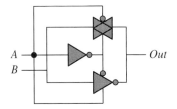

A	B	Out
0	0	0
0	1	1
1	0	1
1	1	0

FIGURE 7-13.

Eight-transistor XOR gate and truth table.

7.8. Power and Activity Coefficient

Power dissipation is an issue for virtually all ICs. High-performance ICs with many transistors can generate 100–200 watts. Most of this power is wasted and has nothing to do with actual computation. Some low power battery ICs generate microwatts and must operate for a few years without replacement. The estimated power dissipation for the combinational logic in this section assumes that all power loss is by the charge and discharge of a load capacitance. This ignores the through current and leakage current power, but does allow a look at the primary variables reducing circuit power. This section is intended to deepen our knowledge and sensitivity to power loss properties. The power loss of an IC to charge or discharge load capacitance is modeled as

$$P_{IC} = \frac{1}{2}\alpha C_L V_{DD}^2 f_{clk} \tag{7-4}$$

where P_{IC} is the power dissipated on the active edge of the clock when load capacitance is charged or discharged, α is the average IC switching activity coefficient, C_L is a lump sum capacitance for all signal switching nodes in the IC, and f_{clk} is the clock frequency. During each clock pulse, only a fraction α of the combinational logic gates undergo a transition, and the load capacitance is either charged or discharged but not both. α is typically on the order of 5–30% for combinational logic gates in various ICs.

The power required to charge and discharge a logic gate capacitance in a *clock network* is

$$P_{IC} = \alpha C_L V_{DD}^2 f_{clk} \tag{7-5}$$

where $\alpha = 1$ since the clock pulse is deterministic and happens with certainty on each period. The $1/2$ factor disappears for a clock circuit since charge and discharge are involved in the same clock period. Some clock networks gate the clock for subcircuits that are not activated during an operation. This allows de-powering of that subcircuit.

We will compute α at the gate level to compare the power efficiency of different logic gates to better understand power reduction. Products may be high-performance high-power or low-frequency low-power ICs, but all products must minimize power dissipation.

Assume that the major power dissipation occurs when the clock makes an active transition. The probability of an output signal transition from zero to one is $\alpha_{0\to1}$ and is

$$\alpha_{0\to1} = p_0\,p_1 = p_0(1 - p_0) \tag{7-6}$$

where p_0 is the probability of an output logic-0, and p_1 is the probability of an output logic-1. The probability of an output signal transition from one to zero is $\alpha_{1\to0}$ and is

$$\alpha_{1\to0} = p_1\,p_0 = p_0 p_1 = p_1(1 - p_1) \tag{7-7}$$

so that

$$\alpha_{1\to0} = \alpha_{0\to1}$$

If the input logic states are random, then we can determine p_0 and p_1. Let N_0 be the number of possible logic-0 states in the output of a logic gate, N_1 is the number of possible logic-1 states in the output of the same logic gate, and N is the number of input signals feeding the logic circuit. The total number of input logic states is 2^N. Then

$$p_0 = \frac{N_0}{2^N}$$

$$p_1 = \frac{N_1}{2^N}$$

The transition activity coefficient is

$$\alpha_{0\to1} = p_0 p_1 = \frac{N_0}{2^N}\frac{N_1}{2^N} = \frac{N_0 N_1}{2^{2N}} = \frac{N_0(2^N - N_0)}{2^{2N}} \tag{7-8}$$

where $N_0 + N_1 = 2^N$

We treat the inputs to combinational logic as random. A random input to an inverter considers that two consecutive zeros or ones could occur that would not cause a logic transition. If the inputs to an inverter are random then the probability of a $0\to1$ or a $1\to0$ transition is

$$\alpha_{Inv} = \alpha_{0\to1} + \alpha_{1\to0} = p_0 p_1 + p_1 p_0 = \frac{1}{2}\frac{1}{2} + \frac{1}{2}\frac{1}{2} = 0.5$$

The inverter has a 0.5 probability of changing state for a random input that we also know from intuition.

EXAMPLE 7-10

(a) Calculate the switching activity coefficient $\alpha_{0\to1}$ for a 2NAND gate using a truth table to identify the number of states.

(b) If the 2NAND gate is clocked at 1 GHZ, has a power supply of 1.5 V, and a load of 200 fF, what is the expected power dissipation of that gate for the $0\to1$ transition?

(c) If an IC has one million such gates each with an activity coefficient of 20%, what is the operating power of the IC due to the combinational logic gates?

A	B	C
0	0	1
0	1	1
1	0	1
1	1	0

(a) $N_0 = 1$, $N_1 = 3$, So

$$\alpha_{0\to1} = \frac{N_0 N_1}{2^{2N}} = \frac{1}{2^{2\cdot2}} \times 3 = \frac{3}{16} = 0.1875$$

(b) $P = \frac{1}{2}\alpha_{0-1}C_L V_{DD}^2 f_{clk} = \frac{1}{2}\frac{3}{16} \times 200 \times 10^{-15} \times 1.5^2 \times 10^9 = 42.2\ \mu W$

(c) Each gate can transition $0\to1$ or $1\to0$. Therefore $\alpha = 2(0.1875) = 0.375$, and 20% Of the one million gates are switching.

$$P_{IC} = 0.20 \times 42.2\ \mu W \times 10^6 = 8.44\ W$$

Notice how quickly power loss increases when dealing with millions or hundreds of millions of gates and the moderating effect of a lower activity coefficient.

EXAMPLE 7-11

Let $C_L = 50$ fF for each signal output node, $V_{DD} = 2.0$ V, and $f_{Clk} = 1.5$ GHz. Make a truth table for the circuit and calculate

(a) The switching activity coefficient $\alpha_{0 \to 1}$ for the circuit including the activity coefficients for all gate output nodes.

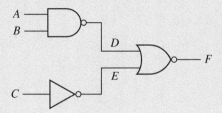

Hint: You must calculate and add the transition probabilities for each of the signal nodes D, E, and F.

(b) Calculate the expected power dissipation for the $0 \to 1$ transition.

(a)

A	B	C	D	E	F
0	0	0	1	1	0
0	0	1	1	0	0
0	1	0	1	1	0
0	1	1	1	0	0
1	0	0	1	1	0
1	0	1	1	0	0
1	1	0	0	1	0
1	1	1	0	0	1

$$(\alpha_{0 \to 1})_D = (p_0 p_1)_D = \left(\frac{2}{8}\right)\left(\frac{6}{8}\right) = \frac{12}{64} = 0.1875$$

$$(\alpha_{0 \to 1})_E = (p_0 p_1)_E = \left(\frac{4}{8}\right)\left(\frac{4}{8}\right) = \frac{16}{64} = 0.25$$

$$(\alpha_{0 \to 1})_F = (p_0 p_1)_F = \left(\frac{7}{8}\right)\left(\frac{1}{8}\right) = \frac{7}{64} = 0.109$$

The sum of the $\alpha_{0 \to 1}$ transition probabilities is

$$0.1875 + 0.25 + 0.109 = 0.5469$$

(b) $P_{DEF} = \dfrac{1}{2}[0.1875 + 0.25 + 0.109]\, 50 \text{ fF } (2)^2(1.5 \text{ GHz}) = 82.0\,\mu\text{W}$

EXAMPLE 7-12

Given that $F = A(B + CD)$ and $V_{DD} = 1.2$ V, $C_L = 100$ fF, and $f_{Clk} = 1.5$ GHz:

(a) Draw the transistor schematic for a static CMOS gate.
(b) Compute the activity coefficients for both output transitions using a truth table.
(c) Calculate the expected power dissipated.

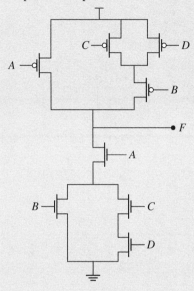

(a) The circuit is to the right.
(b) The truth table is

A	B	C	D	$F = A(B + CD)$
0	0	0	0	0
0	0	0	1	0
0	0	1	0	0
0	0	1	1	0
0	1	0	0	0
0	1	0	1	0
0	1	1	0	0
0	1	1	1	0
1	0	0	0	0
1	0	0	1	0
1	0	1	0	0
1	0	1	1	1
1	1	0	0	1
1	1	0	1	1
1	1	1	0	1
1	1	1	1	1

and

$$\alpha_{Total} = \alpha_{o\to1} + \alpha_{1\to0} = p_0 p_1 + p_1 p_0 = \left(\frac{11}{16}\right)\left(\frac{5}{16}\right) + \left(\frac{5}{16}\right)\left(\frac{11}{16}\right) = 0.430$$

(c) The power is

$$P = \frac{1}{2}\alpha\, C_L V_{DD}^2 f_{clk} = \frac{1}{2}0.430(100 \times 10^{-15})(1.2)^2 \times (1.5 \times 10^9) = 46.4\,\mu W$$

Self-Exercise 7-16
Verify that the 2AND and 2NOR have the same activity coefficient $\alpha_{0\to1}$.

Self-Exercise 7-17
Make a truth table for the circuit and calculate the switching activity coefficient

(a) $\alpha_{0\to1}$ for node-F
(b) $\alpha_{1\to0}$ for node-F
(c) $\alpha_{0\to1}$ for node-D
(d) $\alpha_{0\to1}$ for node-E

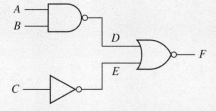

Answers:

(a) $\alpha_{0\to1})_F = 0.234$
(b) $\alpha_{1\to0})_F = 0.234$
(c) $\alpha_{0\to1})_D = 0.1875$
(d) $\alpha_{1\to0})_E = 0.25$

7.9. Summary

This chapter showed how complex CMOS combinational logic gate designs are constructed from Boolean algebra equations. More compact circuits with different power dissipation and speed properties illustrate the popular dynamic logic, and pass transistor logic gate designs. All versions appear in modern CMOS IC design. Lower power design theory was illustrated by calculating activity coefficients and subsequent power dissipation.

Reference

[1] R. H. Krambeck, et al. "High Speed compact circuits with CMOS", *IEEE Journal of Solid State Circuits*, Vol. 17, No. 3, June 1982.

Exercises

Complex Combinational Logic Gates

7-1. Given the Boolean function $F = Z[\overline{X}YZ + X\overline{Z}]$ draw the static CMOS transistor schematic in reduced form.

7-2. Write the Boolean expression F for the input signals in the following circuit:

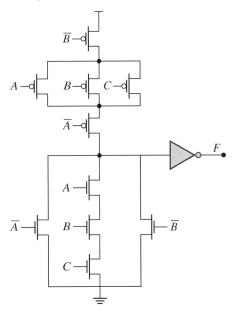

7-3. Draw the static CMOS transistor schematic that performs the Boolean function $F = \overline{(G+F)(M+N)}$

7-4. Draw the CMOS transistor schematic that fulfills the function $F = \overline{(AB + C)D}$ for a static CMOS logic gate.

7-5. Given the schematic in the figure.
  (a) If it corresponds to the CMOS pull-up network of a static circuit, what is the resultant Boolean expression F?
  (b) If the p- and the n-channel transistors are sized for individual equal drive currents

discuss whether the pull-up will be faster than the pull-down network, or they will they be the same.

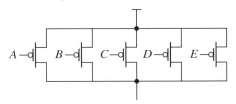

7-6. Determine the logic function of the circuit using a truth table.

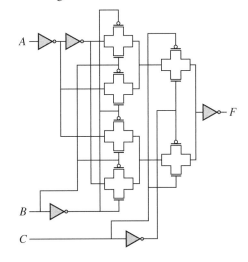

A	B	C	F'	F
0	0	0		
0	0	1		
0	1	0		
0	1	1		
1	0	0		
1	0	1		
1	1	0		
1	1	1		

7-7. Determine the Boolean function of the circuit. *A* and *B* are input signals, and *C* is the output.

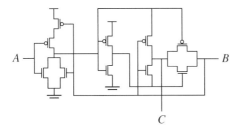

The truth table shows (fill in)

A	B	C
0	0	
0	1	
1	0	
1	1	

Synthesis of DeMorgan Circuits

7-8. Use the DeMorgan equivalent circuit to implement the transistor circuit for $F = XY + Z$ without an output inverter.

7-9. Use the DeMorgan equivalent circuit to implement $F = X\overline{Y}\,\overline{Z} + \overline{X}\,YZ$ without an output inverter.

7-10. Use the DeMorgan equivalent expression to draw the transistor schematic for a CMOS combinational logic circuit for the function $F = MNO + (\overline{P} + Q)\overline{D}$.

7-11. Use the DeMorgan equivalent circuit to implement $F = (X + Y)\overline{Z}$ without an output inverter.

7-12. Use the DeMorgan equivalent expression to draw the transistor schematic for a CMOS combinational logic circuit for the function: $F = (A\overline{B} + C)(\overline{D} + E)$ without using an inverter output stage.

7-13. A switching network has three inputs as shown. *A* and *B* and *C* represent the first and second and third bits of a binary number N1. The output of the circuit (*F*) is to be 1 only if the square root of N1 is greater than or equal to 1 and less than or equal to 2.

(a) Write the truth table.

(b) Draw a CMOS transistor schematic to implement this function.

A	B	C
0	0	0
0	0	1
0	1	0
0	1	1
1	0	0
1	0	1
1	1	0
1	1	1

Sizing Combinational Logic

7-14. $F = AB + CD + E$. Draw the transistor schematic not using an inverter, and size the transistors with respect to current strengths shown for the inverter.

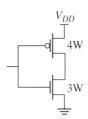

7-15. Consider the circuit and

(a) Provide the logic function that is implemented by the circuit.

Hint: You can provide the truth table or analyze the operation of each transistor.

(b) Size the transistors to have the same delay as a balanced CMOS inverter with $W_p = 3W_n$.

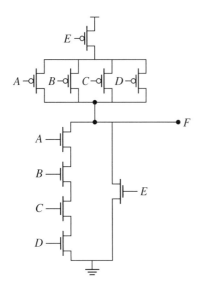

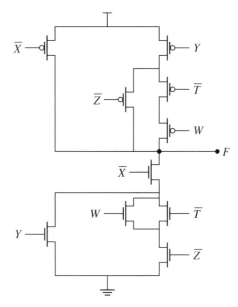

7-16. Given the function: $F = (\overline{C} + \overline{A})(B + AC) + AC$:

   (a) Reduce and express this function to its minimum Boolean statement.

   (b) Draw the transistor schematic and size the transistors for worst-case pull-up and pull-down with respect to the following inverter.

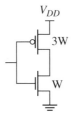

7-17. Size the transistors in the circuit with respect to a symmetrical inverter whose pMOS width is three times the nMOS width of $W_n = 1 \, \mu$m.

7-18. Size the transistors in the circuit with respect to a symmetrical inverter whose pMOS to nMOS width-to-length ratio is 3.

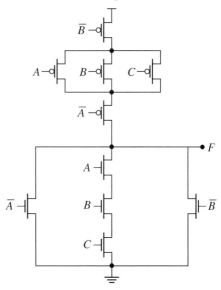

7-19. Size the transistors in the circuit with respect to a symmetrical inverter whose pMOS to nMOS width is three times the nMOS width of $W_n = 1\,\mu$m.

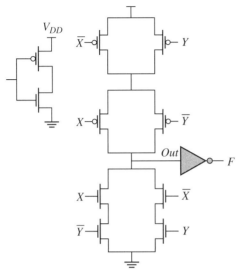

7-20. Given a symmetrical inverter on the left. Size the transistors in the circuit to the right for worst-case pull-up and pull-down that match the current drive of the inverter.

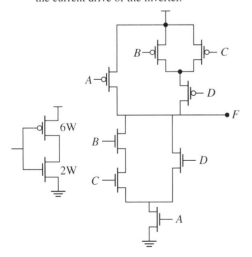

7-21. (a) Draw the pull-up circuit schematic.
   (b) Label every transistor for minimum size that matches the inverter pull-up and pull-down strengths. Use your new schematic to label the width of all transistors.

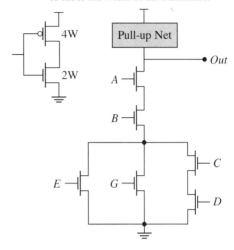

Charge Sharing in Dynamic Circuits

7-22. The dynamic 3NAND gate is precharged and then the A and B n-transistors turn on. $V_{tn} = 0.4$ V. What is the final voltage?

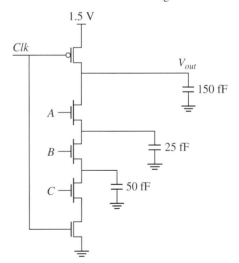

7-23. If the capacitance that absorbs charge after the switching occurs is small enough, then the analysis changes. $V_{tn} = 0.4$ V. Calculate the final voltage V_{final}.

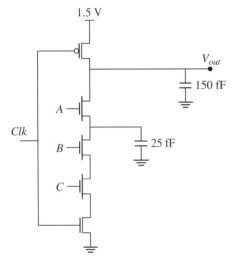

7-24. The dynamic circuit has $V_{tn} = 0.5$ V. Node-C is constant at logic-0. What is V_O during the evaluate phase of the circuit.

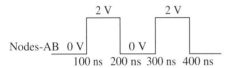

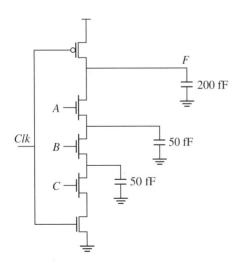

7-25. Given the dynamic circuit with $V_{tn} = 0.6$ V, when the input signals ABC go to 010 during the evaluation, what is the final voltage at the node feeding the inverter?

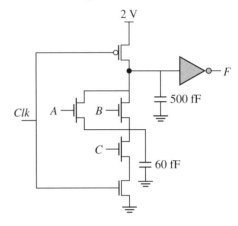

7-26. The dynamic circuit has a vector $ABCDE = 10001$ and $V_{tn} = 0.5$ V. What is the final voltage after charge sharing occurs? The drain capacitance is 30 fF per transistor.

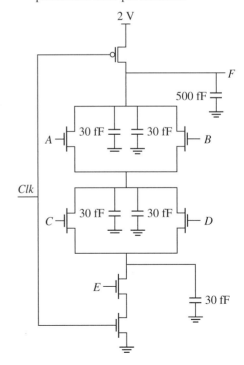

7-27. Repeat Problem 7.26 for the vector $ABCDE$ = 11010.

Domino CMOS Logic

7-28. What Boolean function will the circuit perform?

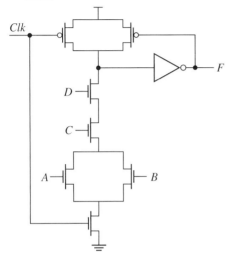

7-29. Draw the transistor schematic for a CMOS domino circuit that satisfies $F = (X + Y)\overline{Z}$.

7-30. Draw the transistor schematic for a CMOS domino circuit that satisfies $F = (X + \overline{Y})(\overline{X} + Y)$.

7-31. Draw a domino logic circuit for the function: $F = [(ABC) + (\overline{D}E)]Z$.

7-32. Draw the transistor schematic for a CMOS domino circuitry that has a 2AND output gate driven by a 2AND and a 2OR gate. The 2AND gate has A and B as input signals, and the 2OR gate has signals C and D. The domino input gates are separate structures. What is the need for the inverters?

7-33. Draw the domino CMOS transistor schematic that fulfills the function $F = (AB + C)D$ for a domino CMOS logic gate.

Pass Transistor Logic

7-34. Given $V_{tn} = 0.5$ V, $V_{tp} = -0.5$ V, and $\gamma = 0$, what is V_{out}?

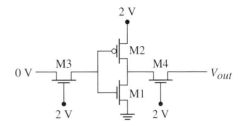

7-35. Determine the node voltages for the circuit where $V_{tn} = 0.5$ V.

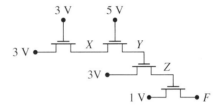

7-36. Write in the voltages at the intermediate and output nodes. $V_{tn} = 0.5$ V.

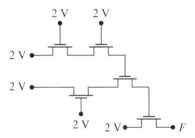

Power and Activity Coefficient

7-37. The probabilities of transitions are $\alpha_{0 \to 1}$ = 0.25 and $\alpha_{1 \to 0} = 0.25$. If $V_{DD} = 1.5$ V, C_L = 10 nF, and clock frequency is 1 GHZ, what is the power dissipated?

7-38. Given $V_{DD} = 1.2$ V and $f_{clk} = 500$ MHz, calculate the activity coefficients and power dissipation at node-D and node-F.

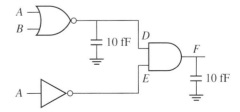

A	B	C	D	E	F

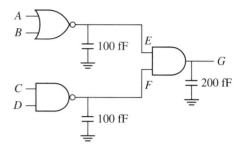

7-42. Given $F = (A + B)C$, $V_{DD} = 2$ V, and $f_{clk} = 2$ GHz, what is the expected power dissipation if the inputs are random?

7-39. What is the probability p_{nt} of no transition at node-F in Problem 7.38.

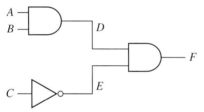

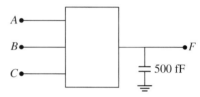

7-40. Calculate the probability of a logic-0 to logic-1 transition and logic-1 to logic-0 transition at node-G. Use a truth table.

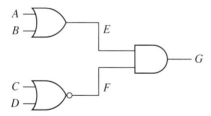

7-43. Calculate the probability of a logic-0 to logic-1 transition and logic-1 to logic-0 transition at node F. Use a truth table. If $V_{DD} = 1.2$ V, $f_{clk} = 500$ MHz. $C_D = 10$ fF, and $C_F = 15$ fF, what is the power dissipated at nodes-D and -F?

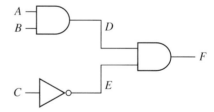

7-41. Estimate the total power consumption at the three capacitive nodes when $V_{DD} = 1.5$ V and $f_{clk} = 100$ kHz.

CHAPTER 8

Sequential Logic Gate Design and Timing

A computer is like an Old Testament God with a lot of rules and no mercy.

Joseph Campbell

Boolean operations define what a computer must do. These operations occur in the combinational logic circuits, but something more is needed to organize the mass of Boolean networks. Simplistically, we could have one giant combinational logic circuit, but many circuit path delays would be hopelessly long, forcing a slow, unworkable design. A better procedure breaks the overall Boolean operation into pieces where portions of the operation are temporarily stored between smaller Boolean networks. This technique is called *pipelining*, and we need a special type of storage circuit to coordinate the passage of combinational logic data with the clock. Circuits that coordinate data processing with a clock are one form of sequential circuits. The most common timing design style is called *synchronous design* where a master clock frequency controls all operations much as a symphony conductor acting as the master clock controls the exact movements of a music piece. The sections of the orchestra are the combinational logic networks.

Sequential circuits in a synchronous design have two or more inputs and one or two outputs (Figure 8-1). The output logic states are dependent on input logic states loaded on the previous clock period. The logical outputs are Q and often its complement $\overline{Q}$. Inputs can be several types, such as data input, clock, set, and reset. Typically, the minimum three signals are a data input (D), a data output (Q), and a clock (Clk). The input signals must follow a precise sequence of timing rules that are described later.

A simple circuit was patented in 1918 by two Englishmen, William Eccles and F. W. Jordan [1]. It was the first circuit to store a bit of information. This Eccles–Jordan vacuum tube circuit marked the beginning of the electronic computer age. It was the first sequential circuit, but it was a few years before its application to computers. It was referred

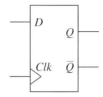

FIGURE 8-1.

Symbol of a sequential circuit.

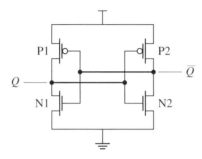

FIGURE 8-2.

A modern CMOS latch version of the Eccles–Jordan flip-flop. [1]

to then and now as a *flip-flop*. One trigger input would flip it into one logic output state while the next trigger might flop it into the other logic state. The flip and flop refer to this back-and-forth transition in logic states. We will use this simple, original definition for the flip-flops (FF). There are several forms of flip-flops, and we will select a gated latch and a D-FF (Delay-FF) to introduce the principles of sequential circuits and timing properties. The D-FF is the dominant sequential circuit in digital integrated circuits (ICs).

Figure 8-2 shows a modern CMOS version of the Eccles–Jordan flip-flop. It is a simple circuit joining two inverters. It is now called a *latch*. In operation, one side of the circuit is logic-1 and the other is logic-0. If a strong logic-1 is applied to the Q node, transistor N2 turns on and P2 turns off dropping $\overline{Q}$ to ground or logic-0. The zero voltage at $\overline{Q}$ is fed back to the N1-P1 inverter input turning on P1 and turning off N1. This positive feedback loop holds the logic states until a different logic value overdrives the latch to its other logic state. When a circuit loop connects an output signal that is in phase with a signal on an input node, then that circuit is said to have positive feedback. Positive feedback is a key to stable memory circuits.

We cannot stress enough the importance of this old circuit to modern computing ICs. The FF is in all of our digital circuits from personal computers, auto electronics, games, coffee percolators, elevator controls, cell phones, and watches to cardiac pacemakers.

The D-FF contains two latches in series. The D-FF temporarily stores bit data from a combinational block on one clock pulse and releases that data to a downstream combinational block of circuitry on the next clock pulse. We will build from this basic latch to larger assemblies that require careful attention to timing rules.

The chapter emphasizes qualitative understanding of sequential circuit operation and its relation to the fundamental timing parameters that designers use. Numerical problems are presented dealing with timing parameters and timing rules. Timing theory is challenging.

8.1. CMOS Latches

The flip-flop in Figure 8-3 is the same circuit as in Figure 8-2 but uses inverter symbols. A latch is a memory circuit that is widely used by itself or as a building block for more complex memory circuits. Latches retain their logic state as long as power is applied or until the latch logic state is overridden by a different external logic state. When a strong overdriving data input signal D is applied, then the latch output Q will acquire and hold logic state $Q = \overline{D}$. If D is noisy, then the Q logic state is also noisy. A latch output looks directly at the input, and this can present a problem in the presence of noise. This is the *transparency* property of latches. Transparency is normally undesirable.

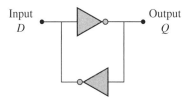

FIGURE 8-3.

Two-inverter latch circuit with input D and output Q.

8.1.1. Clocked Latch

Figure 8-4a shows a gate level design of a clocked latch using 2NOR gates. When $Clk = C = 1$, the outputs of the first set of 2NOR gates are logic-0. This is the

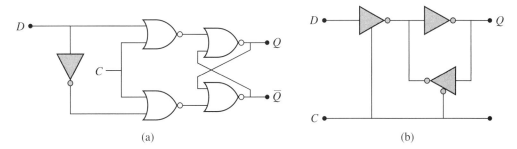

(a)   (b)

FIGURE 8-4.

(a) Basic CMOS latch design with data set on $C = 0$. (b) Gate level design tristate inverter level schematic with data set on $C = 1$. The *Clk* symbol is abbreviated as C for clarity.

non-controlling logic state input to the two output stage 2NOR gates. Therefore, the Q and $\overline{Q}$ signals fed back to the inputs of the two output 2NOR gates set a stable logic condition. If $Q = 1$, then the bottom output 2NOR gate is driven to $\overline{Q} = 0$. The $\overline{Q}$ signal feeds a logic 0 to the upper 2NOR gate setting and holding $Q = 1$ (and $\overline{Q} = 0$). The latch holds its logic state indefinitely unless input signals change or the power is lost.

When $C = 0$ (non-controlling logic state to the input 2NOR gates), the D signal (and $\overline{D}$) has a clear path to set the output NOR latch. When C returns to $C = 1$, then the new data are stored until the next clock change. The latch is transparent when $C = 0$.

Figure 8-4b shows a tristate inverter design of a compact CMOS latch with two tristate inverters and one regular inverter. A clock controls the tristate input of two inverters, so when one inverter is in the tristate (high-Z), and the other is not. When the output of the first tristate inverter stage is active ($C = 1$), the feedback inverter is in the high-Z state, and the latch output is equal to and transparent to D. Data enter the latch when $C = 1$. When C is low the output of the first inverter is in the high-Z state, and the feedback tristate inverter latches Q to D. When $C = 0$, the latch is in its memory state. Reduction of transistor count is often a design goal.

Self-Exercise 8-1

Compare the number of transistors in the latch of Figure 8-4a with a D-latch designed with tristate inverters (Figure 8-4b). Include the inverter that sets the correct level for the clock signal C to the inverters.

8.1.2. Gated Latches

Transmission gates (T-gate) make good elements to control data flow in a latch. Figure 8-5 shows a gated latch with CMOS transmission gates T1 and T2. When T1 is on, D is driven into inverter-I_1 making the output logic state $Q = \overline{D}$. T2 is off when T1 is on. When data are securely settled in the latch, then C can change state to $C = 0$ turning T1 off and

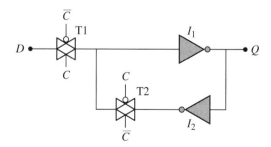

FIGURE 8-5.

Gated latch with two transmission gates. The *Clk* symbol is abbreviated as C for clarity.

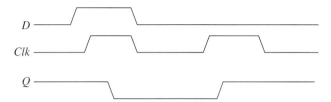

FIGURE 8-6.

Waveform of gated latch operation. Q lags the Clk leading edge due to propagation delay in T1 and I_1.

T2 on. Q is now isolated from D, and the latch will hold its state since the two inverters are connected through T2. A new driving signal D must be strong enough to override the inverters and store a new bit. This is one reason why latches are typically designed with weak transistors having small width-to-length ratios. Transparency exists between D and Q as long as the control signal on T1 is active.

The second transmission gate T2 in Figure 8-5 must remain off until the signal at the I_2 output settles to a steady state. This ensures charge equality across T2 so that T1 can be turned off and T2 turned on without degrading the signal. The output Q is isolated from the input D, and Q is held in a permanent memory condition by the positive feedback loop through T2. This is our first exposure to exact timing sequences necessary to control data flow.

Figure 8-6 shows D, Clk (C), and Q timing waveforms for the gated latch in Figure 8-5. The design sequence is for D to precede the clock signal Clk in the gated latch of our example. When $Clk = 0$, then transmission gate T1 is off (and T2 is on). When the Clk goes from 0 to 1, T1 turns on letting new data enter the latch. T2 turns off as data enter the latch thus preventing any contention between transistors in inverter I_2 and those in the input circuit driving D through T1. When the signal transient settles and the output of I_2 is equal to the input of I_1, then T1 can be turned off and T2 turned on. The node voltages across T2 are now equal so when it switches, no charge transfers across T2. This is important because a transient charge transfer across T2 would affect the overall time delay from D to Q. The response time of Q referenced to the Clk leading edge is due to delay in the electrical path.

8.2. Edge-Triggered Storage Element

Most digital designs clock the sequence of Boolean operations from one stage to another through a memory element called the *edge-triggered* FF. Clocks determine when new data are accepted by the flip-flop and when that data are released to downstream combinational logic circuits. The D-FF is a common edge-triggered design that stores data from the previous clock cycle and then drives that data to an output node on the next clock cycle. This data transfer on two clock periods eliminates transparency. The Clk transition time or

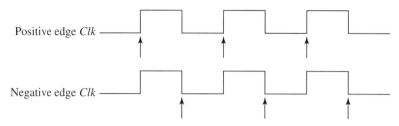

FIGURE 8-7.

Edge-triggered response for D-FF symbol.

edge is the timing reference point for data acceptance and storage. This circuit is called an *edge-sensitive* memory element as opposed to the level sensitive latch.

Figure 8-7 shows two clock waveforms indicating a positive edge (a) and negative edge (b) triggers. Data will transfer from a previous clock pulse to the FF output on the active transition of the clock. The edges also become the points of reference for establishing timing rules. Designs may use either positive or negative edge-triggered FFs, but designers most often choose positive edge triggers.

8.2.1. The D-FF

FFs have several designs, such as the D-FF, T-FF, JK-FF, and SR-FF that you studied in your introductory logic course. We will study the edge-triggered D-FF since it is typically the most common FF in digital ICs. Figure 8-8 shows an edge-triggered D-FF made of two-gated latches in series. The apparent simplicity of connecting two latches is deceptive.

The first latch on the left in Figure 8-8 is called the master, and the second latch on the right is called the slave. Inverters I_1 and I_2 form the master latch with T2. T3 separates the two latches and controls when data pass from the master to the slave. Inverters I_3 and I_4 form the slave latch with T4. Each T-gate has a unique function depending on whether the *Clock* is logic high or low. The master latch accepts input data D on the clock negative state and transfers that data to the slave latch output Q on the positive edge of the next

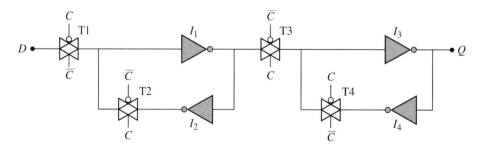

FIGURE 8-8.

A positive edge triggered flip-flop uses two-gated latches in series.

clock pulse. This isolates D from Q during consecutive logic states of the clock removing transparency from D to Q. Notice that $\overline{Q}$ is available from the output of I_4.

The operation begins with $Clk = C = 0$. This turns on T1, and data D pass through T1 into I_1 and I_2. T2 is off. This first phase of operation must wait until the signals settle. Then the Clk can change from logic-0 to logic-1 turning off T1 and turning on T2 and T3. If T2 turns on before its I/O nodes are settled, then a transient charge movement occurs that affects timing accuracy. The action now goes to the slave latch whose T-gates have also changed their respective on-off states.

The positive Clk edge turns on transmission gate T3 and turns off T4. Data from the master latch pass through T3 into I_3 and I_4 where it settles at the output of I_4. When the output of I_4 equals the input of I_3, then the Clk can again change polarity isolating the master and slave latches with T3 now off. The Clk pulse width must be long enough to satisfy this delay. T1 and T3 are never simultaneously on nor are T2 and T4. Each T-gate has a role in steering data through the memory element under direction of the clock polarity.

Clk levels are important, but the release of data to output Q begins at the Clk edge on T3. A short delay occurs from the clock edge at T3 until Q assumes the new logic state. This is known as the Clk-Q delay. The Clk edge is a timing reference mark for the circuit timing parameters that we will discuss shortly.

The Clk signal polarity on transmission gate T3 defines whether the master latch data release is sensitive to a positive or negative transition of the clock. If the T3 nMOS transistor is driven with $\overline{Clk}$ and the pMOS driven by Clk, then data would pass on a negative edge. The choice of positive or negative edge trigger timing belongs to the designer. Either polarity will work. Positive edges are probably more common while negative edges may sometimes offer a local timing advantage by activating a circuit on the next down transition instead of waiting for the next positive edge.

Detailed operation of the edge-triggered FF is important for understanding the sequence of timing rules that digital ICs must follow. The master-slave edge triggered circuit allows us to precisely control the moment when all the flip-flops change state (Figure 8-8). These mechanisms are important so we will review the process thus far for a positive edge-triggered flip-flop clock.

8.2.2. Clock Logic States

Clock = 0 State

In Figure 8-8, T1 passes data when the Clk is logic-0. Simultaneously, T2 and T3 are *off*. Since T3 is *off* the data in the two latches are isolated. Data can be loaded into the master latch independent of the data stored as Q in the slave latch. This isolation gives the non-transparency desired for a register storage cell.

By keeping T2 *off* during the data load time period, we prevent any signal (transistor) contention occurring at the input to I_1. Data can be loaded more rapidly into the master without interference from the feedback loop. Simultaneously, the slave latch is isolated, and the T-gate Clk signals are such that a memory state is held constant at output Q.

Q drives a downstream combinational logic sub-circuit node. When the clock is low, it must remain low long enough for the signal through T1 to settle at the input of T2.

Clock = 1 State

T1 turns *off*, and the master is isolated from new incoming data. T2 turns *on* holding the master memory logic state as it drives the slave. T3 is *on*, and data pass from the master to the slave. With *Clk* = 1, T4 is *off*, and similar to the master operation, this allows data to enter I_3 and I_4 without contention problems at the input to I_3. The *Clk* = 1 pulse width time must allow for the signal to settle at the output of I_4. T3 can then be closed, T4 opened, and the new data are stored as *Q*. *Clk* returns to its low state to accept new data into the master latch.

Clock Active Edge

The *Clk* edge on T3 is the time mark when the data transfer action begins in the FF. A positive *Clk* edge on T3 would pass data on the rising edge of the *Clk*. The T3 *Clk* polarity controls whether the flip-flop is a positive or negative edge triggered memory element. If we reversed all *Clk* polarities on the T-gates in Figure 8-8, then the signal would pass on the falling or negative *Clk* edge. There is a small, finite moment when T1 and T3 are simultaneously on as they transition. For example, if *C* and $\overline{C}$ were momentarily at $V_{DD}/2$ then all T-gates would be weakly on. Since rise and fall times are short, we assume that there is negligible transparency through T1 and T3.

8.2.3. A Tristate D-FF Design

Figure 8-9 shows a positive edge CMOS design of a *D* flip-flop combining tristate inverters and transmission gate design. The slave cell is only half of the master latch design. Data are loaded into the first latch on the negative clock state, and data are read by the output

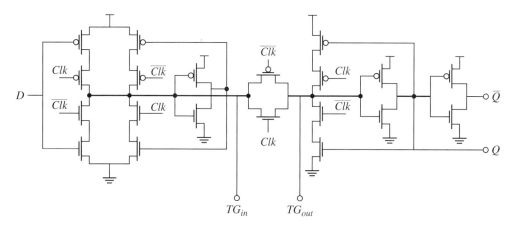

FIGURE 8-9.

CMOS design of a positive edge *D* flip-flop combining tristate inverters and transmission gate design.

Q on the rising clock edge. Then data are stored in the slave latch, when the clock returns to logic-0.

EXAMPLE 8-1

The D and Clk waveforms are given for the FF in Figure 8-9. Sketch the waveforms for the node in front of the controlling CMOS T-gate (TG_{in}), the node after the T-gate (TG_{out}) and Q. Initially, $Q = 0$. Assume no delay in the signal path elements.

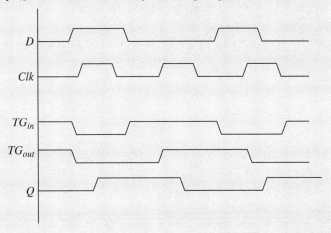

The TG_{in} inverts D when Clk is low. TG_{out} follows the TG_{in} node when the clock is high. The TG_{out} node holds its value until Clk goes high again. Q is the complement of the TG_{out} node.

Self-Exercise 8-2

The D and Clk waveforms are given for the positive D-FF in Figure 8-9. Sketch the waveforms for the node in front of the CMOS T-gate, the node after the T-gate and Q. Assume no delay in the signal path elements.

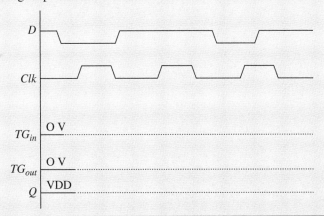

8.3. Timing Rules for Edge-Triggered Flip-Flops

The sequencing of data through both latches of the FFs in Figure 8-8 and the relative arrival times of the data and clock signals require precise timing rules. These rules depend on specific internal path delays. We will define seven timing parameters for the edge-triggered flip-flop. These parameters evolved from early experience with computer circuits. Most timing parameters are referenced to the *Clk* edge. Each parameter is subject to significant statistical variation due to process manufacturing imperfection.

The delays defined by these parameters limit the maximum clock frequency of the chip. Violation of these timing rules leads to failure without mercy. The *D*, *Clk*, and *Q* waveforms for the flip-flop are shown in Figure 8-10 to illustrate these timing parameters. The timing marks are taken at the 50% amplitude points for each waveform. We will adopt the timing symbol convention of [2].

8.3.1. Timing Measurements

Setup Time

t_{su} is the minimal time that a data input signal (D) must precede the clock input (Clk). t_{su} is the circuit settling time for data entering the first latch. It is a function of the delay in T1, I_1, and I_2. t_{su} must be larger than the propagation delay from the input-D through T_1, I_1, and I_2.

Minimum t_{su}

$t_{su,cd}$ is the statistical minimum time that a data input (D) must precede the clock input (Clk). This is a function of manufacturing statistical variation (cd is called the contamination or minimum delay). It is the shortest expected delay of T1, I_1, and I_2 based on engineering estimates of process variation.

Hold Time

t_{hold} is the time that the data input (D) must hold its state after the clock edge. It is the delay between the active *Clk* edge and turning off the input T-gate (T1). When T1 is on,

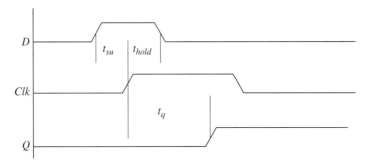

FIGURE 8-10.

Timing definitions for *D*, *Clk*, and *Q* waveforms.

then the incoming D signal can change. t_{hold} is a function of any element delays from the clock edge to its arrival at T1 and T3.

Clock to Q Time

t_{cq} is the maximum delay between the Clk edge and arrival of the signal at the output node Q. This is a function of propagation delays in T3 and I_3.

Minimum t_{cq}

$t_{cq,cd}$ is the statistical minimum delay between the Clk edge and arrival of the signal at the output node Q. $t_{cq,cd}$ is usually based on engineering analysis of the variables that randomly or systematically affect delay in the signal path.

Clock pulse width

t_{cw0} is the minimum clock width when the Clk is low. This is the signal node settling time for the master latch. It is the time to let the I_2 output equal the I_1 input.

t_{cw1} is the minimum clock width when the Clk is high. This is the signal node settling time for the slave latch. It is the time to allow the I_4 output equal to the I_3 input.

The data must precede the clock edge by at least the setup time t_{su}. After the Clk edge, the data must hold stable for an extended time t_{hold}. The hold time is necessary to ensure that any delays between the clock transition time and turning off the input T1 transmission gate are accounted for. It would corrupt the data if T1 was not turned off when new data arrive. Never change D until you are sure that the clock has time to turn off the input transmission gate (T1). The sum of the setup and hold times specify the minimum time that incoming D must remain stable.

EXAMPLE 8-2

The D and Clk waveforms are given for a positive edge D-FF. The initial portion of the Q waveform is drawn to show t_{cq}. Complete a sketch of Q.

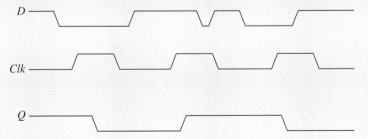

New data transfer from the master to the slave on the rising edge of the Clk pulse. There is a t_{cq} delay before the new data arrive at Q. The glitch in D in the middle of the second Clk pulse is ignored since the first T-gate is off shielding the master latch from accepting the new data.

EXAMPLE 8-3

Given an edge-triggered D-FF.

    (a) Mark the proper clock signals on the T-gate control nodes for a negative edge FF.
    (b) Calculate t_{su}, t_{hold}, t_{cq}, and the minimum clock pulse width for $Clk = 1$ and $Clk = 0$.

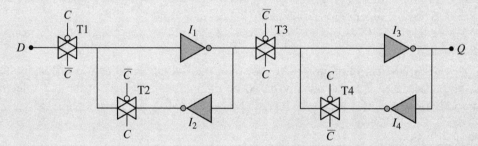

Given:

Ckt	t_{PD} (ns)
T1	2
T2	0
T3	2
T4	0
I_1	1
I_2	2
I_3	1
I_4	2

Answer:
    (a) *Clk* polarities are shown in the schematic.
    (b) $t_{su} = (T1 + I_1 + I_2)t_{PD} = (2 + 1 + 2)\text{ns} = 5$ ns

          $t_{hold} = 0$ ns since the arrival of the *Clk* pulse to T3 and T1 is simultaneous. There is no delay from *Clk* edge in turning off T1, so the new data can be entered as soon as the *Clk* makes its negative edge transition.

        $t_{cq} = (T3 + I_3)t_{PD} = (2 + 1)\text{ns} = 3$ ns

        $t_{pw}(Clk = 0) = (T3 + I_3 + I_4)t_{PD} = (2 + 1 + 2)\text{ns} = 5$ ns

        $t_{pw}(Clk = 1) = (T1 + I_1 + I_2)t_{PD} = (2 + 1 + 2)\text{ns} = 5$ ns

8.3.2. Timing Rule Violation Effect

The simulated *Clk-to-Q delay time* versus *Data-to-Clk* time interval graph in Figure 8-11 shows how the t_{cq} delay of the flip-flop increases as data change occurs closer to the clock

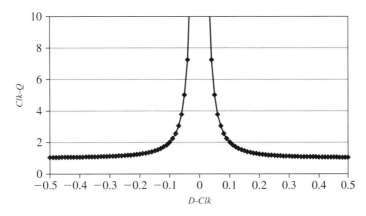

FIGURE 8-11.

Clk-Q delay as a function of *D-Clk* when setup time (left curve) and hold time (right curve) are reduced. The curves are simulated.

edge. The x-axis of the left curve shows changes in setup time violation while the right curve shows hold time violations. The y-axis shows the increase in t_{cq} delay time.

Violation of t_{su} and t_{hold} times does not cause an immediate failure, but an exponentially increasing delay that itself will cause failure. The lower left curve in Figure 8-11 shows the sensitivity to t_{cq} delay as the *D-to-Clk* interval goes below the normalized setup time of -0.5. The right curve shows the t_{cq} delay increase as the delay between the *Clk* edge and data hold time is reduced below 0.5. The setup time violation increase in t_{cq} delay is from the charge contention at the inverter input nodes at the master latch. The increase in t_{hold} is caused by delay at T1 and the subsequent contention of input logic states in transition in the master latch. The conclusion is: let all latch nodes settle before allowing another signal change to enter the latches.

Self-Exercise 8-3

(a) Use Figure 8-12 to estimate the percent increase in t_{cq} if the setup time is decreased from 0.4 to 0.05?

(b) Use Figure 8-12 to estimate the percent increase in t_{cq} if the hold time is decreased from 0.5 to 0.1?

Answers: **(a)** 700%. **(b)** 200%

8.4. Application of D-FFs in ICs

We will describe a brief application of the D-FF before developing deeper insights into the timing process. D-FFs are heavily used to build storage registers (Figure 8-12). Parallel-in

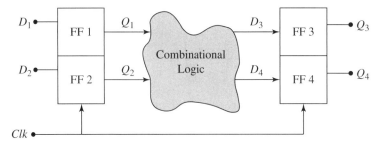

FIGURE 8-12.

Positive edge flip-flops forming a 2-bit parallel-in parallel-out register with D and Clk signals to each FF. The clock goes to each FF.

parallel-out data registers are made of D-FFs. Figure 8-12 shows two positive edge-triggered 2-bit registers with a Clk signal and data inputs going to each FF, and a combinational logic block between the registers.

When Clk is low, input data D_1, D_2, D_3, and D_4 are accepted by their FF master latches. Data D_3 and D_4 are data from the previous Clk pulse. On the next positive Clk transition, each FF delivers data to the slave latches, and their outputs Q_1, Q_2, Q_3, and Q_4 drive that bit data to the next combinational block for processing. The Clk transition (edge) on each register controls the exact time at which this occurs. The timing of D, the Clk pulse edge, and the internal propagation delays is an exact procedure.

8.5. t_{su} and t_{hold} with Delay Elements

Thus far, the setup and hold times were defined for edge-triggered FFs when the D and Clk lines drive directly into the circuit. Typically, combinational logic gates or other elements may lie in these data and clock paths, and their delay can influence t_{su} and t_{hold}. This section will deepen our understanding of these fundamental timing concepts by showing how t_{su} and t_{hold} are altered by line delay elements.

Delay Elements in D-*Line:*   Figure 8-13 shows a positive edge triggered FF with a delay element in the D-line causing a delay $t_{D\text{-}delay}$. How does that affect t_{su} and t_{hold}? The setup time is still defined as the time delay between the edge of the D-waveform and stabilizing of the I_2 output, so that

$$t_{su} = (t_{D\text{-}delay} + \text{T1} + I_1 + I_2) \tag{8-1}$$

The D-signal must start earlier to avoid setup time violation.

Hold time may be analyzed as the sum of two observations. First, D cannot change until T1 receives a Clk signal to turn off. For the circuit in Figure 8-13, T1 turns off instantly with the active edge of the clock, so that D can also change with the clock edge. But a second observation looks for any delays in the D-line and makes an adjustment. If D was

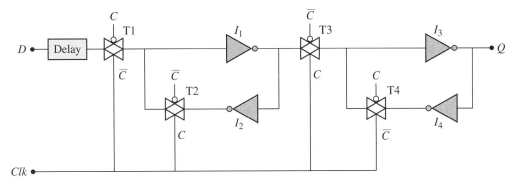

FIGURE 8-13.

Positive edge-triggered FF with a delay element in the *D*-line.

allowed to change with the clock edge, then it would not violate the t_{hold} specification, but new data from *D* would arrive with a delay after T1 was turned off. That is wasted time. Why not initiate an earlier change in *D* equal to that delay, and let the signal on *D* arrive at T1 exactly when the *Clk* turned off T1. The altered hold time in the presence of a delay in the *D*-line is

$$t_{hold} = 0 - t_{D\text{-}delay} = -t_{D\text{-}delay} \qquad (8\text{-}2)$$

Figure 8-14 shows the dotted line waveform adjustment in t_{su} and t_{hold} in the presence of a delay element in the *D*-line. The clock arrival waveform on T1 is shown to illustrate when *D* can change its hold time. The setup time increased, and the hold time decreased as shown by the dotted line waveform.

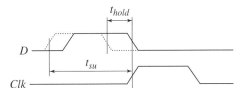

FIGURE 8-14.

Adjusted waveforms for positive edge triggered FF with a delay element in the *D*-line. The dotted line shows alteration of the *D* waveform.

Delay Elements in Clk-*Line:* Figure 8-15 shows a delay element in the *Clk*-line ($t_{Clk\text{-}delay}$). The hold time must adjust for this delay. The *D*-signal must wait for that *Clk* delay before being released to change state.

$$t_{hold} = t_{Clk\text{-}delay}$$

The new setup time ($t_{su\text{-}delay}$) must account for the master latch delays as well as the *Clk*-line delay. The *D*-signal can delay change of state by the amount of the clock line delay so it

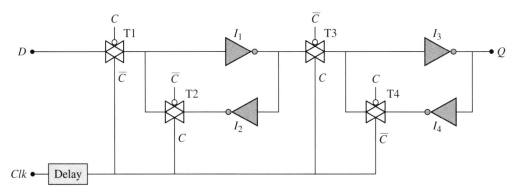

FIGURE 8-15.

Positive edge triggered FF with a delay element in the *Clk*-line.

can shorten by

$$t_{su\text{-}delay} = (-t_{Clk\text{-}delay} + T1 + I_1 + I_2) \text{ delays} \tag{8-3}$$

and the new hold time is

$$t_{hold} = t_{Clk\text{-}delay} \tag{8-4}$$

Figure 8-16 shows the waveform adjustment in t_{su} and t_{hold} in the presence of a delay element in the *D*-line. *D* can release its hold time when the positive T1 waveform turns off T1. The setup time decreased and the hold time increased.

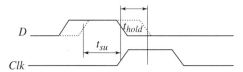

FIGURE 8-16.

Adjusted waveforms for positive edge triggered FF with a delay element in the *Clk*-line. The dotted line shows alteration on the *D* waveform.

Self-Exercise 8-4

The edge-triggered *D* flip-flop circuit with normal timing parameters given as follows has defect-related delays built into certain paths:

(a) If a defect occurred in I_2 and increased its delay time to 6 ns, what would be the effect on operation of the FF?

(b) If a delay defect of 4 ns occurred in the line feeding the *Clk* line, what would be the effect on operation of the FF?

(c) If a delay defect of 5 ns occurred in the line at the input of I_3 what would be the effect on operation of the FF?

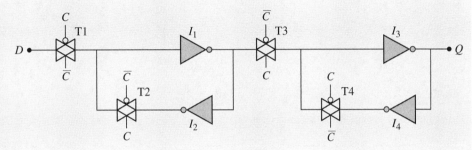

Given:

Ckt	t_{PD} (ns)
T1	2
T2	0
T3	2
T4	0
I_1	1
I_2	2
I_3	1
I_4	2

Answers:

(a) $t_{su} = 9$ ns; an increase of 4 ns. t_{hold} decreases to –4 ns.

(b) t_{hold} increases by the delay in the *Clk* line to 4 ns. t_{su} decreases by 4 ns.

(c) The t_{cq} and clock width times would increase to 8 ns. The clock width must accommodate the extra 5 ns of delay, or we expect failure.

8.6. Edge-Triggered Flip-Flop with Set and Reset

Figure 8-17 shows a positive edge-triggered master-slave FF that uses 2NOR gates in the slave section instead of inverters. When the Reset terminal goes to logic-1, then the Q output goes to logic-0 after the delay through the NOR gate. The Reset function is not necessarily synchronized to the clock. When Reset is set to logic-0, that is the noncontrolling logic state for a NOR gate, and that state then has no influence on Q. Similarly, when the Set terminal at the lower right goes to logic-1, then Q goes to logic-1 when $C = 0$.

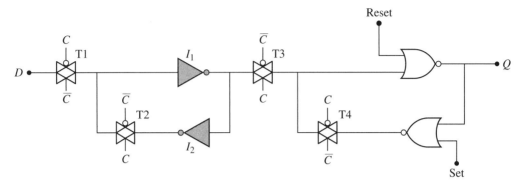

FIGURE 8-17.

Edge-triggered FF with Set and Reset control.

8.7. Clock Generation Circuitry

One more timing subcircuit is needed before we put all of the pieces together, and that circuitry deals with clock generation and its distribution networks. A master clock oscillator generates an off-chip clock pulse typically from a crystal oscillator mounted on the printed circuit board (PCB). A printed circuit board is an assembly of many ICs and discrete components on a thin fiberglass laminate that performs a dedicated function. The PCB has large pF load capacitances and inductances on the long board traces (interconnect lines). The board is not a good environment for high frequency signals (>200 MHz). One solution uses a PCB crystal oscillator circuit at a low frequency (50–166 MHz) and transports that low-frequency clock across the board traces to each chip. Each chip then has a special circuit called a phase locked loop (PLL) that converts the low board master clock to a higher (or lower) synchronized clock frequency suited to the specifications of the chip. Virtually all microprocessor type chips have a PLL system.

Figure 8-18 shows the major blocks of a clock generation system: the phase locked loop (PLL) and its frequency divider, the metal interconnections, the buffer inverters on the interconnect lines, and the timing distribution network to the FF registers. The board master clock f_{osc} enters the PLL where its frequency is multiplied by a constant N, and its phase is compared with a clock signal taken from a timing node deep in the circuit. The $\div N$ digital block feeds a lower frequency clock f_o/N to the PLL, but the PLL has an unusual property that by dividing the timing node frequency by N, the output of the PLL actually multiplies the PCB reference frequency f_{osc} by N and the lower f_o/N input frequency is phase synchronized to f_{osc}. The frequency at the PLL output is a multiple $f_o = N f_{osc}$. That frequency is distributed to many FFs across the IC. The buffer inverters in the interconnect lines are necessary to minimize path delay. The whole assembly is a negative feedback circuit that locks in the frequency f_o. The next question is how the PLL works.

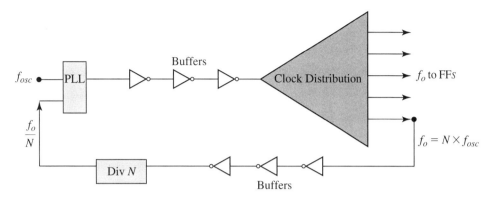

FIGURE 8-18.

PLL and clock distribution.

8.7.1. Phase Locked Loop Circuit

The PLL and the clock distribution circuitry: (1) convert a lower frequency board clock to a higher frequency for the chip; (2) distribute that chip clock signal to thousands or millions of edge-triggered memory elements; and (3) ensure that these distributed clock signals are sufficiently synchronized so that stable, high-frequency operation can occur. Managing the clock arrival times is challenging where violations of local setup and hold times can cause chip failure.

Figure 8-19 shows the signal flow and subcircuit blocks of a PLL driven by an off-chip oscillator. The phase detector senses when f_{osc} and f_o/N are not aligned. Misaligned signals drive a charge pump that either turns on a pull-up current source or a pull-down current source depending on the misalignment. The third stage is a simple low-pass filter that smooths the rapid transitions of the charge pumps and converts them to a slow quasi-DC signal for the V_{DD} of the voltage-controlled oscillator (VCO). Next we will examine the details of the PLL blocks.

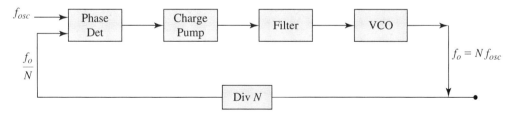

FIGURE 8-19.

The major blocks of the PLL.

Figure 8-20a shows one version of a phase detector. It uses two positive edge D-FFs, an AND gate, and Reset signals to each D-FF. The reset signals instantly drive the FF outputs to logic-0 when the 2AND output is logic-0, and this control is independent of the

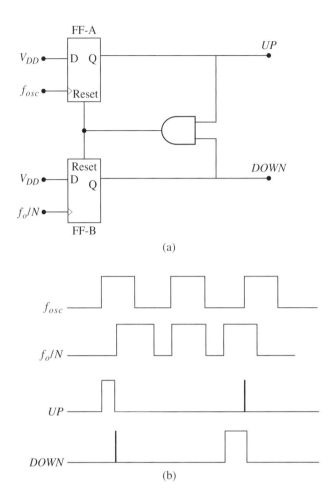

FIGURE 8-20.

(a) Phase detector circuit. (b) Waveforms.

input clocks. The master clock, f_{osc} drives the clock input of FF-A, and the chip feedback frequency f_o/N drives the clock node of FF-B. V_{DD} is tied to the D-input of both FFs.

The *UP* signal in Figure 8-20 is taken from FF-A and the DOWN signal from FF-B. When both FFs are logic-1, the AND gate responds and drives both FFs to logic-0 through the Reset nodes. The FFs respond to the rising clock edge, so if both clock input signals are logic-1 the *UP* and *DOWN* signals remain at logic-0, and no error correction signal is driven to the charge pump. Error correction begins at the time at which the signal phases differ and the length of time that they differ.

Figure 8-20b shows the waveforms of these signal responses. The two glitches in the waveforms correspond to delay time in the circuit before the Reset terminal is activated. Focus on the two clock's *rising edges* to help interpret the *UP/DOWN* waveforms.

Self-Exercise 8-5

Sketch the *UP* and *DOWN* waveforms

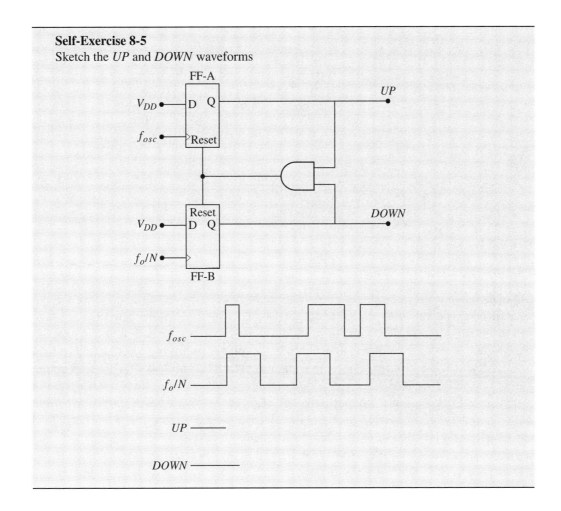

Charge pumps can be thought of as modified CMOS inverters with special biasing and signal steering transistors added. *p*MOS and *n*MOS transistors act as constant current source pull-up and pull-down devices when they are in the saturation bias mode. The pull-up and pull-down signals control the charge pump output node voltage.

The RC *filter* in Figure 8-19 eliminates the high-frequency components from the charge pump output. The filtered signal contains the necessary phase and frequency information to drive the V_{DD} node of the VCO. The VCO output is at the required chip frequency. Changes in V_{DD} affect the VCO output frequency and phase that are described next.

A common VCO design uses a ring oscillator (RO) that connects an odd number of inverters in series (Figure 8-21) with the output of the last inverter connected to the input of the first inverter. When power is applied to the RO, noise in the system initiates a logic signal at the input of one of the inverters such as the first inverter. That signal is continuously inverted and amplified to a logic level as it passes through the chain. After

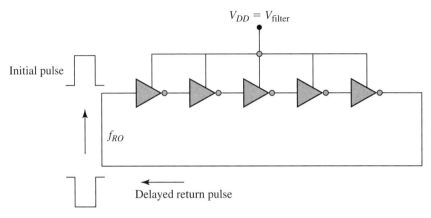

FIGURE 8-21.

Ring oscillator.

a total path delay, the last inverter output arrives back at the input of the first inverter, but its logic level is opposite to that of the original signal. The first inverter now changes state and the process repeats. The signal ripples through the chain with only three inverters in the On-state at a single moment; an inverter that is full on, its driving inverter, and its load inverter. Typical RO inverter counts are from 5 to over 100. The lower the inverter count, the higher the RO frequency.

The signal delay around the loop sets the period of the VCO. Since transistor current drive strength is dependent on V_{DD}, the ability of each inverter to charge and discharge load capacitance is a function of V_{DD}. The RO frequency increases when V_{DD} increases and decrease when V_{DD} decreases. Any noise in V_{DD} or V_{SS} is sensitively transmitted to the RO frequency. RO *jitter noise* is a major concern and discussed shortly in system timing properties. The overall function of the PLL is complete when the VCO output drives a local or regional clock node through line buffers.

EXAMPLE 8-4

(a) The ring oscillator has $V_{DD} = 1.5$ V and an average inverter propagation delay of 500 ps. What is the RO frequency?

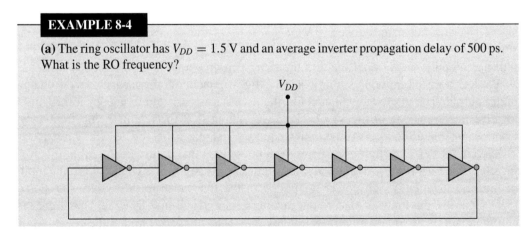

$$f_{RO} = \frac{1}{\sum T_{pd}} = \frac{1}{7 \times 500 \text{ ps}} = 285.7 \text{ MHz}$$

(b) If V_{DD} drops by 10 mV, the inverter propagation delays increase by 1%. What is the new RO frequency?
The adjusted propagation delay is 1.01×500 ps $= 505$ ps, and the RO frequency is

$$f_{RO} = \frac{1}{\sum T_{pd}} = \frac{1}{7 \times 505 \text{ ps}} = 282.9 \text{ MHz}$$

The difference seems small, but the PLL will sense the change in misalignment, and increase the strength of the *UP* signal in the phase detector. The charge pump reacts by increasing the V_{DD} drive to the RO, and its frequency will return to 285.7 MHz.

8.8. Metal Interconnect Parasitic Effects

Metal interconnect lines of the clock generation circuitry have a complexity described in Chapter 4. Their R, C, and L properties are especially relevant, since interconnect lines can be long and form many parallel paths. Their long length increases the chance for random interaction with adjacent signal lines. We will briefly review the complexity of exact timing placement before addressing system timing.

Clock arrival times at local flip-flop clock nodes depend on many variables. The total load capacitance of the clock interconnect lines can be large and in the nF range. Line resistance can degrade the clock logic strength by an Ohm's law voltage drop. Reduced signal strength slows a circuit. Cross-talk coupling with adjacent signal lines can slow or speed up a pulse. Inductive voltages (*Ldi/dt*) reduce V_{DD} and V_{SS} strengths. These uncertainties confuse the prediction of when a clock pulse will actually arrive. The rise and fall times of modern ICs are in the range of 20–100 ps making inductive voltages a problem in stabilizing signal and power voltages. Long interconnect lines have distributed RC loads, and that is a problem since delay times are proportional to the square of the interconnect line length (Chapter 4). Line buffers help solve some problems.

8.9. Timing Skew and Jitter

Clock circuitry has two other significant inaccuracies that challenge accurate placement of clock signals at the D-FF registers. The first is called *timing skew* (δ). Skew is the difference in arrival times of two clock signals with respect to each other or with respect to a reference node. Skew is typically a constant offset caused by differences in interconnect path length and buffer circuit delay.

Jitter is the second placement inaccuracy. It is a random variable that alters successive clock periods, and jitter's major source is from V_{DD} and V_{SS} noise variations in the PLL.

The output stage of a PLL is a voltage controlled digital oscillator (VCO) whose frequency is sensitive to the instantaneous changes in the power rails. Other jitter sources include differential temperatures across the die and cross-coupling effects. In the next section, skew and jitter will be incorporated into an overall timing budget for a chip. Jitter is usually the more difficult timing uncertainty to control in modern ICs.

8.10. Overall System Timing in Chip Designs

All timing properties studied to this point come together in a description of the whole chip system timing. Figure 8-22 illustrates the IC system timing components. The primary inputs (PI), the pipeline structure, the clock generation circuitry, and the primary outputs (PO) make a system. We must consider

- Combinational logic maximum and minimum propagation delays
- The setup, hold, and *Clk-Q* times of the FFs in the registers
- The skew and jitter inaccuracies of the clock generator circuit

Two major timing constraints must be obeyed. The first is the *clock period constraint* stating that all relevant propagation delays must be completed before the next clock edge can occur. The second is the *hold time constraint* of the register FFs that states that you cannot generate a subsequent downstream clock pulse until the FF hold times are satisfied. If the next clock pulse arrives too soon, then erroneous data will pass early through the

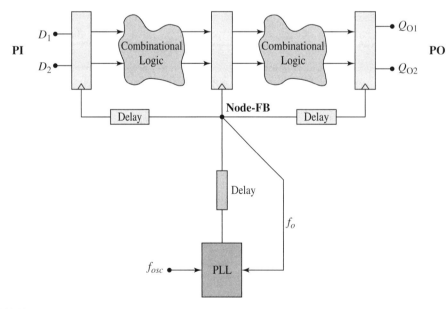

FIGURE 8-22.

System blocks and timing connections.

combinational logic block arriving at the next register too soon. That can violate hold times, and data from the wrong clock edge will enter in the next register.

8.10.1. Period Constraint

Figure 8-23 shows a pipeline structure with A and B clock waveforms. The clocks are perfectly aligned. When clock-A initiates data transfer in the first register, the signal undergoes a *Clk-Q* path delay followed by a delay in the combinational logic block and setup time of the second register. This is simple common sense. All the relevant events must take place before the next clock edge can occur. The period constraint equation is

$$T_{period} > t_{cq} + t_{logic} + t_{su} \tag{8-5}$$

The maximum operating frequency is

$$F_{MAX} = \frac{1}{T_{period}} \tag{8-6}$$

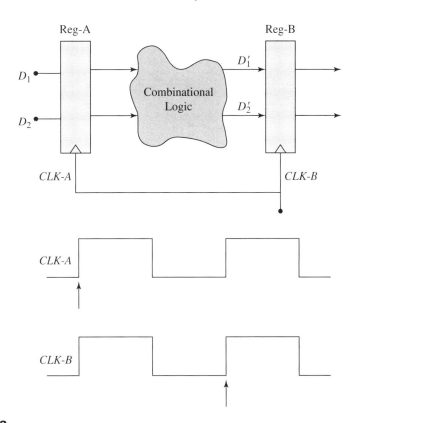

FIGURE 8-23.

Period constraints.

EXAMPLE 8-5

How fast can the circuit in Figure 8-23 be clocked if $t_{su} = 2$ ns, $t_{hold} = 3$ ns, $t_{logic} = 15$ ns, and $t_{cq} = 4$ ns.

$$T > t_{cq} + t_{logic} + t_{su} = (4 + 15 + 2)\ \text{ns} = 21\ \text{ns}$$

$$F_{MAX} = \frac{1}{T} = \frac{1}{21\ \text{ns}} = 47.62\ \text{MHz}$$

If the clock frequency increases, then false data can be entered in the downstream load register.

8.10.2. Period Constraint and Skew

We will illustrate the hold time constraint introducing skew offset. Figure 8-24 shows clocks A and B again, but clock-B occurs later and is positively offset by a skew amount δ.

The period constraint is modified as

$$T_{period} > t_{cq} + t_{logic} + t_{su} - \delta \qquad (8\text{-}7)$$

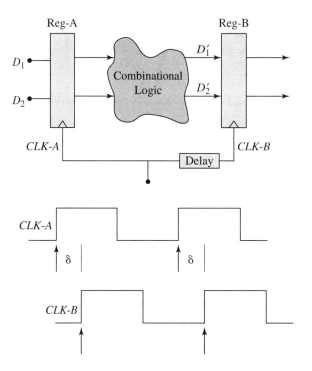

FIGURE 8-24.

Clock waveforms with positive skew.

Since clock-B occurs later, there is more time available for the path delays. That means that the clock period can be shortened by an amount δ with a corresponding increase in clock frequency. That is good, but positive skew causes a corresponding problem for the hold time that is explained in the next section.

If the delay element were now placed in the clock-A line, then clock-B occurs before clock-A. The period must increase by the negative skew as

$$T_{period} > t_{cq} + t_{logic} + t_{su} + \delta \qquad (8\text{-}8)$$

EXAMPLE 8-6

How fast can the circuit in Figure 8-23 be clocked if $t_{su} = 40$ ps, $t_{hold} = 60$ ps, $t_{logic} = 600$ ps, $t_{logic,cd} = 400$ ps, $t_{cq} = 50$ ps, and a positive clock skew of $\delta = 100$ ps.

$$t_{period} > 50 \text{ ps} + 600 \text{ ps} + 40 \text{ ps} - 100 \text{ ps} = 590 \text{ ps}$$

$$F_{MAX} = \frac{1}{590 \text{ ps}} = 1.695 \text{ GHz}$$

8.10.3. Hold Time Constraint

If a short data path in the combinational logic had a delay less than δ, then it is possible that data would reach the load registers before clock-B. When clock-B arrived shortly after, this short path data would be loaded in the second register instead of the previous data that should have gone there. That error is a *hold time violation*. Hold time is a common problem for designers. The hold time constraint can be expressed for positive skew as

$$t_{hold} < t_{cq,cd} + t_{logic,cd} - \delta \qquad (8\text{-}9)$$

The contamination delays (cd) are the shortest delays expected. A requirement that t_{hold} be made smaller is a challenge to the designer since it must reduce the propagation delay between a clock edge and turning off the master latch input transmission gate. The minimum propagation delay time parameters of the combinational logic and the *Clk-Q* time are used. In the hold time constraint, speed in these two paths can kill functionality.

EXAMPLE 8-7

What is the maximum hold time allowed for the circuit of Figure 8-24 if $t_{su} = 40$ ps, $t_{logic} = 600$ ps, $t_{logic,cd} = 400$ ps, $t_{cq} = 50$ ps, $t_{cq,cd} = 40$ ps, $\delta = 75$ ps?

$$t_{hold} < 40 \text{ ps} + 400 \text{ ps} - 75 \text{ ps} = 365 \text{ ps}$$

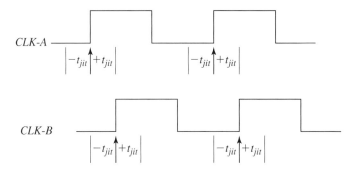

FIGURE 8-25.

Clock waveforms with jitter.

8.10.4. Period Constraint with Skew and Jitter

Jitter is a random distribution of clock edges about their expected mean value. Jitter adversely affects the period and hold time constraints. Figure 8-25 illustrates the uncertainty range of the clock edges and includes a positive skew. The full range of jitter uncertainty is $2t_{jitter}$. The period constraint is

$$T_{period} > t_{cq} + t_{logic} + t_{su} - \delta + 2t_{jitter} \qquad (8\text{-}10)$$

EXAMPLE 8-8

Given $\delta = 75$ ps, $t_{jitter} = 100$ ps, $t_{su} = 50$ ps, $t_{logic} = 550$ ps, and $t_{cq} = 80$ ps, how fast can this IC be clocked (F_{MAX})?

$$T_{period} > t_{cq} + t_{logic} + t_{su} - \delta + 2t_{jitter}$$

$$T_{period} > 80\text{ ps} + 550\text{ ps} + 50\text{ ps} - 75\text{ ps} + (2)(100)\text{ ps}$$

$$T_{period} > 805\text{ ps}$$

$$F_{MAX} = \frac{1}{805\text{ ps}} = 1.24\text{ GHz}$$

The constraints include a $2t_{jitter}$ term since this is the total range of uncertainty. The clock period must be increased by this amount, thus lowering the maximum operating frequency. Jitter comes primarily from the noise in the VCO power rail. The hold time constraint in the presence of jitter and positive skew is

$$t_{hold} < t_{cq,cd} + t_{logic,cd} - \delta - 2t_{jitter} \qquad (8\text{-}11)$$

EXAMPLE 8-9

Given $\delta = 90$ ps, $t_{jitter} = 100$ ps, $t_{su} = 50$ ps, $t_{logic} = 550$ ps, $t_{logic,cd} = 450$ ps, $t_{cq} = 80$ ps, and $t_{cq,cd} = 60$ ps, what is the maximum hold time allowed for this circuit to function?

$$t_{hold} < t_{cq,cd} + t_{logic,cd} - \delta - 2t_{jitter}$$

$$t_{hold} < 60 \text{ ps} + 450 \text{ ps} - 90 \text{ ps} - (2)(100) \text{ ps}$$

$$t_{hold} < 220 \text{ ps}$$

EXAMPLE 8-10

Given the system D-FF with designated delay elements at the D, Clk, and Q nodes. The FF element delays are T-gate delays are 65 ps, the inverter delays are 110 ps. The system parameters are $\delta = +75$ ps, $t_{jitter} = 60$ ps, $t_{logic} = 500$ ps, $t_{logic,cd} = 425$ ps, and $t_{cq,cd} = 210$ ps.

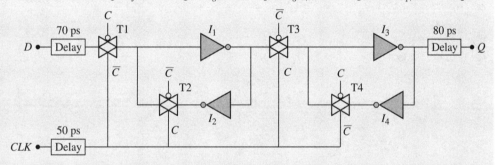

What is the maximum frequency of operation?
Calculate the system hold time constraint.

(a) $T_{period} > t_{cq} + t_{logic} + t_{su} - \delta + 2t_{jitter}$
Calculate the unknown elements of the period expression.

$$t_{su} = [T1 + I_1 + I_2] \text{ delays } + D\text{-delay} - Clk\text{-delay}$$

$$= 65 \text{ ps} + 110 \text{ ps} + 110 \text{ ps} + 70 \text{ ps} - 50 \text{ ps} = 305 \text{ ps}$$

$$t_{hold} = 0 \text{ ns} + 50 \text{ ps} - 70 \text{ ps} = -20 \text{ ps}$$

$$t_{cq} = [T3 + I_3 + Q\text{-delay}] \text{ delays } = 65 \text{ ps} + 110 \text{ ps} + 80 \text{ ps} = 255 \text{ ps}$$

The system period constraint is

$$T_{period} > 255 \text{ ps} + 500 \text{ ps} + 305 \text{ ps} - 75 \text{ ps} + 2(60 \text{ ps})$$

$$T_{period} > 1.105 \text{ ns}$$

$$F_{MAX} = \frac{1}{1.105 \text{ ns}} = 905 \text{ MHz}$$

(b) The hold time constraint is

$$t_{hold} < t_{cq,cd} + t_{logic,cd} - \delta - 2t_{jitter}$$

$$t_{hold} < 210 \text{ ps} + 425 \text{ ps} - 75 \text{ ps} - 2(60 \text{ ps}) + 50 \text{ ps} - 70 \text{ ps} = 420 \text{ ps}$$

$$t_{hold} < 420 \text{ ps}$$

Self-Exercise 8-6

This self-exercise illustrates how the maximum attainable clock frequency of an IC is affected by the system timing variables. The timing parameters are

$t_{su} = 125$ ps

$t_{logic} = 1000$ ps

$t_{logic,cd} = 900$ ps

$t_{cq} = 150$ ps

$t_{cq,cd} = 100$ ps

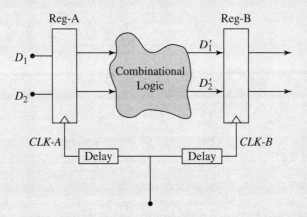

Fill in the table for the three values of skew, $\delta = 0$, $\delta = +200$ ps, and $\delta = -200$ ps.

	F_{MAX} (MHz)	t_{hold} (ns)
$\delta = 0$		
$\delta = 200$ ps		
$\delta = -200$ ps		

Observe the spread and consider the difficulty that a designer faces when trying to meet clock period and a hold time specification.

Self-Exercise 8-7

This self-exercise illustrates how the maximum attainable clock frequency of an IC is affected by the system timing variables. The timing parameters are

$t_{su} = 125$ ps

$t_{logic} = 1000$ ps

$t_{logic,cd} = 900$ ps

$t_{cq} = 150$ ps

$t_{cq,cd} = 100$ ps

$t_{jitter} = 150$ ps

Solve for skew = +200 ps and −200 ps

 Jitter is zero in the first row. Fill F_{MAX} and t_{hold} in the next two rows of the table for the values of skew and jitter.

Conditions	F_{MAX} (MHz)	t_{hold} (ns)
$\delta = 0$		
δ^+, t_{jitter}		
δ^-, t_{jitter}		

Observe the results and think of the difficulty that a designer faces when trying to meet the clock period and a hold time specifications.

Jitter and positive skew negatively impact the hold time, and designers must work within these time constraints. Circuits cannot be driven faster than the period constraint allows nor can hold time be violated. These rules apply to all millions of flip-flops in a circuit. We illustrated positive skew in the previous examples. Negative skew has opposite effects on the period and hold time constraints. Negative skew increases the clock period requirement reducing the maximum operating frequency of the chip. But negative skew causes no hold time problem, since Clock-B arrives before data are launched by Clock-A.

Statistical variation in the transistor and interconnect line parameters is a final timing accuracy challenge. IC manufacturers cannot precisely replicate the small transistor and interconnect dimensions and doping concentrations. This causes transistor-to-transistor variation in parameters such as L_{eff}, W/L, R_{sheet}, V_t, and $I_D sat$. The parameter variations modulate the transistor and interconnect propagation delays. Designers must have good estimates of the variables to allow safety margins in clock settings. Too much margin means that the chip highest frequency, F_{MAX}, is compromised. Too little margin means more failed chips. Clock generation circuits are difficult to analyze in the presence of tens of picoseconds of uncontrolled variations.

8.11. Timing and Environmental Noise

At this point, the reader has progressed through the basics of digital electronics. This section will advance that level to briefly describe real environmental effects on IC

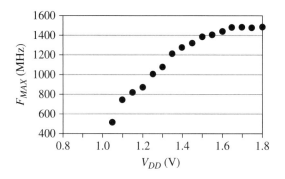

FIGURE 8-26.

F_{MAX} versus V_{DD} for a 150 nm IC [3]. (Reproduced with permission of Intel Corp.)

timing. Environmental effect is a general term that refers to variables such as V_{DD} and temperature.

Figure 8-26 plots the maximum operating frequency F_{MAX} versus V_{DD} of an Intel 1.0 GHz, 6.6 million transistor router chip built in a 150 nm technology. As V_{DD} increases, F_{MAX} increases and then saturates just above 1.6 V. Noise-induced millivolt changes in V_{DD} can significantly influence the chip maximum frequency. Self-Exercise 8-8 measures that sensitivity in a real IC for maximum operating frequency (F_{MAX}) versus V_{DD}. Designers must have characterization data to anticipate power supply fluctuations from noise and from ±5% voltage regulators on the printed circuit boards upon which this chip might be mounted.

Self-Exercise 8-8
Refer to Figure 8-26:

(a) Use a pencil and ruler to estimate the slope and change in F_{MAX} for a 1 mV change in V_{DD}.

(b) A 1 GHz chip in inserted on a printed wire board that has a voltage regulator that is 5% low of its intended value of 1.3 V. Estimate the maximum operating frequency that the chip will have on that board?

Answers: (a) Approximately 1.67 MHz per mV. (b) Approximately 892 MHz.

There is another source of timing uncertainty that originates in the environment outside of the chip. Two major noise variables in the power supply lines are temperature on the chip induced by the outside room temperature and self-heat generated by the chip during operation. Low power chips may generate microwatts of power while high-performance microprocessors can generate 150–200 watts. In addition, significant temperature gradients

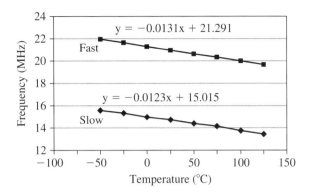

FIGURE 8-27.

Maximum operating frequency capability (F_{MAX}) versus temperature [3]. (Reproduced with permission of Intel Corp.)

exist on the high power chips causing uncertainty in the exact delay times of different interconnect paths.

Figure 8-27 plots F_{MAX} versus average temperature of a 20,000 transistor test chip. The two curves represent a chip design fabricated with two different transistor channel lengths. The "slow" chip had the longer channel lengths. The regression coefficient of each curve gives the temperature sensitivity of each design. The fast chips had a change of –13.3 kHz per degree centigrade and the slow chips –12.3 kHz per degree centigrade. As expected, circuits slow as temperature rises. The change with frequency derives from the transistor and the metal responses to temperature change. IC hot spots of 150°C can significantly affect the expected chip performance.

Temperature can change in microseconds on high power chips following a rapid increase in chip computation demand, and these changes can be local affecting only a portion of the transistors. Or, average temperature can rise affecting the speed of all transistors. Generally, a rise in temperature will cause carrier mobility to decrease thus decreasing the saturated current driving strength ($I_D sat$). Charge and discharge of load capacitance then takes longer. Compare this high performance environment with a very low power, low frequency implanted cardiac pacemaker. The thermal environment is a constant 37°C in the human body, and a slow 33 kHz clock virtually eliminates chip self-generated temperature rise.

8.12. Summary

The rules of designing circuits for a designated frequency require knowledge of timing properties of flip-flops, combinational logic, clock generation circuitry, clock period and hold time constraints, and the uncertainties imposed by jitter, skew, and manufacturing variation. Whether the circuit has fast or slow frequency specifications does not affect the need to understand all variables.

References

[1] W. Eccles and F. Jordan, "A trigger relay utilizing three-electrode thermionic vacuum tubes," *Radio Review*, **1**, 143, 1919.

[2] J. Rabaey, A. Chandrakasan, and B. Nikolic, *Digital Integrated Circuits*, 2nd Ed., Prentice-Hall, 2003. Saddle River, New Jersey, USA.

[3] J. Segura and C. Hawkins, "The nature of nanometer timing failures," *Electron Device Failure Analysis Magazine*, 2nd Quarter, May 2005.

Exercises

8-1. The D and Clk waveforms are given for the negative edge-triggered MS FF in the figure. Sketch the waveforms for the node in front of the CMOS T-gate and after the T-gate and sketch Q. Assume no transistor delay and that TG_{in} and TG_{out} are initially at logic-1.

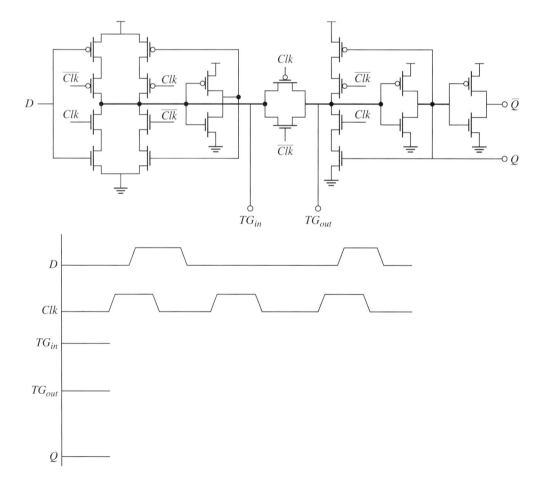

8-2. The *D* and *Clk* waveforms are given for the negative edge-triggered MS FF in the figure. Sketch the waveforms for the node in front of the CMOS T-gate and after the T-gate and sketch *Q*. Assume no transistor delay and that TG_{in} and TG_{out} are initially at logic-0.

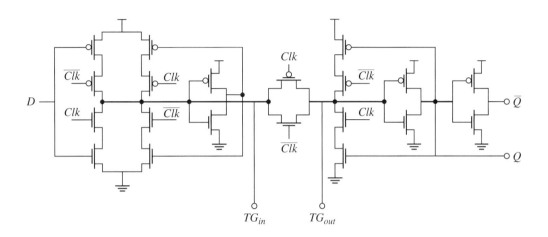

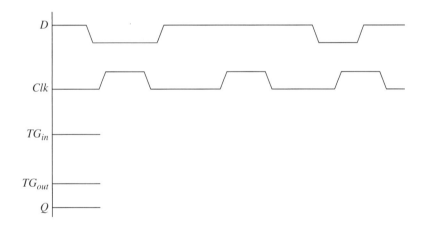

8-3. Sketch the TG_{in}, TG_{out}, and Q waveforms.

<u>Initial conditions</u>:
$TG_{in} = TG_{out} = 0$ V
$Q = V_{DD}$

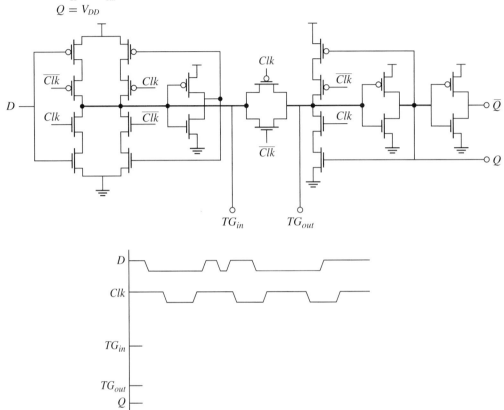

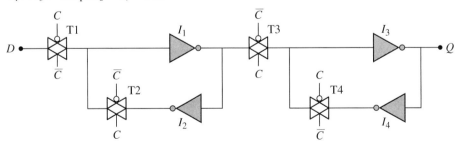

8-4. Given the positive edge MS FF whose delay elements are: T1 = T2 = T3 = T4 = 500 ps and $I_1 = I_3 = 700$ ps. $I_2 = I_4 = 0$ ns.

(a) Calculate t_{su}, t_{hold}, and t_{cq}.
(b) What are the minimum clock widths (CW) for clock = logic-0 and logic-1?
(c) What is the effect if T2 = T4 = 800 ps?

8-5. The circuit of the figure has the same function as a basic block used in sequential circuits. Identify the circuit type and the conventional names given to the inputs and outputs.

Hint: Analyze the equivalent circuit for $Y = 0$ and $Y = 1$.

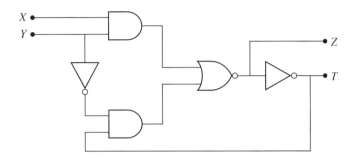

X	Y	Z	T	Comments
0	0			
0	1			
1	0			
1	1			

8-6. Given a positive edge MS FF. If T1 = T2 = T3 = T4 and $I_1 = I_2 = I_3 = I_4$. If $t_{su} = 4.5$ ns and $t_{cq} = 3$ ns, what are the T-gate and inverter propagation delays?

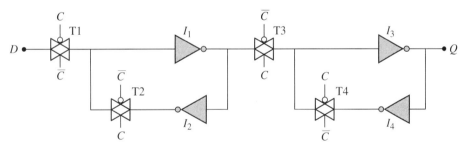

8-7. Without the delay element in the data line, $t_{su} = 1.5$ ns, $t_{hold} = 0$ ns, and $t_{cq} = 1$ ns. If the delay element has a delay of 1.8 ns, what are the new values for t_{su}, t_{hold}, and t_{cq}? Draw the waveforms of the signals D, Clk, T1 (clock arrival), and Q with these new timing parameters. Initially, D and Q are at logic-0.

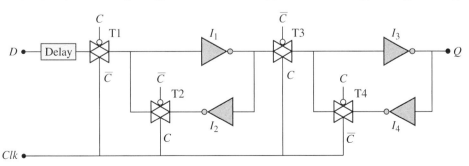

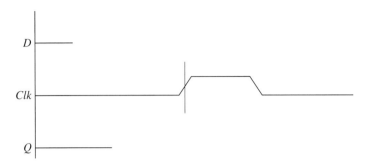

8-8. Without the delay element in the *clock* line, $t_{su} = 1.5$ ns, $t_{hold} = 0$ ns, and $t_{cq} = 1$ ns. If the clock delay element has a delay of 1.8 ns, what are the new values for t_{su}, t_{hold}, and t_{cq}? Draw the waveforms of the signals D, *Clk*, T1 (clock arrival), and Q with these new timing parameters. Initially, D and Q are at logic-0.

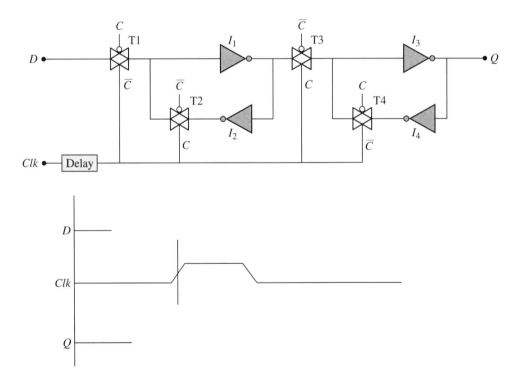

8-9. Defect delay elements are in the *D*-path and the *Clk*-path. If a defect-free circuit has $t_{su} = 3$ ns and $t_{hold} = 2$ ns, what are the setup and hold times when these delay elements are present?

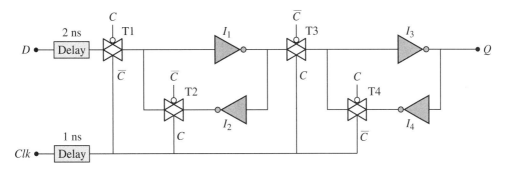

8-10. An IC has a clock frequency specification of 1.5 GHz. The FF setup and clock to *Q* times are 190 ps and 100 ps. What combinational logic delay must not be exceeded to meet the 1.5 GHZ specification?

Clock Generation Circuitry

8-11. Sketch the *UP* and *DOWN* waveforms for the phase splitter.

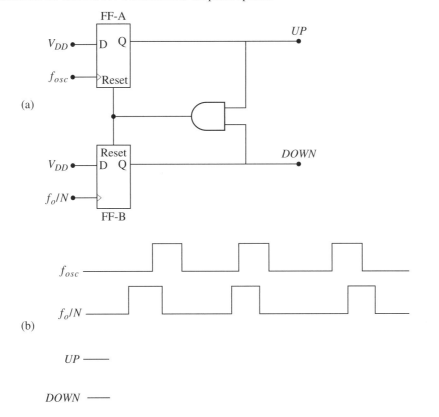

System Timing

8-12. A pipelined circuit has a clock frequency of 2 GHZ, $t_{su} = 60$ ps, $t_{hold} = -40$ ps, $t_{cq} = 70$ ps, $t_{logic} = 550$ ps, and $t_{logic,cd} = 500$ ps. How much positive skew will cause a timing error?

8-13. A pipeline design has a negative skew of 70 ps, $t_{su} = 90$ ps, $t_{hold} = -30$ ps, $t_{cq} = 110$ ps, $t_{cq,cd} = 90$ ps, $t_{logic} = 650$ ps, and $t_{logic,cd} = 580$ ps. What is the maximum frequency of operation?

8-14. Repeat Problem 8-10 if the skew changes to +70 ps.

8-15. Given a skew of +70 ps, $t_{su} = 90$ ps, $t_{hold} = -30$ ps, $t_{cq} = 110$ ps, $t_{cq,cd} = 90$ ps, $t_{logic} = 650$ ps, and $t_{logic,cd} = 580$ ps:

(a) What is the hold time maximum value that allows functionality in the design?
(b) Repeat the calculation for a negative skew of 100 ps.

8-16. Given that $t_{jitter} = 120$ ps, $\delta = +130$ ps, $t_{su} = 80$ ps, $t_{hold} = -30$ ps, $t_{cq} = 110$ ps, $t_{cq,cd} = 80$ ps, $t_{logic} = 580$ ps, and $t_{logic,cd} = 520$ ps, what is the maximum safe operating frequency (F_{MAX})?

8-17. A low-frequency, low-power medical electronics IC has a frequency specification of 33 kHz. It has $t_{jitter} = 120$ ps, $\delta = +130$ ps, $t_{su} = 80$ ps, $t_{hold} = -30$ ps, $t_{cq} = 110$ ps, $t_{cq,cd} = 80$ ps, $t_{logic} = 580$ ps, and $t_{logic,cd} = 500$ ps. What period margin does the IC have for combined positive skew and jitter effects?

8-18. Jitter in a pipeline circuit is $t_{jitter} = 85$ ps, $\delta = +100$ ps, $t_{su} = 100$ ps, $t_{cq} = 80$ ps, $t_{cq,cd} = 70$ ps, and $F_{MAX} = 2.2$ GHz. What delay must the pipeline combinational logic circuits be kept under?

8-19. The flip-flop and system parameters are $t_{jitter} = 160$ ps, $\delta = +145$ ps, $t_{su} = 150$ ps, $t_{hold} = -30$ ps, $t_{cq} = 130$ ps, $t_{cq,cd} = 105$ ps, $t_{logic} = 620$ ps, and $t_{logic,cd} = 575$ ps.

(a) Calculate the hold time specification to allow functionality.
(b) Calculate the maximum operating frequency (F_{MAX}).

8-20. Given $t_{jitter} = 105$ ps, $t_{hold} = 130$ ps, $t_{cq} = 125$ ps, $t_{cq,cd} = 110$ ps, $t_{su} = 140$ ps, $t_{logic} = 550$ ps, and $t_{logic,cd} = 450$ ps, find F_{MAX}.

The following problems first calculate the basic FF parameters, then estimate system period and hold time constraints.

8-21. Given a pipeline structure with positive edge triggered FFs. The FF element delays are T-gate delays are 600 ps, $I_1 = I_3 = 900$ ps, and $I_2 = I_4 = 800$ ps. The system parameters are $\delta = +150$ ps, $t_{jitter} = 100$ ps, $t_{logic} = 2$ ns, $t_{logic,cd} = 1.7$ ns, and $t_{cq,cd} = 1.1$ ns.

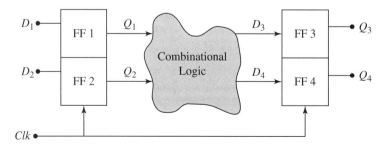

(a) What is the maximum frequency of operation?
(b) Calculate the system hold time constraint.

8-22. Given a pipeline structure with negative edge-triggered FFs and a delay element of 250 ps in the D-signal line. The FF element delays are T-gate delays are 150 ps, and the inverter delays are 200 ps. The other timing parameters are $\delta = -65$ ps, $t_{jitter} = 45$ ps, $t_{logic} = 600$ ps, $t_{logic,cd} = 550$ ps, and $t_{cq,cd} = 300$ ps.

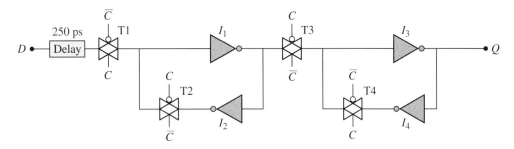

    (a) What is the maximum frequency of operation?
    (b) Calculate the system hold time constraint.

8-23. Given a pipeline structure with positive edge-triggered FFs and a delay element of 125 ps at the Q-signal line. The FF element delays are T-gate delays are 75 ps, the inverter delays are 150 ps. The other timing parameters are $\delta = +55$ ps, $t_{jitter} = 45$ ps, $t_{logic} = 700$ ps, $t_{logic,cd} = 550$ ps, and $t_{cq,cd} = 295$ ps.

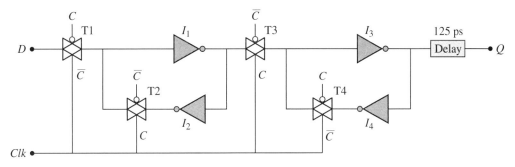

    (a) What is the maximum frequency of operation?
    (b) Calculate the system hold time constraint.

CHAPTER 9

IC Memory Circuits

Somewhere, something is waiting to be known.

Carl Sagan

Computers store information in a variety of ways. The flip-flops studied in Chapter 8 perform temporary storage that synchronizes data movement to each clock cycle. The memory circuits described here deal with longer-term storage of large amounts of data. Data are stored as single bit information, and collections of bits can be grouped in parallel as a digital word. Computer performance is enhanced when more memory is available, especially when those data are stored on the chip itself.

There are many types of memory that have the acronyms SRAM, DRAM, ROM, EPROM, E²PROM, and FLASH. SRAM stands for static random access memory, and DRAM stands for dynamic random access memory. The word *random* describes some memory types and is not meant in the literal sense, but that any memory location can be addressed independently of any other location. This contrasts with memories whose information retrieval depends on the order that data were written. Examples are FIFO (first-in, first-out) or LIFO (last-in, first-out) memories. Random access is a loose, inaccurate phrase that stuck for most memory types.

ROMs are read-only memories that store permanent information with no capability to change the content once the data are written. ROMs retain their data when power is turned off. EPROMs are nonvolatile electrically programmable read-only memories. E²PROMs can be electrically programmed and electrically erased during the life of the product. FLASH is a high-density nonvolatile memory that can electrically read and write new data. FLASH has many high volume applications including cell phones, digital cameras, and the small portable flash memory drives.

While these various memory designs are important, we will emphasize two memory designs. The first is the static random access memory (SRAM). Modern microprocessor chips may dedicate more than 65% of their total transistor count to embedded SRAM circuits. Systems can operate much faster if memory access is on the chip rather than on a separate memory chip on a printed circuit board. The second memory design we will study is the DRAM.

The SRAM is the design of choice for embedding memory within a computing chip. An SRAM retains its information as long as power is applied to the circuit making SRAM data volatile when power is removed. However, SRAMs are relatively fast and easy to merge with the CMOS fabrication technology.

The dynamic random access memory (DRAM) memory ICs are typically stand-alone chips wholly dedicated to storing data that link to a host processor IC through the wire traces on the circuit board. DRAMs (dynamic random access memory) have small dynamic cells that require constant refresh clocking, but they have gigabit memory capacity. DRAMs are comparatively slow and volatile, but large memories are obtained for a low cost. DRAM memory cells are small providing very dense memory chips. The IC technologies with the smallest metal and transistor dimensions are typically DRAMs.

This chapter focuses on SRAMs because of their use in embedded memories in modern ICs, and DRAMs to illustrate a radically different high volume design. Memories have three states: standby, reading, and writing. Typically a memory cell spends most of its time in standby with its contents undisturbed by the digital signals whirling around it. The Read and Write operations and their transistor sizing requirements are detailed in this chapter.

9.1. Memory Circuit Organization

Figure 9-1 shows a memory diagram. It may be a whole IC, or the memory could be embedded as a subcircuit within an IC such as a microprocessor, game, DSP, or controller chip. Each memory cell in the matrix stores one bit of data as a logic-0 or logic-1. The figure shows a small 16-bit memory IC. The ordered row and vertical column lines control reading or writing to individual bit cells or may group some core cells as digital words.

Most memories use an address word to select a specific bit. An n-bit address word is divided into two sections, one section contains the m-row address line bits and the other the k-column bits (Figure 9-2). There are 2^m rows and 2^k columns. The total number of cells in the memory are $2^{m+k} = 2^n$. The 4-bit address word has 16 values each corresponding to one of the 16 memory cells. The 16-bit memory in Figure 9-1 is typically too small for practical use, but this size allows us to see the whole operation. Megabits and gigabits of memory size are common. The horizontal word line connects to many cell transistors through a long, doped polysilicon interconnect line. It has a large capacitance on the order of pFs or hundreds of fFs and relatively high resistance. The vertical column lines have a metal resistance that is lower and also carry large capacitance. Both column address lines and the word lines require buffer drivers.

Decode circuits design for minimum die area since the goal is to pack as much memory circuitry as possible into the die. The row address bits are decoded to activate individual *word lines* $W_0 \ldots W_3$. The column address bits are decoded to individual column lines

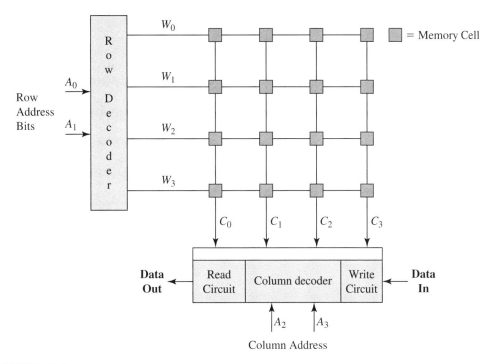

FIGURE 9-1.

System diagram of a circuit with address word $A_3 A_2 A_1 A_0$.

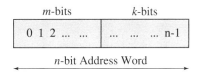

FIGURE 9-2.

Bit allocation for m-row bits and k-column bits in an n-bit address word.

$C_0 \dots C_3$ called *bit lines*. It is the intersection of particular activated column and row lines that target a memory cell for a Read or Write operation (Figure 9-1).

The column interconnect lines also connect to the write and read circuitry (Figure 9-1). Write and Read circuits are different. New incoming data are routed to a specific cell activated by the address decoders and then written into the designated cell through selected column bit lines. A Read command is coordinated with the address decoders to select a cell row and column address, and to drive the cell logic value to the data output node through the bit lines to a sense amplifier. While this memory shows selection of a single bit cell, the column decoders can activate several columns in parallel to select 2-, 4-, or 8-bit words. We will examine these subsystems in more detail.

9.2. Memory Cell

Figure 9-3 shows an SRAM cell. The two inverters form a latch whose data entry and exit are controlled by two nMOS pass transistors called memory access transistors (M5, M6). Each column has two bit lines, bl and $\overline{\text{bl}}$. When the word line is activated ($wl = 1$), the access transistors turn on and connect the latch nodes Q and $\overline{Q}$ to the bit lines, bl and $\overline{\text{bl}}$. When activated, the bit lines carry data to and from the cell supporting the Write and Read operations.

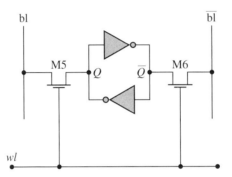

FIGURE 9-3.

SRAM cell using a latch and two access transistors.

Figure 9-4 shows a full 6-transistor schematic for the SRAM cell and bit lines. This cell is used extensively in embedded memory applications, and the operation is the same as described in Figure 9-3. The positive feedback connection from one output drain node to the input gate node of the other inverter assures that a logic state is held. The W/L sizing of the transistors is set by the Read and Write operations, and that will be discussed shortly. We will use the transistor schematic in Figure 9-4 to analyze properties of the SRAM cell.

Assume a 2 V memory power supply with $Q = 0$ V and $\overline{Q} = 2$ V in Figure 9-4. Simplistically, if a new Write operation sets the bit line drivers to bl $= 2$ V, $\overline{\text{bl}} = 0$ V,

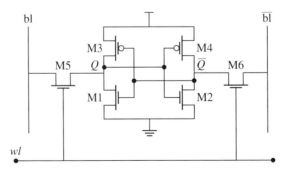

FIGURE 9-4.

Full 6-transistor schematic for the SRAM cell.

and $wl = 2$ V turns on M5 and M6, then the latch responds to the new bl and $\overline{bl}$ data setting $Q = 2$ V and $\overline{Q} = 0$ V. The transistors are designed so that the bit line drivers will overpower the weak transistors of the cell. When the word line goes to $wl = 0$, access transistors M5 and M6 turn off and the latch holds the new logic value until power is turned off or a new logic value is written. A Read operation also activates the access transistors, but the bit lines would be connected to a sensing circuit that can detect the logic state of the latch. The Read operation does not alter the logic state of the memory cell.

One last feature must be introduced before analyzing the W/L ratios of the access, pull-up, and pull-down transistors. Both bit lines are *precharged* to a high voltage, typically V_{DD} or close to it, before a Read or a Write operation (Figure 9-5). After the precharge pulse (PC) is turned off and the bit lines float, the access transistors M5 and M6 turn on and charge flows between the bit line and the latch node that is in the logic-0 state. A difference voltage develops between the charged bit lines allowing the Read (sense) amplifier to respond to a difference in bit line voltages. This method uses the noise reduction properties of the sense amp achieving better stability and sensitivity. The Write operation also precharges both bit lines and then drives the lines with the desired write data. One of the bit lines drivers is at logic-0 and that pulls down its attached cell node voltage thus changing the state. A pull-down bit line driver uses an nMOS transistor that is more efficient than a pMOS pull-up.

Figure 9-5 shows the precharged circuitry. Assume $V_{DD} = 2$ V and $Q = 0$ V. The precharge pulse (PC = 0) turns on transistors M7, M8, and M9 prior to a Write or Read

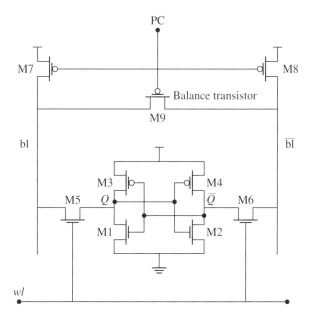

FIGURE 9-5.

SRAM cell with precharge transistors M7, M8, and M9.

operation. M9 is a balance transistor that ensures bl = $\overline{bl}$ during precharge. The M9 path to a low bit line allows two pull-up paths during precharge. For example, if bl = 0 V, then the two pull-up paths are through M7 and through M8–M9.

When PC = 2 V, the bit lines float in a high impedance state at the high voltage. When the access transistors turn on, charge will move from the bit line into the logic-0 cell node disturbing that bit line voltage. The Read circuitry senses this bit line voltage disturbance and makes a logic-1 or logic-0 assessment. By design the memory cell latch is not sufficiently disturbed to change the logic state during a Read. A Write operation is designed such that the bit line drivers overpower the latch and reset its contents to the new value.

9.3. Memory Decoders

9.3.1. Row Decoders

Figure 9-6 shows a 2NAND 2-bit row decoder and the memory cells for a 16-bit memory. The goal is to take an address word, decode it, and select one of the individual memory rows and one column. The address word is 4 bits, two for row decoding and two for the column decode block. Decoders require the address bits and their complements. Each row uses one 2NAND and one inverter to decode. The inverter delivers a logic-1 to the whole row and serves as a buffer driver for the high capacitance word lines. Each 2NAND gate

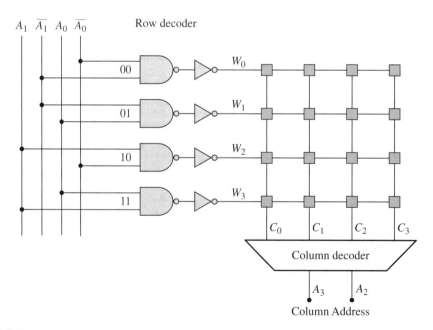

FIGURE 9-6.

2-Bit row decoder using NAND logic.

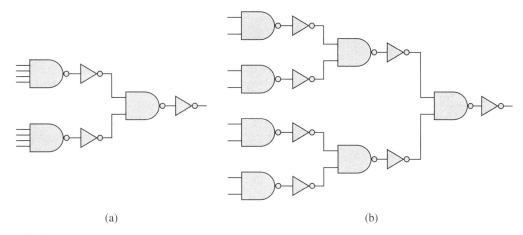

(a)   (b)

FIGURE 9-7.

4-Bit row decoder designs: (a) 4NAND. (b) 2NAND.

picks two of the four possible logic signals $A_1 \overline{A_1} A_0 \overline{A_0}$ to decode that row. The first row decoder for w_0 uses $A_1 A_0 = 00$ for selection, the second row uses $A_1 \overline{A_0} = 01$, and so forth.

This small 2-bit row and 2-bit column example uses 2NAND decoders for the 16-bit memory. Typical embedded memories are much larger on the order of 256 k to several megabits bits. An 8-bit (256 rows) decoder design cannot use 256 8NAND gates. The pull down strength of eight nMOS transistors in series is prohibitively slow, and the decoder physical size must reside within the layout pitch of the row lines. So, the row and column decoders use extensive pre-predecoding to break up the logic gate size of a large memory. This requires smaller logic gates, but more of them to physically fit within the row pitch. Figure 9-7 shows two decoder designs for a 4-bit row. There are four bits for the address and four bits for address bit complement.

The 4NAND gates (Figure 9-7a) can be broken into 2NAND gates (Figure 9-7b) if the 4NAND was too slow or too large for the row pitch dimension. The 4-bit decoders in Figure 9-7 will support 16 word lines. Each circuit in Figure 9-7 is replicated 16 times, once for each row.

Row decoders use a line buffer to drive a long highly capacitive polysilicon word line that connects to the nMOS access transistor gates. This structure can generate pFs. The polysilicon word line connects directly to the transistor poly gate eliminating the need for a contact.

9.3.2. Column Decoders

Column decoders are similar to row decoders but have a different load circuit. Column decoders drive relatively low impedance CMOS transmission gates that connect the bit lines to the Write or Read circuit, but the bit lines have hundreds of fF to pF capacitance for the bit line drivers to overcome. The bit lines use a contact to connect to the access transistor drain.

Figure 9-8 shows column decoders driving the CMOS transmission gates and bit lines. The transmission gates (T0 – T3) connect the bit lines with either the Write circuit to drive new data into the memory cells, or a Read (sense) circuit that interprets the logic state of a given memory cell. There are, in fact, two transmission gates per column since there are two bit lines per column, but only one bit line is shown for figure clarity. A column has a bit line (*bl*) and its complement ($\overline{bl}$). Each transmission gate is driven by one of the

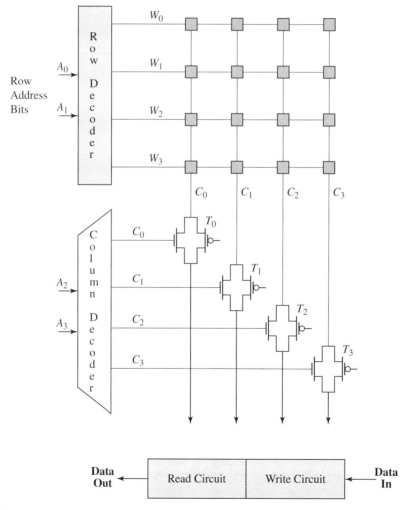

FIGURE 9-8.

Column decoder driving CMOS transmission gates that connect to Read or Write circuit.

column decoded outputs and its complement (the complement connection is not shown for the pMOS transistors). The steering circuitry to the Read and Write circuits is not shown.

A single pass transistor in the column decoders would also work instead of the CMOS transmission gates, but it has a weakness. An nMOS pass transistor will pass a strong logic-0, but a weak logic-1. A pMOS transistor will pass a strong logic-1, but a weak logic-0. Weak logic voltages slow the response time of the memory. A single pass transistor performs the access function in the 6-transistor SRAM cell, but the increase in cell size would not allow using a CMOS T-gate, and the access transistors don't need to pass full logic voltages to operate the memory cell. The cell will flip at much less than full logic voltage. The memory core transistor design (sizing) for the Read and Write operations will be discussed next and then the peripheral circuits that support column reading and writing.

9.4. The Read Operation

The Read operation uses five elements: (1) the memory cell; (2) word and bit line decoders; (3) word and bit lines; (4) bit line precharge transistors; and (5) a sense amp. The Read operation begins with a bit line precharge followed by a control line $wl = V_{DD}$ pulse that turns on the access transistors. Data in a cell are then exposed to floating bl and $\overline{bl}$ lines that were precharged to V_{DD}. The cell node that is logic-0 will pull down the bit line connected to it while the cell node that is logic-1 will have no effect on its bit line. This causes the voltages on bl and $\overline{bl}$ to differ. This difference in bit line voltages is fed to a special circuit called a sense amplifier.

The *sense amp* is an analog circuit more generally called a *differential amplifier* (described later) that interprets whether a memory cell stores a logic-0 or logic-1. The sense amp amplifies the difference of two voltages and suppresses the average (DC) voltage that is common to each bit line. The small difference voltage $\Delta V_{in} = V_{bl} - \overline{V_{bl}}$ is amplified to give a strong logic-0 or logic-1.

The difference amplifier raises the question of why we just don't read the cell contents of one bit line through a simple inverter. We could do that and the memory would work. However, the difference amplifier is more sensitive to small bit line difference changes allowing faster readings, and it is less sensitive to noise. We also save power since the bit line rails don't traverse a full V_{DD} voltage. The basics of difference amplifiers are described later, but the details of difference amplifier design are left to a course in analog electronics.

A portion of the memory cell schematic in Figure 9-4 is repeated in Figure 9-9 showing the bl line on the left side. As long as the word line $wl = 0$, the access transistors are off and the cell remains undisturbed. The Read operation begins with a precharge signal PC that drives both bit lines to $V_{DD} = 2$ V in our example. Next, PC is turned off, and the bit lines float at 2 V. The word line then asserts $wl = 2$ V turning on the access transistors and exposing node-Q to bl and node-$\overline{Q}$ to $\overline{bl}$. Assume that initially $V_Q = 0$ V representing a cell logic-0. Since the bit lines were floating, charge will flow to ground through transistors M5 and M1, and V_Q rises. If $\overline{Q}$ is at logic-1 during and after precharge, then no charge transfers occur on the right-hand side transistors of the memory cell when $wl = 2$ V.

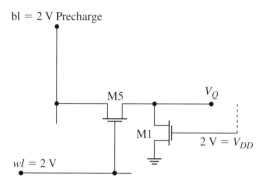

FIGURE 9-9.

Relevant memory cell elements of the Read operation.

After the bit lines precharge, transistors M5 and M1 are on with their gate voltages at 2 V. M5 is in saturation since $V_{G5} = 2$ V, $V_{D5} \approx 2$ V, and $V_{S5} = V_Q \approx 0$ V. M1 is in the ohmic state since initially $V_{D1} = V_Q = 0$ V and $V_{G1} = 2$ V. Charge passes through the transistors charging the node capacitance at V_Q. V_Q rises and V_{bl} drops slightly buffered by its large capacitance. However, if V_Q was unchecked and continued to rise, then it would turn on M2 on the opposite side of the cell causing a drop in $\overline{V_Q}$. The latch could erroneously flip its logic state if the rise in V_Q was not limited by the design. The W/L ratios are critical for correct operation, and are discussed next.

9.5. Sizing Transistor Width to Length Ratio for Read Operation

The width-to-length ratios of the six-cell transistors are critical for failure free and rapid Read and Write operations. M5 and M1 in Figure 9-9 act as nonlinear voltage dividers for V_Q. The W/L ratios of M1 and M5 must be adjusted so that node Q does not rise above a certain voltage that would flip the latch during a Read operation.

bl is initially at its precharge of V_{DD} but drops slowly due to the current through M5 and M1. V_Q will then be at its highest voltage before it begins to drop with the discharge of bl. If the circuit design is safe for bl $= V_{DD}$, then it will be safe as bl drops. When the word line is asserted, M5 is in saturation and M1 is in nonsaturation, so we can equate $I_{D5} = I_{D1}$ as

$$K_5 \left(\frac{W_5}{L_5}\right)[V_{G5} - V_{S5} - V_{t5}]^2 = K_1 \left(\frac{W_1}{L_1}\right)[2(V_{GS1} - V_{t1})V_{DS1} - V_{DS1}^2] \qquad (9\text{-}1)$$

$$K_5 \left(\frac{W_5}{L_5}\right)[V_{DD} - V_Q - V_{t5}]^2 = K_1 \left(\frac{W_1}{L_1}\right)[2(V_{DD} - V_{t1})V_Q - V_Q^2] \qquad (9\text{-}2)$$

and

$$\left(\frac{W_1/L_1}{W_5/L_5}\right) = \frac{K_5}{K_1} \frac{[V_{DD} - V_Q - V_{t5}]^2}{[2(V_{DD} - V_{t1})V_Q - V_Q^2]} \qquad (9\text{-}3)$$

This ratio of $\left(\frac{W_1}{L_1}\right)/\left(\frac{W_5}{L_5}\right)$ will ensure that V_Q is never higher than a specified low voltage. The bl and V_Q voltages will continue to drop if the wl line is left activated. A further design task is to relate capacitive charging at node Q to the current required to charge 100–200 mV in a given small time period. We will leave that topic to another course.

An example will illustrate the design approach for setting W/L ratios that support a stable Read operation.

EXAMPLE 9-1

Use the parameters $V_{DD} = 1.5$ V, $V_{tn} = 0.5$ V, and $K_n = 100\,\mu A/V^2$. Restrict V_Q to a 10% rise equal to 0.15 V, and find the W/L ratio needed to do this. Assume a precharge of $V_{bl} = 1.5$ V.

$$100\ \mu A \left(\frac{W_5}{L_5}\right)[1.5 - 0.15 - 0.5]^2 = 100\ \mu A \left(\frac{W_1}{L_1}\right)[2(1.5 - 0.5)0.15 - 0.15^2]$$

and

$$\left(\frac{W_1}{L_1}\right)\bigg/\left(\frac{W_5}{L_5}\right) = 2.6$$

Since $L_1 = L_5$, then $W_1/W_5 = 2.6$.

This transistor β-ratio will ensure that V_Q will not go higher than 150 mV during the read.

Since $\left(\frac{W_1}{L_1}\right)/\left(\frac{W_5}{L_5}\right)$ is a large clumsy expression, the β symbol is often used to indicate $\beta = W/L$. The transistor ratio of the previous example becomes $\beta_1/\beta_5 = 2.6$. An exercise problem at the end of the chapter asks you to plot the β-ratio over a continuous range of V_Q and form conclusions.

Self-Exercise 9-1

Use $V_{DD} = 1.5$ V, $V_{tn} = 0.5$ V, and $K_n = 100\,\mu A/V^2$. Find the W/L ratios required to restrict the rise in $V_Q = 100$ mV, 150 mV, and 200 mV. Assume $V_{bl} = 1.5$ V for the analysis.

Answers:

(a) $\dfrac{\beta_1}{\beta_5} = 4.26$

(b) $\dfrac{\beta_1}{\beta_5} = 2.60$

(c) $\dfrac{\beta_1}{\beta_5} = 1.78$

9.6. Memory Write Operation

A meaningful data Write operation occurs when an opposite logic state is written to a memory cell. The access transistors expose the latch to the bit lines logic states that must overdrive and flip the cell logic state. However, the access and pMOS pull-up transistors must be accurately sized to ensure safe Write operation. The sequence is that the bit lines are first precharged, then exposed to the Write drivers with their new logic states. Precharging sets both bit lines to a known state before the Write operation begins. The access transistors turn on, and the bit line drivers cause the cell to flip logic states. The bit line driven to logic-0 initiates the transition action.

9.6.1. Cell Write Operation

Assume a Write operation with a new data bit logic-0 placed on the $\overline{bl}$ bit line. This will flip a logic-1 stored on the $\overline{Q}$ node to a logic-0. Figure 9-10 shows M4, M6, and the right side bit line. Initially $\overline{Q} = \overline{V}_Q = $ logic-1, and $\overline{bl}$ is driven low by the bit line driver. The word line activates, and charge flows from M4 to node $\overline{Q}$ then to the bit line $\overline{bl}$. $\overline{V}_Q$ drops sufficiently to initiate the feedback action that flips the cell logic state. The details of that operation are next.

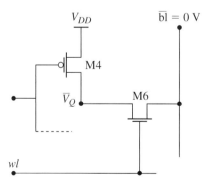

FIGURE 9-10.

Relevant elements at steady state for determining W/L ratios for a Write operation. $\overline{Q}$ is initially at $\overline{V}_Q = V_{DD}$ but is pulled sufficiently low to flip the cell.

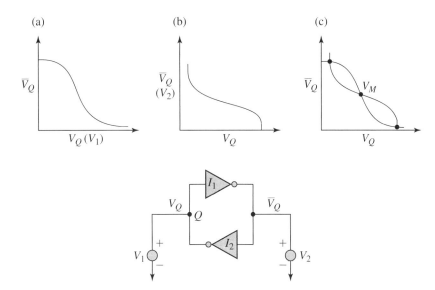

FIGURE 9-11.

Transfer curves for (a) Inverter-1, (b) Inverter-2, (c) Composite curves of (a,b).

9.6.2. Latch Transfer Curve

We studied the inverter transfer curve in Chapter 5, and we will examine the latch transfer curve to better understand the transition properties. Figure 9-11 shows the latch and its transfer curves when driven by a strong signal source at one inverter input. The input signal clamps the node voltage neutralizing the influence of the feedback loop. When V_1 is active, V_2 is disconnected, and vice versa.

Figure 9-11a is a conventional inverter voltage transfer curve when Inverter-1 is driven by the V_1 signal source. Figure 9-11b is the VTC when Inverter-2 is driven by V_2, but plots the transfer curve of $\overline{V_Q}$ versus V_Q.

Figure 9-11c combines the curves to illustrate the action of the latch. Observing $\overline{V_Q}$ and V_Q is like watching a teeter-totter with its two stable positions and its unstable midpoint. There are three important intersections in Figure 9-11c. We see that the cell will change state if either node is taken beyond the transition point V_M in the middle of the curves. A bit line must be pulled low with a transition range greater than V_M to effect a Write operation. A Read operation must not allow the bit line voltages to get near the transition region. The two intersections at the curve's extremes define the logic-0 and logic-1 states and represent the high and low noise margins. This curve is referred to as a butterfly curve relating the two latch signal node voltages over the operating range.

9.7. Sizing Transistor Width to Length Ratio for Write Operation

How low should the $\overline{Q}$ latch node voltage be pulled to ensure a safe transition? We could pull it just beyond V_M, say by 100 mV, and that might work in a perfect world. But a safer

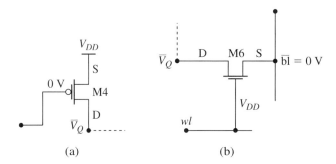

FIGURE 9-12.

Transistor bias states: (a) Initial M4 node voltages. (b) Initial M6 node voltages.

way pulls the node below the threshold voltage of the opposite n-channel transistor (M1 in Figure 9-5). This cuts off M1 ensuring a complete transition and gives a safety margin against statistical variation in V_{tn}. Our examples will assume that the latch node is pulled to V_{tn}.

In sizing the transistors, let us look at the M4 and M6 bias states before equating I_{D4} and I_{D6} and solving for the β ratios. Assume initially that $\overline{V_Q}$ is logic-1 and $\overline{bl} = 0$ V in Figure 9-12. We must calculate the M4–M6 width to length ratios that ensure that $\overline{V_Q}$ is pulled sufficiently low to flip the logic state. We will assume that when M4 and M6 turn on, the initial voltage drop across each transistor is about $V_{DD}/2$. Figure 9-12a shows M4 with its steady state node voltages. M4 is in nonsaturation. M6 in Figure 9-12b also has a V_{GD6} greater than threshold, so M6 is also in nonsaturation. $\overline{V_Q} > V_t$ during the transition so that M4 and M6 remain in the ohmic state until the cell flips.

The W/L ratios must ensure that $\overline{V_Q}$ is pulled down to V_{tn} so that the cell flips its logic state. We analyze the action similar to the Read operation by equating drain currents of M4 and M6 that are both in the nonsaturated state, and we calculate β_4/β_6 as

$$K_{n6}\left(\frac{W_6}{L_6}\right)\left[2(V_{GS6} - V_{tn6})V_{DS6} - V_{DS6}^2\right]$$

$$= K_{p4}\left(\frac{W_4}{L_4}\right)\left[2(V_{GS4} - V_{tp4})V_{DS4} - V_{DS4}^2\right] \tag{9-4}$$

$$\frac{\beta_4}{\beta_6} = \frac{W_4/L_4}{W_6/L_6} = \left(\frac{K_{n6}}{K_{p4}}\right)\frac{\left[2(V_{GS6} - V_{tn6})V_{DS6} - V_{DS6}^2\right]}{\left[2(V_{GS4} - V_{tp4})V_{DS4} - V_{DS4}^2\right]} \tag{9-5}$$

We seek the ratio of widths to lengths for the two transistors as β_4/β_6. The designer sets the low voltage of $\overline{V_Q}$ to ensure a logic state transition by the cell. An example will illustrate the Write operation design of a memory cell.

EXAMPLE 9-2

Use the parameters $V_{DD} = 1.5$ V, $V_{tn} = 0.5$ V, $V_{tp} = -0.5$ V, $K_n = 100\ \mu\text{A/V}^2$, and $K_p = 50\ \mu\text{A/V}^2$. Determine the width to length ratios of M4 and M6 in Figure 9-10 to perform a successful logic-1 to logic-0. Write operation when the initial latch node is $\overline{V_Q} = 1.5\ V$. Assume that $\overline{V_Q} = V_{tn} = 0.5$ V when the cell flips.

Using Eq. (9-5)

$$\frac{\beta_4}{\beta_6} = \frac{100\ \mu\text{A}}{50\ \mu\text{A}}\ \frac{[2(1.5 - 0.5)0.5 - 0.5^2]}{[2(0 - 1.5 - (-0.5))(0.5 - 1.5) - (0.5 - 1.5)^2]}$$

$$\beta_4/\beta_6 = 1.50$$

EXAMPLE 9-3

$V_{DD} = 2.0$ V, $V_{tn} = 0.5$ V, $K_p = 45\ \mu\text{A}$, $K_n = 85\ \mu\text{A}$. If $\beta_4/\beta_6 = 1.45$, what is the nMOS threshold voltage when the change of cell state is triggered by a latch voltage of 0. 6 V?

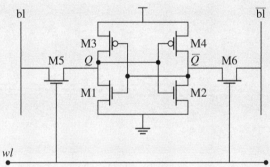

$$K_p \left(\frac{W}{L}\right)_4 \left[2(0 - 2.0 - V_{tp})V_{DS} - V_{DS}^2\right] = K_n \left(\frac{W}{L}\right)_6 \left[2(2 - V_{tn})V_{DS} - V_{DS}^2\right]$$

$$1.45 = \frac{85\ \mu\text{A}}{45\ \mu\text{A}}\ \frac{[2(2 - 0.5)(0.6) - (0.6)^2]}{[2(0 - 2 - V_{tp})(0.6 - 2) - (0.6 - 2)^2]}$$

$$V_{tp} = -0.63\ \text{V}$$

Self-Exercise 9-3

$V_{DD} = 1.2$ V, $V_{tn} = 0.4$ V, $V_{tp} = -0.4$ V, $K_n = 75\ \mu\text{A/V}^2$, and $K_p = 45\ \mu\text{A/V}^2$. What is the β ratio (β_4/β_6) necessary for a Write operation to change a cell node from logic-1 to logic-0?

Answer: $\dfrac{\beta_4}{\beta_6} = 1.25$

Self-Exercise 9-4
Given $V_{tn} = 0.5$ V, $V_{tp} = -0.5$ V, $K_p = 40\ \mu A/V^2$, and $K_n = 95\ \mu A/V^2$, $\beta_4/\beta_6 = 1.99$, and the cell flip occurs at $V_{logic} = 0.4\ V_{DD}$. What is the V_{DD} to allow a Write operation to change a cell from logic-1 to logic-0?

Answer: $V_{DD} = 1.61$ V

Self-Exercise 9-5
Size all transistors for the memory cell given $V_{tn} = 0.45$ V, $V_{tp} = -0.45$ V, $K_p = 65\ \mu A/V^2$, $K_n = 120\ \mu A/V^2$, and $V_{DD} = 1.2$ V. Set V_Q at 10% of V_{DD} for the Read operation and $\overline{V_Q}$ at the switching threshold at 0.5 V for the Write operation.

Answers:
Read operation:
$$\frac{\beta_1}{\beta_5} = \frac{\beta_2}{\beta_6} = 2.40$$

Write operation:
$$\frac{\beta_4}{\beta_6} = \frac{\beta_3}{\beta_5} = 1.65$$

If W_5 and W_6 are unit size $= 1$
$W_1 = W_2 = 2.40$
$W_3 = W_4 = 1.65$

Sizing Summary: We sized the nMOS access and pull-down transistors and the pMOS pull-up transistors. Transistor size is one trade-off of noise sensitivity against the trigger levels that cause change of cell state. The latter concern is called guard banding which reflects the degree of safety from voltage noise. Guard banding is a term related to noise margin that we studied in Chapter 5. The smallest transistors in our sizing examples were the access transistors M5 and M6. The other two transistor sizes are referenced to these small access transistors. Notice that we don't concern ourselves with calculations for a logic-0 to logic-1 transition at the $\overline{Q}$ node since that would be governed by pull-down of M3–M5 that are sized identically to M4–M6.

A major effort goes into memory design and layout. The basic cell must be laid out to a minimum area that satisfies performance and power specifications. These design ratios used the long channel models. A modern design would use the same principles as described here, but would use more accurate computer models for short channel transistor dimension calculations. The ratios calculated with short channel transistor models typically yield β-ratio numbers that are closer to 1.5 for both Read and the Write operations.

9.8. **Column Write Circuits**

Figure 9-13 shows the relevant transistors and interconnects of a Write driver for a selected memory cell [1]. Two AND gates decode the information necessary to determine which bit line must be pulled low to write new data into the cell. The Write signal W is ANDed with Data D and $\overline{D}$ to activate either M10 or M11 to pull down the designated bit line to ground. Transistor M12 is a strong write driver pulling the selected bit line low causing a change of state in the memory cell. The gate signal of M12 is a column address select (CAS). When CAS is high, M12 allows current to pass and when $CAS = 0$, then no current passes and the memory cell is unaffected. There is considerable capacitance in the bit line

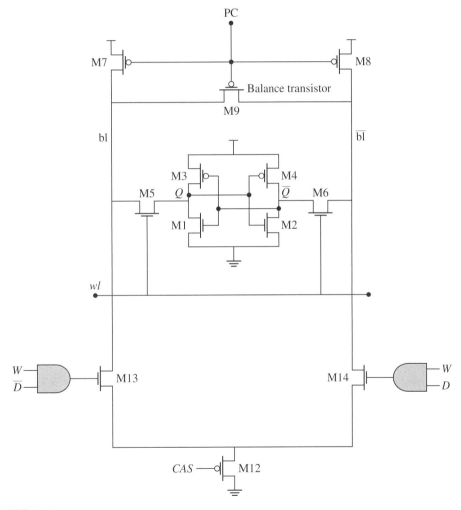

FIGURE 9-13.

Write circuit driver using pass transistors to decode column. W is the write enable signal.

circuitry (pFs) so that M12 must be sized adequately to pull the line low in the specified delay time of the memory.

The simple timing sequence for the Write operation is then

1. The bit lines are precharged.
2. Decoded bit lines are driven to the new logic signals.
3. The word line is asserted and charge flows from the memory cell from its high voltage node to the low bit line.
4. The cell flips its memory state, and the Access and *CAS* control signals return to their Off-state.

9.9. Read Operation and Sense Amplifier

We discussed the Read operation at the cell level in Section 9.4. The bit lines are precharged in a Read operation prior to activating the word lines and their access transistors. The access and pull down transistors must be designed to read a small difference in the two bit line voltages to determine which of the bit lines is reading a logic-0 from the cell. The memory cell must not flip its logic state during a Read operation.

The bl and $\overline{bl}$ lines may have about a 100–200 mV difference after access transistors are turned on. There is a sense amplifier near the end of the bit lines that is designed to amplify the difference of these two voltages. In memory circuits, the diff amp is called a sense amp. It senses the logic state of the bit lines, and delivers a logic state voltage at the *Sense Output* node (Figure 9-14). Each bit line is an input to the diff amp, and the output is a logic-0 or logic-1. The sense enable (SE) signal activates the diff amp for a Read operation.

A diff amp is one of a few analog circuits that appear in digital circuits. Analog circuits deal with signals that use a full range of voltage and current. The recorded music that we hear comes from analog circuits, as well as simple microphone sound. The operational amplifier (op amp) is an analog circuit that finds use in digital interface with analog signals. These interface circuits are called converters; analog-to-digital and digital-to-analog. The diff amp is an integral part of many analog ICs, converters, and all memory circuits.

MOSFET transistors in analog applications have an important distinction. They operate in the saturated bias state where an output signal current is linearly related to an input signal over a small voltage range. A small input sinusoidal voltage of 100 mV might be linearly amplified by a factor 10–50 giving a large sinusoidal output voltage of 1–5 V. Typically an entire course is devoted to analog circuits, and we cannot replicate that information here. However, we must begin the learning process to better understand sense amps. We will analyze the schematic for the amplifier and transistor function describing the purpose of the elements.

Figure 9-15a shows a 2-transistor differential amplifier. It is a symmetrical circuit with M1 = M2, and the load resistors are large and equal. Load resistors are necessary to allow a signal voltage change caused by the current drawn by the transistor. The amount

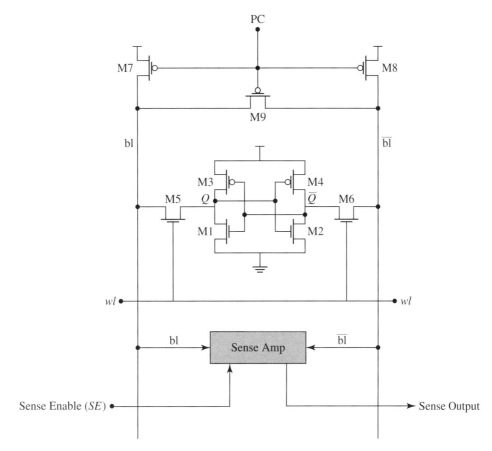

FIGURE 9-14.

Read operation circuitry.

of drain current is modulated by the gate-to-source voltage consistent with our studies in Chapter 3. The bias resistor from the transistor source nodes to the ground path is large. A large resistor in series with a power supply approximates a current source as shown in the Norton equivalent circuit in Figure 9-15b.

Figure 9-15c replaces the large load resistors with pMOS transistors M4 and M5. All transistors are in saturation so the output signal nodes v_1 and v_2 still see a large resistance. A transistor in saturation presents a reverse bias pn junction to signals at its drain node, and that is a large resistance to a small signal change. Transistors allow a better matching of resistance. Also, the constant current source I_{SS} is replaced with an nMOS transistor. Since M3 is in saturation, its drain node acts as a constant current source with a high resistance to ground. MOSFETs are much smaller than large resistors.

The small signal operation lies at the heart of sense amplifier function. The description here is admittedly simplistic, but serves as an introduction. If $v_1 = v_2$ then the drain currents

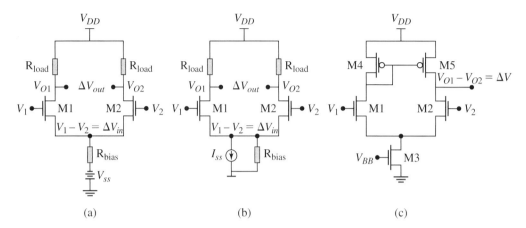

FIGURE 9-15.

(a) Differential amplifiers. (b) Approximate current source biasing. (c) Active load transistors M4 and M5.

are equal in each transistor, and the drain voltages are equal. If V_2 increases, then I_{D2} increases causing the drain voltage V_{O2} of M2 to drop since the IR drop across the load resistance increases. If V_1 decreases, then I_{D1} decreases and V_{O1} rises. The output difference voltage is $\Delta V_O = V_{O1} - V_{O2}$ and the input difference voltage is $\Delta V_{in} = V_1 - V_2$. The voltage amplification or gain factor is $A_v = \Delta V_O / \Delta V_{in}$ where A_v is the amplifier voltage gain.

If v_1 and v_2 change by an equal amount, then I_{D1} and I_{D2} increase by an equal amount and the difference voltage across the drains is zero. Input voltages that are common to both inputs are not amplified, and the effect is called common-mode rejection. In summary, the diff amp amplifies differences in the two input signals while suppressing the effect when input voltages are common.

The sense amp uses the differential input voltages from the bit lines as $\Delta V_1 = V_{bl} - V_{\overline{bl}}$. The amplified output difference voltage may be taken from a single output and fed to an inverter that restores the strong logic state. Figure 9-16 shows the final circuitry. The sense amp control signal V_{Sense} allows the sense amp to pass current and to function. When $V_{Sense} = 0$ V, then M15 and M16 gates can be toggled, but no current passes in M15–M18.

Figure 9-17 shows another sense amp design that incorporates full logic correction from the M1 and M2 diff amp transistors eliminating the inverter output in Figure 9-16. The latch will give a strong logic-1 and a slightly weak logic-0. *SE* is the sense enable signal that activates the Read operation. When *Out* goes high, it is fed by M5 giving a full V_{DD} signal. $\overline{Out}$ is low when *Out* is high, but $\overline{Out}$ has two transistor drain-source voltage drops above ground. These voltage drops are small since the gate to source voltage is at V_{DD} for M3, and slightly lower than V_{DD} for M1. M1 and M2 are driven into the ohmic state.

There are several sense amp designs, so which design is better? The design is primarily driven by the memory specifications on speed (access times), power dissipation, noise immunity, and area of the layout.

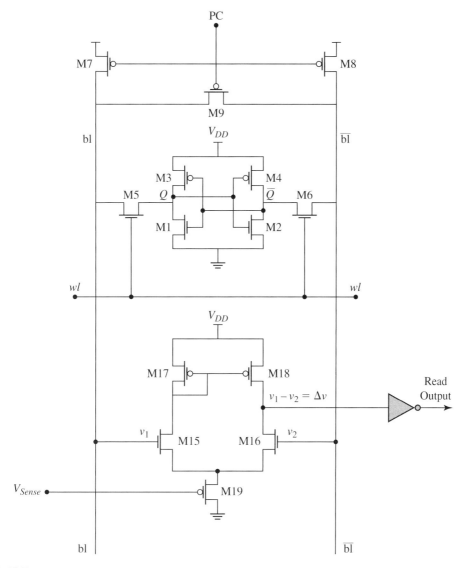

FIGURE 9-16.

Read operation circuitry.

9.10. Dynamic Memories

Robert Dennard of IBM conceived the dynamic memory (DRAM) in 1966. A memory cell with a single, small transistor and capacitor was an attractive alternative to the memory cell designs of the day. And MOSFETs were just beginning to attract the eye of designers. The first commercial DRAM product was the Intel 1103 with 1024 bits of memory. The

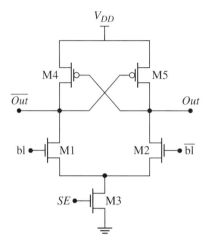

FIGURE 9-17.

A cross-coupled double-ended sense amp.

1103 DRAM was introduced in 1968 using pMOS transistors. The significance of the Robert Dennard invention and the subsequent Intel 1103 was huge for future computing efficiency and size. It allowed the powerful evolution of our modern PCs and laptops.

DRAMs contrast to SRAMS is several ways. Table 9-1 compares these properties. Each memory type has a unique and significant place in the market. DRAMs are more dense and preferred when making large stand-alone memories. DRAMs require a clock driven data refresh operation to constantly assure that the cell data are held true. The refresh requirement is the basis for the word dynamic. SRAMs use a 6-transistor cell that takes more chip area than the 1-transistor or the 3-transistor DRAM cell. The complexity of the DRAM peripheral circuits makes them inherently slower. DRAMs often use a precharge voltage in excess of the normal V_{DD} that is often supplied by an on-chip charge pump (Chapter 8). DRAMs are compatible with CMOS technology.

TABLE 9-1 SRAM and DRAM Properties

SRAM	DRAM
Faster	Slower
Larger cell area (6T cell)	Smaller cell area (1T or 3T cell)
– Smaller size (Mbytes)	– More memory per IC (Gbytes)
Simpler peripheral (I/O) circuit	Complex peripheral circuits and timing
No clocked refresh	Needs clocked refresh
No V_{DD} special needs	Needs precharge > V_{DD} (charge pump)
Requires ratioed transistors	Does not require ratioed transistors

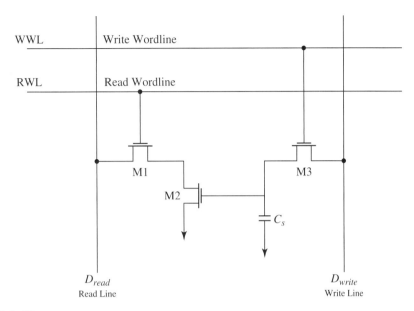

FIGURE 9-18.

3-Transistor DRAM cell.

This section will introduce a 3-transistor DRAM cell to illustrate DRAM basic principles. The 1-transistor cell is preferred in modern DRAMs, but its structural and electronic complexity makes it a better choice of study in the next level of digital electronics course.

9.10.1. 3-Transistor DRAM Cell

A 3-transistor DRAM cell is shown in Figure 9-18. The bit line symbols are changed to D_{read} and D_{write} to differentiate from the SRAM bit lines. The capacitance C_S is the heart of the cell storage. C_S is the parasitic capacitance at the M3 drain and M2 gate. New bit data are written by first precharging the bit lines, then driving the D_{write} bit line to the new data. M3 is turned on and C_S either charges to a logic-1 or logic-0 depending on the new data. The large capacitance of the bit line prevents a charge sharing with C_S sufficient to disturb the bit line precharge voltage.

The Read operation precharges the bit lines and then activates M1. If a logic-0 is stored on C_S, then the read bit line remains at the precharge of logic-1 because M2 is off. If a logic-1 is stored on C_S then M2 turns on providing a pull-down path for the read bit line through M1 and M2. The DRAM does not require fixed transistor ratios to operate as does the SRAM, therefore, minimum size transistors can be used.

When C_S is charged and the M3 access transistor is turned off, the charge on C_S leaks through the M3 drain and the M2 gate oxide. The leakage time constant is typically a few milliseconds at room temperature. If not corrected, this leakage would corrupt data in C_S. The solution is to *refresh* the data at a rate in excess of the leakage time constant. This is typically above 150 kHz. Dynamic means that the IC clock must constantly be active in contrast to static designs. The refresh operation uses circuitry that restores the original value in the cell once it is read, and then reloads that data in the cell. Pure static CMOS designs are easier to test and debug, and may also achieve lower power.

The body effect in M3 in Figure 9-18 lowers the logic-1 voltage on C_S to $V_{DD} - V_{tn}$. This can be compensated by elevating the bit line precharge voltage to $V_{DD} + V_{tn}$. A charge pump (Chapter 8) is one way to do this. We see that the simple DRAM schematic requires more support circuitry that the SRAM. However, DRAMs are economic for very large memory sizes that can be tens of gigabytes. DRAMs are a high-volume product.

9.10.2. 1-Transistor DRAM cell

A single transistor DRAM can be constructed. The 1-transistor cell (Figure 9-19) has design simplicity, but requires more complex support (peripheral) circuitry. Notice that the DRAM cell has a single bit line and a cell storage capacitor connected to one end of the MOSFET. M1 forms a switch connecting C_S to data on the bit line. Data are written into the cell capacitance through the bit line and read back through the same bit line. The capacitor holds the logic value.

Each Read operation discharges a significant amount of charge from C_S so that the cell must be recharged, or refreshed after each operation. In addition, the DRAM capacitor C_S will passively leak charge through adjacent reverse bias *pn* junctions, thin oxide tunneling, and adjacent cell-to-cell leakage. So an automatic refreshing of data must be part of the design. Typically a dummy cell stores the data from the specified cell, and after a read, write, or in standby operation that data are written back into the dynamic cell. This complicates DRAM design, but the design works and is economical. The refresh operation increases

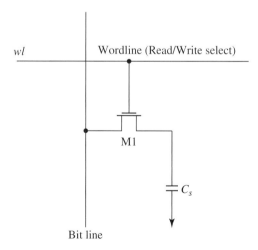

FIGURE 9-19.

1-Transistor DRAM cell.

power consumption and slows DRAM circuit operation. In contrast to the relatively simple SRAM design, DRAMs may have 20 or more internal clocks to control Read, Write, and refresh operations.

9.11. Summary

This chapter emphasized how the SRAM and DRAM work. Design equations allowed practice and deepening of these concepts. However, many memory topics were not described since they involve the details that are best left to advanced texts and books [1,2,3]. These important omitted design topics include: the complex timing signals; the lower signal-to-noise ratio of densely packed cells; coupling noise; the use of column redundancy to increase yield; the design of local power voltage different than V_{DD}; the increase is power dissipation for nm transistors with leakage through their ultra-thin oxides; error correction schemes; and description of the other types of memories listed at the beginning of the chapter.

References

[1] D. Hodges, H. Jackson, and R. Saleh, *Analysis and Design of Digital Integrated Circuits*, 3rd Ed., McGraw-Hill, 2003, Chapters 8 and 9.

[2] J. Rabaey, A. Chandrakasan, and B. Nikolic, *Digital Integrated Circuits,* 2nd Ed., Prentice-Hall, 2003, Chapter 12.

[3] K. Itoh, *VLSI Memory Chip Design,* Springer-Verlag, 2001.

Exercises

Read Operation Sizing

Assume that the maximum node voltage for a read is 10% of V_{DD}.

9-1. Given: $V_{DD} = 1.2$ V, $K_n = 120\ \mu$A/V^2, and $V_{tn} = 0.4$ V. Calculate the β-ratio β_1/β_5 for a Read operation.

bl = 1.2 V Precharge

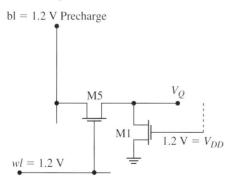

9-2. Ratio the appropriate transistors for a Read operation, given $V_{DD} = 1.8$ V, $i = 130\ \mu$A/V^2, $i = 50\ \mu$A/V^2, $V_{tn} = 0.5$ V, and $V_{tp} = -0.5$ V.

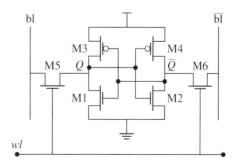

9-3. Plot the β-ratio (β_1/β_5) versus V_Q for a Read operation given $V_{DD} = 2$ V, $K_n = 100\ \mu$A/V^2, and $V_{tn} = 0.6$ V. Plot 100 mV $< V_Q < 400$ mV and comment on the implications of the curve.

9-4. Plot the noise margin versus V_Q for 100 mV $< V_Q <$ 400 mV for the previous problem given $V_{DD} = 2$ V, $K_n = 100$ μA/V^2, and $V_{tn} = 0.6$ V. The noise margin is defined as the difference in voltage between $V_{tn} = 0.6$ V and V_Q. Comment on the implications of the curve.

9-5. An SRAM memory cell with $K_n = 100$ μA/V^2, $K_p = 60$ μA/V^2, $V_{tn} = 0.6$ V, $V_{tp} = -0.6$ V, and the Read operation β-ratio is 3.0. What V_{DD} will satisfy this design?

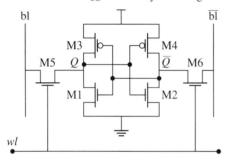

9-6. A low-voltage SRAM is designed for $\beta_1/\beta_5 = 1.35$. If $K_n = 120$ μA/V^2, $i = 85$ μA/V^2, and $V_{DD} = 0.9$ V, what threshold voltage will achieve this for a Read operation.

Write Operation Sizing

Unless otherwise note, assume that the required change in cell node voltage to cause a transition pulls down the node voltage to V_{tn}.

9-7. Given $V_{tn} = 0.5$ V, $V_{tp} = -0.5$ V, $K_p = 40$ μA/V^2, and $K_n = 95$ μA/V^2, and $\beta_4/\beta_6 = 1.99$. What is the V_{DD} to allow a Write operation to change a cell node from logic-1 to logic-0?

9-8. Determine the safe β-ratio to write a logic-1 to one of the memory cell nodes given that $K_n = 100$ μA/V^2, $K_p = 50$ μA/V^2, $V_{tn} = 0.5$ V, $V_{tp} = -0.5$ V, and $V_{DD} = 1.8$ V.

9-9. Given $K_n = 110$ μA/V^2, $K_p = 50$ μA/V^2, $V_{tn} = 0.5$ V, $V_{tp} = -0.5$ V, and $V_{DD} = 1.5$ V, calculate the relevant Write operation β-ratio for the memory cell.

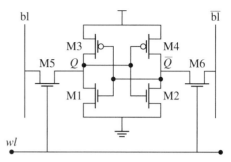

9-10. Plot the Write β-ratio of M4/M6 versus $\overline{V_Q}$ for $i = 100$ μA/V^2, $i = 50$ μA/V^2, $V_{tn} = 0.5$ V, $V_{tp} = -0.5$ V, and $V_{DD} = 1.8$ V. Comment on the result.

9-11. Given $K_n = 105$ μA/V^2, $K_p = 50$ μA/V^2, $V_{tn} = 0.5$ V, and $V_{tp} = -0.5$ V, calculate V_{DD} if $\beta_1/\beta_5 = 1.778$ and $\beta_4/\beta_6 = 1.867$. Verify that the answer is valid for both Read and Write operations, and don't assume that V_Q(write) $= V_{tn}$.

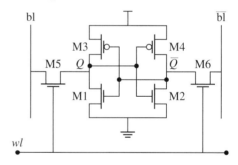

9-12. What is W_4 if $i = 0.25$ μm given $i = 120$ μA/V^2, $K_p = 60$ μA/V^2, $V_{tn} = 0.6$ V, $V_{tp} = -0.5$ V, and $V_{DD} = 1.5$ V?

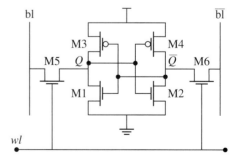

CHAPTER 10

Programmable Logic—FPGAs

Programmable logic circuit (PLC[1]) technology is a powerful alternative to implement digital integrated circuits (ICs) that provides flexibility and high integration. Programmable logic circuits evolved from initial fab mask-programmed gate arrays to more advanced structures that contain both logic and interconnect resources that are programmed by the user. PLCs were introduced in the mid-1980s, and they are based on a reconfigurable interconnect technology to establish IC functionality. In general, users can reprogram PLCs virtually as many times as desired. Its rapid evolution in performance and capabilities provided a quick market acceptance as these circuits allowed the construction of complex digital circuits in very short development times. In addition, programmable circuits are a valid solution for prototyping and in-field applications thus reducing drastically development time, cost and power while providing high performance. The development of flexible and integrated computer-aided design (CAD) tools helped significantly to this rapid market growth.

Unlike previous generation PLCs using input–outputs (I/Os) with programmable logic and interconnects, today's field-programmable gate arrays (FPGAs) use various mixes of configurable embedded SRAM, high-speed transceivers, high-speed I/O, logic blocks, and routing. Specifically, an FPGA contains programmable logic components and a hierarchy of reconfigurable interconnects that allow the logic components to be physically connected. They can be configured to perform complex combinational functions, or merely simple logic gates like AND and XOR. In most FPGAs, the logic blocks also include memory elements, which may be simple flip-flops or more complete blocks of memory.

Some advantages to designing with an FPGA versus full custom (ASICs) are as follows:

- Rapid prototyping
- Shorter time to market

---

[1] We use the acronym PLC for basic programmable logic circuits. Nomenclature is not standard, and other textbooks and papers use PLA (programmable logic array) or PLD (programmable logic devices) to refer these basic structures. Complex programmable logic circuits are more commonly referred to FPGAs.

- The ability to reprogram in the field for debugging
- Long-product life cycle to mitigate obsolescence risk

FPGAs represent the most attractive solution for small or even complex logic solutions requiring low- to mid-volume production since the initial investment is much smaller than full custom (ASIC) alternatives. For high volume productions FPGA cost per chip is too high compared with ASICs.

In this chapter we introduce the fundamentals of this technology, providing insight into the basic programmable device operation. Then we outline its evolution from the basic AND/OR matrix structure to the today's complex and powerful huge FPGAs by showing some examples of products from various vendors. We are not focusing on the latest FPGAs for simplicity; the objective is to provide the reader with an idea of how these circuits work.

10.1. A Simple Programmable Circuit—The PLA

The first user-programmable structures were focused on implementing basic logic functions, and were constructed following the two-level sum-of-products function structure. The capability of reprogramming a circuit to change its logic function was based on implementing *distributed gates*, being basic AND/OR gates whose inputs could be selectively connected and disconnected through programming. The circuit was constructed by connecting an array of distributed ANDs to an array of distributed ORs. In some cases one of the matrixes was fixed, and the other was programmable, while in other circuits both matrix were programmable.

10.1.1. Programmable Logic Gates

PLCs are based on the simple concept that any Boolean function can be implemented as a sum of products. A logic function F1 can be obtained in two steps: first performing product operations, and then summing these product terms. Once the function is physically implemented within a circuit, changing its original behavior to another function F2 implies modifying the product terms that are summed. Assume a simple case where these functions are

$$F1 = xyz + \overline{x}z$$
$$F2 = xyz + x\overline{y}$$

Changing the circuit operation from F1 to F2 requires modifying the second-product term. This can be done either by changing the logic gate that does that product, or changing the inputs arriving to the corresponding 2-input AND gate. PLCs use this second option by changing the inputs driving the gate using programmable logic gates. Programmable logic gates are special logic gates having multiple inputs that can be electrically connected or disconnected through programmable interconnects. These gates were the essential elements of the primary PLCs that implemented basic functions.

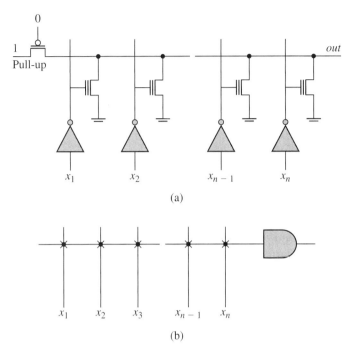

FIGURE 10-1.

(a) Transistor-level structure of a programmable logic AND gate. (b) Schematic representation.

In programmable logic gates, a number n of input signals $x_1, x_2, \ldots, x_n$ are connected to n programmable transistors[2] with their drains shortened to a pull-up line (the output node), with sources grounded (see Figure 10-1a). In this structure, if all the programmable transistors are programmed (i.e. they never turn on), then the output node is always logic 1, independently of the input signals values (the programmable transistors will never pull the line to GND). If the programmable transistors are all unprogrammed (i.e., they behave as normal nMOS transistors) then the output will go low if any of the signal inputs are low. The output will remain high only if all the inputs are high. Note that this structure behaves like an AND gate whose inputs can be "connected" and "disconnected" depending on the programming state of the programming transistor driven by each input. To "connect" a given input to the AND gate, its corresponding programming transistor must be unprogrammed (conducting), and to "disconnect" such an input, the corresponding programming transistor must be programmed. Figure 10-1b shows the schematics used to represent the structure in Figure 10-1a, where the gate inputs are illustrated to be joined through programming interconnects to the AND gate input.

---

[2]Programmable transistors are explained in Section 10.4. These devices can be programmed to work as normal transistors or to never turn on for any gate terminal voltage.

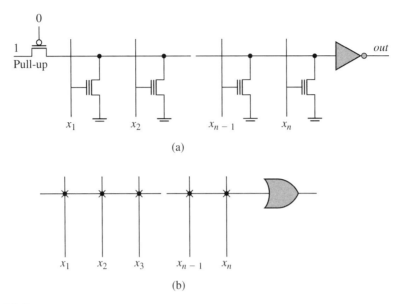

FIGURE 10-2.

(a) Transistor-level structure of a programmable logic OR gate. (b) Schematic representation.

Similarly to the programmable logic AND gate, an OR gate structure is constructed (Figure 10-2a). In this case, if any of the gate inputs whose transistor is unprogrammed (conducting) are high, then the gate output is high, being an OR gate function. Figure 10-2b shows the schematics of this gate.

10.1.2. AND/OR Matrix Gates

A basic programmable circuit is constructed by combining a set of programmable logic AND gates (that implement the product terms) together with a set of programmable logic OR gates (that sum the product terms), in a two-array arrangement as shown in Figure 10-3. This simple array-based structure is typically known as programmable logic array (PLA), being one type of PLC.

The PLA in Figure 10-3 has three external inputs (x, y, and z) that enter the AND matrix both in direct and negated forms. The AND matrix is composed of eight programmable logic AND gates, each one connected through programmable switches to the external inputs. This structure allows implementing all eight possible 3-input *minterms* that can be constructed with the external input variables. The AND gate outputs are inputs to the OR Matrix constructed with four programmable logic OR gates. Each distributed AND gate output is connected to each distributed OR gate input through a programmable switch, thus allowing us to sum the desired *minterm* outputs to implement four of the 256 possible 3-input logic functions. This is done by connecting and disconnecting the appropriated programmable switches.

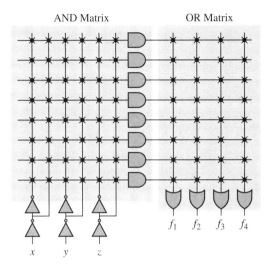

FIGURE 10-3.

Typical AND-OR matrix structure of a basic PLA.

EXAMPLE 10-1

Determine the interconnect map for the circuit of Figure 10-3 to implement a full adder.

   This circuit will use two of the four available circuit outputs. Output f_1 will be used to implement the carry function, and output f_2 for the sum. The canonical sum-of-products for each function is

$$Carry = f_1 = \overline{x}yz + x\overline{y}z + xy\overline{z} + xyz$$
$$Sum = f_2 = \overline{x}\,\overline{y}z + \overline{x}y\overline{z} + x\overline{y}\,\overline{z} + xyz$$

Therefore, the resulting interconnect scheme is

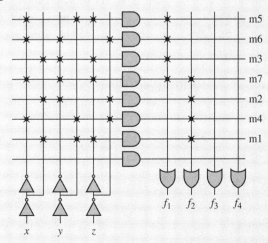

Self-Exercise 10-1
Determine the interconnect scheme within the circuit of Figure 10-3a to obtain a 3-bit binary
to Gray code converter.

The structure of the circuit shown in Figure 10-3 is a particular case of AND/OR matrix
structure where both planes have programmable interconnects. There were variations to
this configuration in which the AND matrix was fixed (all the minterms were generated
by default) and the OR matrix was programmable, as well as the opposite configuration
in which the AND matrix was programmable, and the OR was fixed.

The basic PLA structure was the first to appear, and it had many limitations. The number
of inputs and outputs was fixed, and only functions constructed as sum of products could be
implemented. Various improvements were developed to increase circuit capabilities such
as bidirectional terminals that could be programmed either as circuit inputs or outputs,
and the capability of using intermediate variables by enlarging the AND matrix to allow
feedback from OR matrix outputs. These circuits were known as extended PLAs.

10.2. The Next Step: Implementing Sequential Circuits—The CPLDs

The flexibility of programmable circuits and its rapid completion cycle motivated the
improvement of their capabilities in terms of computational power and device density.
The path to such complexity started with the incorporation of basic memory blocks to
implement sequential circuits.

10.2.1. Incorporating Sequential Blocks—The Complex Programmable Logic Device (CPLD)

Figure 10-4 illustrates one basic building block included in the early family of Altera's
CPLDs [1]. The incorporation of several of these basic blocs (running from eight up to
about 50 depending on the device) provided programmable circuits with enough features
to implement mid-complexity sequential circuits. The left-hand structure of the block in
Figure 10-4 is the switch matrix where primary inputs and outputs from other blocks
can be connected to the multiple-AND, single OR gate structure used to implement the
sum-of-products programmed function. The memory element is a single D-type latch,
whose input may be driven by the programmed function (passed through an XOR gate for
inversion control).

The circuit block has a bidirectional PIN. In this way, the block function can be directed
to the output PIN by appropriate programming of the corresponding bidirectional control
circuitry. The bidirectional circuitry has a tristate buffer that can be programmed either in
low impedance (the PIN is used as a circuit output), or in high impedance (the PIN is used
as an input).

The block in Figure 10-4 can be used either as a combinational block or a sequential
one as illustrated in Figure 10-5a and 10-5b, respectively. This is done by programming

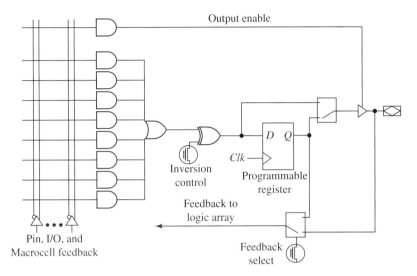

FIGURE 10-4.

Basic block containing one sequential element.

the top-most multiplexer (MUX) shown in the circuit block. The module is used as a combinational block when the output of the memory element is not driven toward the circuit output (Figure 10-5a). If the top-most MUX of the block drives the flip-flop output to the circuit output, then the whole block works as a sequential element (Figure 10-5b). This sequential output can be redirected toward the switch matrix using the bottom-MUX where it can be used in other blocks as an input.

It is also possible not to use the functional block at all, and program its associated PIN as an input by setting the appropriated value to the "Output Enable" signal in the figure that puts the output buffer in tristate.

The first CPLDs were oriented to synchronous circuit applications and had one particular PIN that could be selected either as a data input, or clock. Figure 10-6 shows the schematic of one CPLD from Cypress, having 10 building blocks similar to those shown in Figure 10-4. The central global interconnect section of the circuit is on left side. Note that primary inputs (located on the left) and programmable input/outputs (on the right side), are merged into the global interconnect section, and can be used as inputs by any of the ten blocks. PIN number 1 can be used either as one additional input or as the circuit clock signal.

The main characteristics of CPLDs are their easy of design, lower development cost, and the relatively small delay of the circuit, which can be predicted before programming the application.

CPLDs evolved from the simple structure shown in Figure 10-6 to more complex architectures based on incorporating a higher number of similar building blocks to the ones shown, but with more complex interconnect topologies. The goal was to increase circuit functional capabilities while keeping delay within bound values.

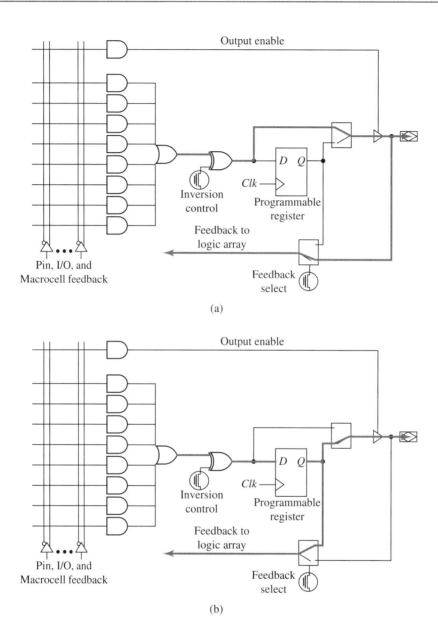

(a)

(b)

FIGURE 10-5.

(a) Using the basic block to implement a combinational function and (b) a sequential one.

10.2.2. Advanced CPLDs

In today's market CPLDs remain the best option for low-power, low-cost programmable solutions, as opposed to the high-end FPGAs (explained in the next section), representing a remarkable advance from the initial architectures in terms of integration and computation

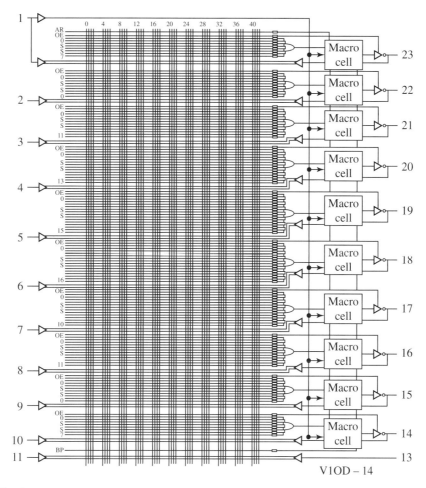

FIGURE 10-6.

Schematic of a CPLD (22V10 from Cypress) [2].

power. We present two main architectures as CPLDs remain in the market and have their target applications.

Altera Max CPLDs

The Altera family of CPLs started with the Classic family to evolve to the MAX architecture composed successively of the MAX5000, MAX7000, and MAX9000 families, until the actual Max®II family (that remains together with the MAX3000 family). We detail the architecture and organization of the MAX7000 family being a representative architecture.

Figure 10-7 shows the architecture of the MAX7000 device block diagram based in interconnecting logic array blocks (LAB) through a programmable interconnect array (PIA)

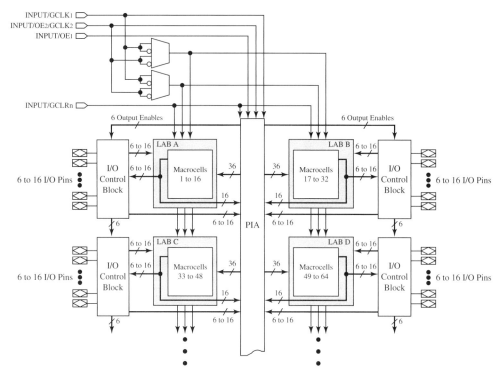

FIGURE 10-7.

Altera's MAX7000 device block diagram.

and I/O control blocks. Each LAB has 16 macrocells that are an evolution of the basic blocks in Figure 10-4 and detailed next. A given device may contain from 32 up to 256 macrocells and has up to two global clock signals.

Figure 10-8 shows the internal structure of the macrocells composing each LAB in groups of 16. Each macrocell can be configured either for combinational or sequential logic operation and they have three functional blocks: the logic array, the product-term select matrix, and the programmable register.

The logic array provides five product-terms per macrocell similar to the structure shown in Figure 10-4. But it adds two product-term expanders: shareable expanders and parallel expanders. Each LAB has 16 shareable expanders that take one product term from each macrocell in the LAB that are feedback into the logic array. Parallel expanders are unused product terms that can be allocated to a neighboring macrocell to implement fast, complex logic functions. They allow up to 20 product terms to directly feed the macrocell OR logic with 5 product terms provided by the macrocell, and 15 parallel expanders provided by neighboring macrocells in the LAB. The software complier automatically allocates up to three sets and up to five parallel expanders to the macrocells that require additional product terms. These connections cause an incremental delay due to the additional routing. Each

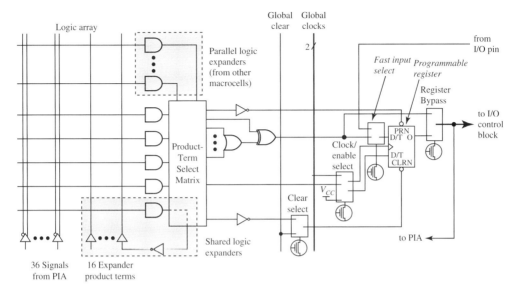

FIGURE 10-8.

Altera's MAX7000 Device macrocell.

macrocell can take parallel expanders from lower-numbered macrocells. The macrocell has a configurable register with an independently programmable clock, clock enable, clear and preset functions, and the flip-flop can be individually programmed to implement D, T, JK, or SR operations.

The connections between LABs are routed through the interconnect array PIA. It is a programmable path that connects any signal source to any destination on the device. All the circuit dedicated inputs, I/O pins, and macrocell outputs feed the PIA making them available through the whole device.

The I/O control block allows each I/O pin to be configured individually as input, output, or bidirectional operation. All I/O pins have tristate capability through a tristate buffer that is controlled by one of the global output enable signals or directly connected to ground or V_{DD}.

Once a given application is to be programmed into a CPLD, it is desirable to determine the overall circuit delay. The mid-complexity of a CPLD and its internal structure allows simulating the device timing, once the specific resources have been allocated and connected internally. Software tools are available that analyze how the device has been routed and compute the timing point-to-point delay prediction. The pin-to-pin delay is computed by summing the delay of the internal elements through which the worst-case path is routed.

The MAX7000 Altera family is an intermediate CPLD family. The higher end CPLDs are based on an evolution of the macrocell shown here—called logic elements (LE) with advanced possibilities. Additionally, the architecture is more complex including both local and global interconnects.

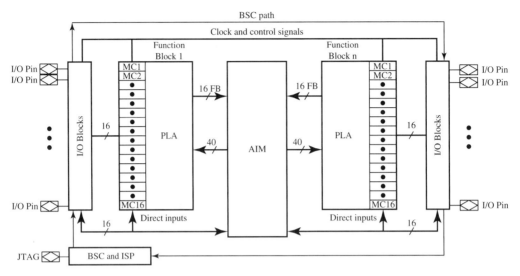

FIGURE 10-9.

CoolRunner-II CPLD architecture.

Xilinx CoolRunner CPLD Series

The underlying architecture is a traditional CPLD architecture combining macrocells with up to 32 function blocks (FBs) interconnected with a global routing matrix, called the Xilinx Advanced Interconnect Matrix (AIM) [3]. FBs use a PLA configuration that allows all product terms to be routed and shared among any of the macrocells of the FB.

Figure 10-9 shows the high-level architecture and how FBs attach to pins and interconnect to each other within the internal interconnect matrix. Each FB contains 16 macrocells. The BSC path is the JTAG boundary scan control (BSC) path. The BSC and in-system programming (ISP) block has the JTAG controller and ISP circuits.

The CoolRunner-II CPLD FBs contain 16 macrocells, with 40 entry sites for signals to arrive for logic creation and connection. The internal logic is a 56-product term PLA. All FBs are identical regardless of the number contained in the device.

Figure 10-10 shows a schematic of the CoolRunner-II CPLD macrocell. It allows a direct manage of up to 40-input sum of products (SOP) with up to 56 product terms within a single function block whose polarity can be controlled through an XOR gate. The 56 product terms have 49 generic product terms, four control terms (CTC, CTR, CTs, and CTE) made available to all macrocells within a given FB. The three individual product terms (PTA, PTB, and PTC) are locally available within the macrocell. The control term signals can control the macrocell clock (CTC), the set (CTS), reset (CTR), or I/O control (CTE) with the property that these signals can be shared among various macrocells within the FB. The individual product terms are available at the macrocell level only, and they can control the set/reset signal (PTA), the I/O control (PTB), and clock:

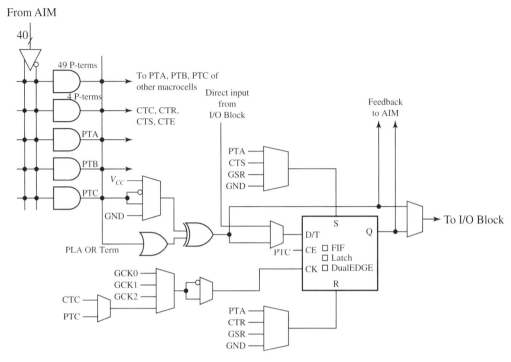

FIGURE 10-10.

Macrocell structure for the Xilinx CoolRunner-II.

The memory element can be configured either as a D or T flip-flop or as a transparent latch, and clocking can occur on either edge or dual edge clocking. The set and reset circuits are similar and have various options depending on their driving multiplexer configuration. They have the option of no set/reset condition (GND input), a product term set signal (PTA), a control term set (CTS), or the global set/reset condition (GSR). Note that global signals are available through the entire chip, control term signals are common within a given FB, while product terms are available within each macrocell.

Figure 10-11 represents a simplified structure of the macrocell in Figure 10-10 illustrating control term set and control term clock usage. Control-term signals are used when a given clock, reset, or set signal must be synchronized across various macrocells within the same FB. Control terms are the same as product terms except that they connect to the same mux sites on every macrocell within the same FB. Figure 10-11 illustrates how two control-term signals are used for set and clock signals while being available for other macrocells in the same FB.

To compute the overall delay once the circuit has been programmed, each block through which the signal goes is a time delay that adds to a signal that passes through such resource. Timing reports are created by adding up the incremental signal delays as signals progress

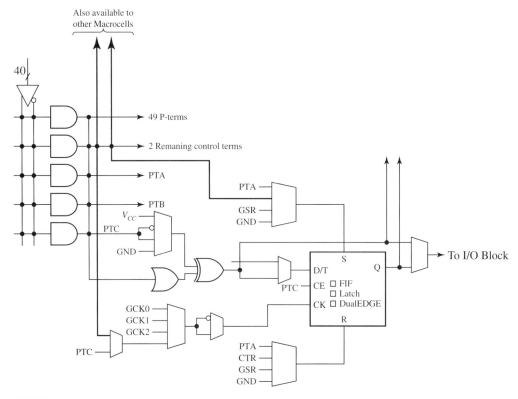

FIGURE 10-11.

Simplified structure of the Xilinx CoolRunner-II macrocell to illustrate control-term set and control-term clock.

within the CPLD. Software creates the timing reports after a design has been mapped onto the specific part and knows the specific delay values for a given speed grade.

10.3. Advanced Programmable Logic Circuits—The FPGA

Field-programmable gate arrays (FPGAs) can be seen as the evolution of programmable technology toward the most complex and powerful programmable circuits. Although FPGAs and CPLDs coexist in today's market (as opposed to CPLDs vs. PLAs, CPLDs replaced PLAs that are no longer in the market) as they target different applications. Form an application standpoint FPGAs are much bigger than CPLDs. CPLDs are "coarse-grain" devices containing relatively few large blocks of logic with flip-flops, and are used for simple glue logic applications. They are cheap and nonvolatile. FPGAs are "fine-grain" devices and contain a huge number of tiny blocks of logic with flip-flops and can implement very complex systems. They are more expensive on a chip basis than CPLDs, but cheaper on a per gate basis and are volatile in general. Typically, FPGAs require some support

logic and must be configured at each power-up, while CPLDs are active at power-up
if they have been programmed. Another difference is that CPLDs have faster and more
predictable input-to-output timing than FPGAs due to their coarse-grain architecture and
relative simplicity.

Although there are many kinds of FPGAs, all have some elements in common. FPGAs
have a regular array of block cells that are configured to perform a given function and are
based on the principle of integrating gates and memory together with a rich interconnect
structure. Four main internal structures can be distinguished:

– Configurable logic blocks (CLB) or configurable logic elements (CLE)
– Programmable interconnect or programmable switch matrix (PSM)
– Input/output buffers (IOB)
– Other elements (e.g., memory, arithmetic units, clock trees, timers, processors)

Figure 10-12 illustrates the generic structure of an FPGA as a circuit composed of logic
blocks, interconnect subsystem, and input/output blocks. It also contains specific blocks
that perform a specific function like multipliers.

Logic blocks are commonly referred to as CLBs or Logic elements (LEs) and are an
evolution of the sequential blocks developed for CPLDs (Figure 10-4). These elements are

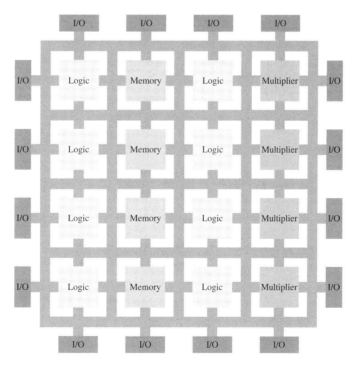

FIGURE 10-12.

Illustration of the FPGA concept.

constructed to provide high flexibility in terms of the function that can be implemented while providing an acceptable speed.

Typically, both the LBs and the interconnect system are programmable, thus allowing us to implement specific functions. The particular architecture and capabilities of each particular FPGA depends highly on the manufacturer. We will cover the most popular low-level FPGA structures looking separately to the logic blocks, circuit architecture, interconnect system, and I/O resources.

There are multiple ways of implementing programmable logic elements, and there are various alternatives in the market depending on the particular vendor. These different approaches are essentially the evolution of the logic blocks used in CPLD applications, and each vendor offers a family of devices that cover a variety of I/O Pins and internal elements. We will cover the basic structures of three major vendors: Altera [1], Xlinx [3], and Actel [4].

10.3.1. Actel ACT FGPAs

Logic Blocks

The early FPGA family from this vendor was the Actel ACT whose basic logic blocks were called Logic Modules. The ACT1 family used just one type of Logic Module (Figure 10-13), while the ACT2 and ACT3 FPGA families used two different types of Logic Modules (Figure 10-14) [4].

The ACT1 module (Figure 10-13) uses three multiplexers and one OR gate constituting a fairly simple, fine-grained logic module with low delay, small area, and flexibility. Using Shannon's expansion theorem, the ACT1 module can implement many logic functions: any 2-input combinational function, almost any 3-input function, many 4-input functions and some functions up to 8 inputs. It can also implement latch gates. These logic functions

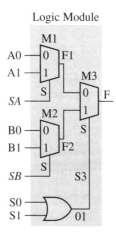

FIGURE 10-13.

ACT1 Actel Logic Module.

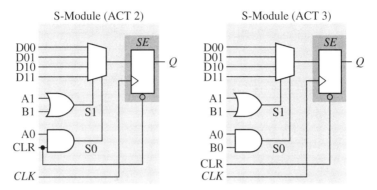

FIGURE 10-14.

Actel ACT2 and ACT3 S-Modules.

are implemented by interconnecting signals from the routing tracks to the data inputs and select lines of the multiplexers. Inputs can be also tied to a logic one or logic zero, as these signals are always available in the routing.

ACT1 module is also referred to as a C-Module as it is purely a combinational block. To expand these module possibilities, Actel introduced the sequential S-Modules ACT2 and ACT3 (Figure 10-14). These modules basically added a flip-flop to the MUX based C-Module.

Architecture

Actel ACT FPGAs are based on a channeled gate array architecture that combines rows of logic blocks and routing channels (Figure 10-15). The logic modules are therefore interconnected with segmented tracks whose length are predefined and can be connected with low-impedance switching elements to create the required routing length. Other wiring segments run vertically through the modules and across the channels. Surrounding the logic core is the interface to the I/O pads. Interconnect is based on antifuse technology (explained later in this chapter).

The ACT 1 interconnect architecture uses 22 horizontal tracks per channel for signal routing with three tracks dedicated to V_{DD}, GND, and the global clock (GCLK), making a total of 25 tracks per channel. Horizontal segments vary in length from four columns of logic modules to the entire row of modules. Four logic module inputs are available to the channel below the logic module and four inputs to the channel above the logic module. Thus eight vertical tracks per logic module are available for inputs (four from the logic module above the channel and four from the logic module below). These connections are the input stubs. The single logic module output connects to a vertical track that extends across the two channels above the module and across the two channels below the module. This is the output stub. Thus module outputs use four vertical tracks per module (counting two tracks from the modules below and two tracks from the modules above each channel).

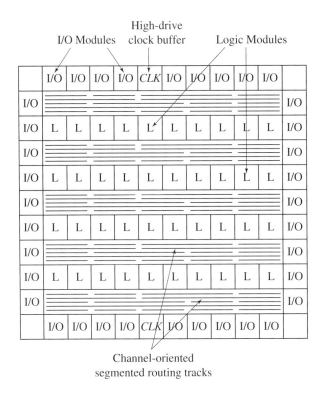

FIGURE 10-15.

Actel ACT FPGA architecture.

10.3.2. Xilinx Spartan FPGAs

We briefly describe the structure and characteristics of the Spartan-3 Xilinx generation being a currently available FGPA. This FPGA uses five fundamental programmable functional elements:

- CLBs contain flexible look-up tables (LUTs) that implement logic plus storage elements used as flip-flops or latches. CLBs perform a wide variety of logical functions as well as store data.
- IOBs control data flow between the I/O pins and the internal circuitry. IOBs support bidirectional data flow plus 3-state operation and various signal standards, including several high-performance differential signal standards.
- Block RAM provides data storage in the form of 18-Kbit dual-port blocks.
- Multiplier Blocks accept two 18-bit binary numbers as inputs and calculate the product.
- Digital Clock Manager (DCM) Blocks provide self-calibrating, fully digital solutions for distributing, delaying, multiplying, dividing, and phase-shifting clock signals.

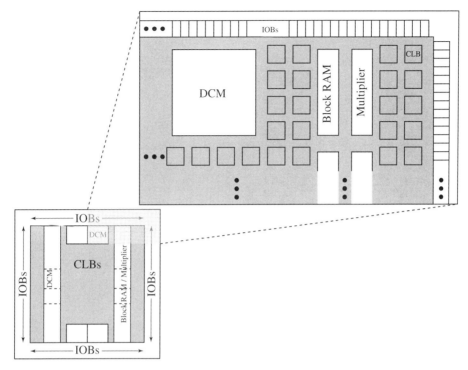

FIGURE 10-16.

Xilinx Spartan platform.

These elements are organized as shown in Figure 10-16. Each RAM column block has several 18K-bit RAM blocks, and each block RAM is associated with a dedicated multiplier. The circuit has a network of traces that interconnect all five functional elements. Each functional element has an associated switch matrix that permits multiple connections to the routing. We describe in more detail the circuit CLBs and the interconnect system.

Spartan Configurable Logic Blocks

The CLBs constitute the main logic resource for implementing combinational and synchronous logic. Each CLB contains four slices, and each slice contains two LUTs to implement logic. The CLB also has two dedicated storage elements that can be used as flip-flops or latches. The LUTs can be used as a 16×1 memory or as a 16-bit shirt register. Additional multiplexers and carry logic simplify wide logic and arithmetic functions. The CLBs are arranged in a regular array of rows and columns. Each CLB has four interconnected slices (Figure 10-17a). These slices are grouped in pairs, and each pair is organized as a column with an independent carry chain. The left pair supports both logic and memory functions, and its slices are called SLICEM. The right pair supports logic only and its slices are called SLICEL. Therefore, half the LUTs support both logic and memory while half support logic only, and the two types alternate throughout the array columns.

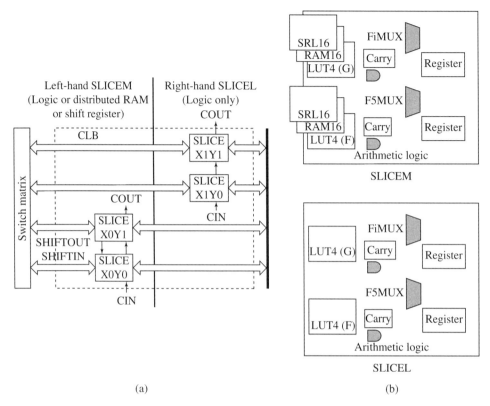

(a)   (b)

FIGURE 10-17.

(a) Arrangement of slices within the CLB. (b) Resources in each slice.

A slice includes two LUT function generators and two storage elements, along with additional logic, as shown in Figure 10-17b. Both SLICEM and SLICEL have the following elements in common to provide logic, arithmetic, and ROM functions:

- Two 4-input LUT function generators, F and G
- Two storage elements
- Two wide-function multiplexers, F5MUX and FiMUX
- Carry and arithmetic logic

The SLICEM pair supports two additional functions:

- Two 16 × 1 distributed RAM blocks, RAM16
- Two 16-bit shift registers, SRL16

The combination of a LUT and a storage element is known as a logic cell. The additional features in a slice, such as the wide multiplexers, carry logic, and arithmetic gates, add to the capacity of a slice, implementing logic that would otherwise require additional LUTs. For a detailed view of the slice circuitry, we refer the reader to the Xilinx literature [5].

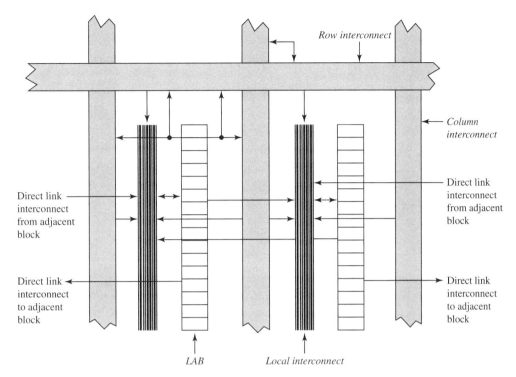

FIGURE 10-18.

Altera Cyclone III LAB structure.

10.3.3. Altera Cyclone III FPGAs

Similar to the FPGAs from other manufacturers Altera Cyclone III FPGAs combine a number of logic blocks (called LE) organized in LAB, memory blocks, embedded multipliers and digital processing (DSP) systems, interconnect structures, and I/O blocks. We describe some of these subsystems in detail.

Logic Blocks

The basic LEs are arranged in LABs that interface through global and local interconnect structures. The LAB structure is shown in Figure 10-18. Each LAB is formed by 16 Logic Elements, and has both a local and a global Interconnect structure. Local interconnects connect signals between LEs in the same LAB, and to adjacent blocks (e.g., either other neighboring memory blocks, embedded multipliers, or LABs) through specific lines called direct link interconnect. Global row and column interconnects can connect any resources within the chip. Each LAB contains dedicated logic to manage specific control signals like clocks, clock enable, clears (synchronous and asynchronous) and load.

Logic elements are the smallest units of logic and are LUT based. They have one programmable register that can be configured as a D, T, JK, or SR flip-flop, while the LE

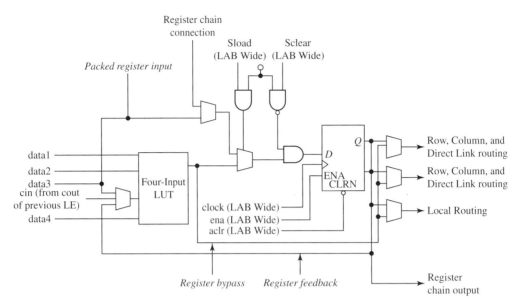

FIGURE 10-19.

Cyclone III LE in normal mode.

outputs can drive either local or global row and column interconnects. Each LE has six inputs available (four data inputs and two chain connections from previous LE) and can operate in two modes: normal mode or arithmetic mode.

The LE normal model is used for general logic applications and combinational functions (Figure 10-19). In this mode, the four data inputs arrive to the LUT module whose output can be directed to the available outputs for combinational logic functions or driven into the flip-flop for sequential logic.

When in arithmetic mode the LE is configured to implement adders, counters, accumulators, and comparators. In this mode, the LE implements 2-bit full adders and basic carry chain (Figure 10-20).

Other elements

The Altera Cyclone III FPGAs implement embedded memory structures organized in columns of memory blocks, called M9K blocks, that can be configured to get RAM, shift registers, ROM, and FIFO buffers. Each M9K memory block has 8.192 memory bits that can be configured either in single bit mode or words of 2, 4, 8, 9, 16, 18, 32, and 36 bits. They allow implementing single or dual port memory blocks. The memory blocks can also work in shift register mode for DSP applications, ROM mode or FIFO buffer mode.

The chip also contains embedded multiplier blocks that can be configured either as one 18×18 multiplier or two 9×9 multipliers. For operations larger than 18×18, multiple

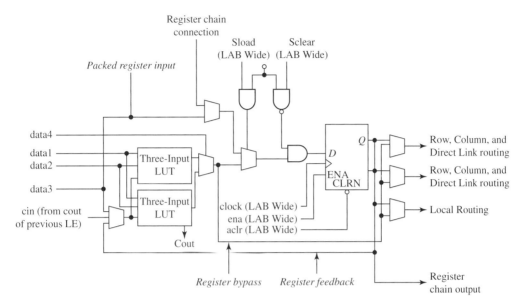

FIGURE 10-20.

Cyclone III LE in arithmetic mode.

embedded multiplier blocks are cascaded together. The number of embedded multipliers in a chip ranges from 23 to 396.

10.3.4. Today's FPGAs

Today's FPGAs are much more advanced than the structures detailed in the previous section. These circuits include hundreds of thousands of logic blocks, more than a thousand user I/Os, memory blocks with capacities beyond 20 Mbits, multi-gigabit high speed serial transceivers, and include various embedded processors blocks as part of their structure. They are manufactured in advanced technologies.

There are also a variety of FPGAs depending on the final application and contain specialized features for high-end, low-power, low-cost, mixed-signal, or radiation-tolerant devices. They have specific markets such as aerospace or automotive applications. Some companies also maintain their own antifuse-based devices.

10.3.5. Working with FPGAs—Design Tools

Implementing a design within an FPGA is done by following a design flow using a set of procedures that allow designers to cover the gap between specifications and final design in a systematic way. A detailed study of design flow techniques is out of the scope of this book, and only a brief comment is given here.

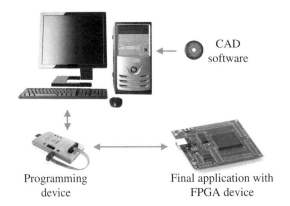

FIGURE 10-21.

FPGA design and programming sequence.

FPGAs are programmed through computer-aided design (CAD) tools that ultimately interface the circuit and configure the internal connections to obtain a given application. Circuit designs are done through the CAD tools, and FPGA design flow shares many steps with full-custom (or ASIC) design practices, although generally FPGA vendors provide their specific tools to design an application and program the device. In some cases, it is possible to interface a given FPGA vendor tool with generic tools.

CAD tools allow designing a circuit through Hardware Description Languages (HDL) that provide either a behavioral (the circuit is described by writing a code that specifies its desired behavior) or a structural (the "physical" structure of the circuit is described) specification. It is also possible to enter the design through a schematic description. The CAD tools allow simulating the behavior of circuits containing a mix of blocs specified by any of the given choices.

Once the designer validates the simulation, the CAD tool "synthesizes" the circuit and programs the FPGA by interfacing the board containing the FPGA (Figure 10-21). Such a board contains a specific bus to interface the FPGA for programming. Once the FPGA is programmed, the circuit is available for its final application. Such a design cycle can be repeated if further modifications to the FPGA function are required, thus providing the flexibility of what is called "in-system programming."

10.4. Understanding the Programmable Technology

There are three main technology options that sustain programmable logic circuits, and they offer two main types of characteristics: reprogrammable or one-time programmable (OTP), and volatile or nonvolatile. We describe the technology basis of the three main devices that support programmable circuits: antifuse, double-gate transistor, and SRAM-based programmable circuits.

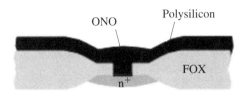

FIGURE 10-22.

Poly-diffusion antifuse structure and cross-section picture.

10.4.1. Antifuse Technology

The concept of antifuses dates back at least to 1957, when they were considered for use in memories [5]. An antifuse device is a normally open circuit that is short-circuit when a programming current is forced through it. It is constructed by inserting an insulator between two conducting layers that melts when a high current density causes a large power dissipation in its small area. It is a one-time programmable technology since the insulating material is destroyed and cannot be restored. There are two main structures used to build antifuses: polysilicon-diffusion and metal-metal antifuse.

Poly-diffusion antifuse

One of the most common poly-diffusion antifuse structure inserts an oxide-nitride-oxide (ONO) layer between a polysilicon and a heavily n^+ doped diffusion (Figure 10-22). This structure is typically known as a programmable low impedance circuit element (PLICE) [5]. It is constructed by growing a silicon dioxide (SiO_2) layer over the n-type antifuse diffusion, a silicon nitride (Si_3N_4) layer, and another thin SiO_2 layer. Such an ONO structure produces a tighter spread of blown antifuse resistance values and also improves both the yield and the reliability compared to simple oxide antifuses. The programming process melts the dielectric structure by applying a high electric field inducing a current that also drives dopant atoms from the polysilicon and diffusion electrodes into the link. The final doping level into the link determines the antifuse resistance.

Typically once the PLICE is programmed, a single link is observed and its radius increases with programming current, thus lowering the resistance. For a minimum area PLICE is programmed with 5 mA. The resistance is distributed as shown in Figure 10-23 [5] with a mode of 600 ohms. With a programming current of 15 mA, the mode of the distribution is shifted down to about 100 ohms. The reliability of these structures has been shown to be equivalent to the typical CMOS process reliability.

Metal-to-Metal Antifuse

These structures use amorphous silicon sandwiched between two metal interconnect layers, and have the advantage that they don't need to be close to the silicon crystal and can be placed on the top conducting layers. The programming mechanism is similar to

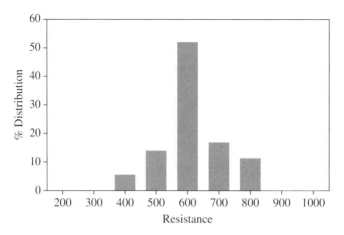

FIGURE 10-23.

Resistance distribution (in ohms) for a 5 mA programmed PLICE [5].

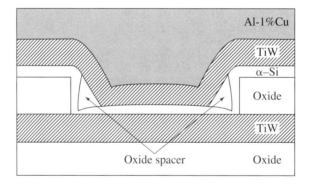

FIGURE 10-24.

Cross-section drawing of the metal electrode amorphous silicon antifuse [8].

the poly-diffusion structure, although it was initially observed that the antifuse, once broken down in the programming stage, could be switched back to the OFF-state during the operation when the direct current (DC) Read current became comparable or larger than the programming current [6]. This phenomenon, known as Read disturb mechanism, is avoided by keeping the read current below a critical level. Such a reliability problem was shown not to be a concern for product lifetime of 20 years [7].

The antifuse structure is shown in Figure 10-24 illustrating a cross section of the structure consisting of an amorphous silicon oxide on top of a dielectric material.

Metal-to-metal antifuses have some advantages over polysilicon-diffusion structures, and are related to their smaller size limited by lithography resolution, and their smaller parasitic capacitance. Figure 10-25 shows the programmed antifuse resistance distribution

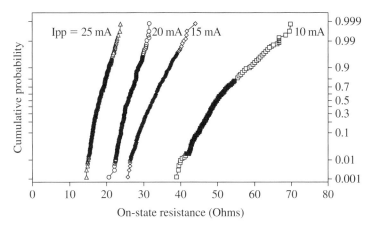

FIGURE 10-25.

On-state antifuse resistance distribution for various programming currents [7].

for various programming currents. It was shown that the programmed antifuses resistance is stable with temperature varying less than 15% per 100°C.

10.4.2. EEPROM Technology

The main drawback of antifuse technology resides in its one-time programmable property. Double gate (or floating-gate) devices are a special kind of transistors having one floating gate inserted between the control gate and the transistor channel that can be programmed and unprogrammed being nonvolatile. Flash memory devices contain the last step of floating-gate devices evolution referred to as EEPROM devices. EEPROM transistors are in turn the evolution of UVEPROM devices being transistors that could be electrically programmed (i.e., under a specific bias conditions) and where unprogrammed by using ultraviolet light (therefore they are called UV from ultraviolet for erasing, and E for electrically programming).

UVEPROM devices were not practical since they must be removed from the system to be erased before reprogramming, and required expensive packaging with a window through which the UV light could pass. EEPROM transistors used an electrical bias condition to program an erase (and therefore the EE nomenclature). They are more flexible since they can be reprogrammed in field.

The EEPROM transistor structure is shown in Figure 10-26, being a modification of an nMOS transistor structure. The control gate acts as the normal gate of the transistor, while the floating gate is surrounded by oxide and cannot be contacted directly. Programming the device consists in driving electrons into the floating gate where negative charge is accumulated. This charge in the floating gate shields the device channel from the control gate electric field in a way that a conducting channel is never created even if the control gate

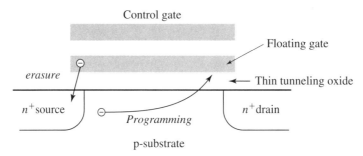

FIGURE 10-26.

Structure and schematics programming/unprogramming of an EEPROM device.

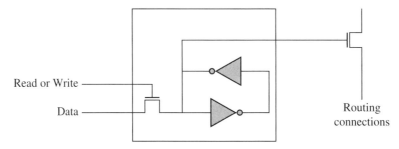

FIGURE 10-27.

SRAM-based programming technology structure.

is driven at VDD. Therefore, a programmed device never turns on while unprogrammed devices act as normal transistors.

Programming is done by driving the device in heavy saturation so that hot electrons accelerated in the channel collide with the substrate atoms and tunnel through the thin oxide to the floating gate. Unprogramming is done by reversing the voltage in the control gate and tunneling the electrons back from the floating gate to the source.

These floating-gate transistors are the double gate devices shown earlier in Figures 10-1 and 10-2.

10.4.3. Static RAM Switch Technology

This technology uses a typical cross-coupled inverter SRAM cell structure to store the conducting state of a pass transistor or transmission gate embedded within the FPGA structure as shown in Figure 10-27.

The advantage of this technology resides fundamentally in its short programming time compared to floating gate devices. The negative aspects relate to its relatively larger area since transistor count is higher and the need to reprogram the circuit after power-up. Early FPGAs had to be programmed at each power-up from external ROM or flash circuitry.

Today, many FPGAs based on SRAM technology include flash memory blocks that retain the circuit configuration and internally program the circuit at power-up.

References

[1] www.altera.com.

[2] www.cypress.com.

[3] www.xilinx.com.

[4] www.actel.com.

[5] J. Greene, E. Hamddy, and S. Beal, "Antifuse field programmable gate arrays," *Proceedings of the IEEE*, Vol. 81, No. 7, pp. 1042–1056, July 1993.

[6] S. Chiang, R. Forouhi, W. Chen, F. Hawley, J. McCollum, E. Hamdy, and C. Hu, "Antifuse Structure Comparison for Field Programmable Gate Arrays," *IEDM Tech. Dig.*, pp. 1–4. 1992.

[7] C. Shih, R. Lambertson, F. Hawley, F. Issaq, J. McCollum, E. Hamdy, H. Sakurai, H. Yuasa, H. Honda, T. Yamaoka, T. Wada, and C. Hu, "Characterization and modeling of a highly reliable metal-to-metal antifuse for high-performance and high-density field-programmable gate arrays," *IEEE International Reliability Physics Symposium*, pp. 25–33, 1997.

[8] R. Wong, K. Gordon, and A. Chan, "Time dependent reliability of the programmed metal electrode antifuse," *IEEE International Reliability Physics Symposium*, pp. 22–26, 1996.

CHAPTER 11

CMOS Circuit Layout

Integrated circuits use a photochemical process and physical implant techniques to manufacture the metal lines, drains, sources, gates, wells, polysilicon, and dielectric structures. Designers specify these structural dimensions, such as the transistor gate width and length or the metal width. But the process engineer specifies the minimum spacing between the elements where minimum spacing means smaller die size. These minimum dimensions form a list of design rules (DR) for a given technology. We cannot arbitrarily let two metal lines get too close, or they can merge or bridge to each other during manufacturing. The process engineer guarantees that if the design rules are not violated, then the circuit can be manufactured without isolated structures touching into each other. Structures violating DRs may merge causing electrical shorts or create other problems such as charge leakage.

The integrated circuit (IC) layout exercises in this chapter use PowerPoint for simplicity in learning about design rules. However, other layout tools exist and can easily be used as an instructor's preference. L-edit is a relatively easy to learn tool and is used by many instructors. The layout figures use a typical industry color code scheme. Color images of the layout are shown in the inside front and back covers.

11.1. Layout and Design Rules

The plan view (top-down view) of a portion of an nMOS transistor is shown in Figure 11-1 with color-coded elements. The intersection of poly and diffusion layout regions defines a transistor. Notice how the red polygate overlaps the green active regions of the n-diffusion drain and source. The gate poly line extends beyond the active drain-source diffusion regions in a minimum distance called the endcap. The poly must extend at least this distance beyond the diffusion regions to eliminate charge leakage between the drain and source. This minimum distance specification is a design rule. Other design rules include minimum poly dimensions (transistor minimum width and length), minimum metal and contact sizes, overlap of diffusion on contacts, and spacing of contacts to poly.

FIGURE 11-1.

Partial layout of an *n*MOS transistor showing the vertical polygate (red), *n*-diffusion region (green), contacts (gray), and metal-1 (blue). The colors refer to images on the inside front cover.

The (blue) metals must overlap the (gray) contacts by a specified minimum distance (DR). Design rules form a list of minimum allowed dimensions. A CMOS fabrication process has a long list of DRs, and we later present a representative reduced set of DRs to guide layouts of various logic gates. The layout plan view allows DR and circuit density examination.

Figure 11-2 shows the layout of a CMOS inverter. For illustration, it is laid out without strict adherence to minimum DR dimensions. One key to reading a layout is to identify structural markers, such as the poly signal input line-A (red), the V_{DD} and GND, and the transistor types whose sources are tied to V_{DD} and GND. The two power lines (V_{DD}, GND) are the larger flat horizontal (blue) lines at the top and bottom. They are usually wider than the signal lines and the poly lines since they carry more current and line resistance

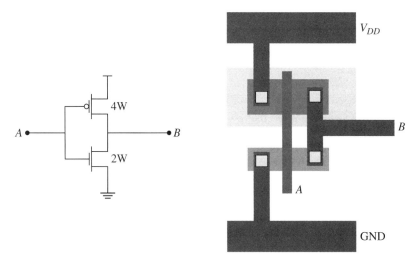

FIGURE 11-2.

Color-coded layout of CMOS inverter in *n*-well technology. The poly is red, the *n*-diff is green, the *p*-diff is brown, the metal-1 is blue, the *n*-well is yellow, and the contacts are metal grain gray. The colors refer to images on the inside front cover.

must be kept low. The vertical polysilicon line (red) is the logic gate input. Typically, all transistors on the IC have the same minimum gate length dimension.

The poly line-A short dimension between the drain and source active regions defines the gate length of the two transistors. It is the narrow dimension of the red poly line. The diffusion region height along the poly line defines the transistor width.

A third clue is to identify pMOS and nMOS transistors by their source contacts to V_{DD} and GND. Notice the p-channel transistor source region is connected to V_{DD} by a contact to metal-1. The source of the nMOS transistor is connected to GND by a contact to the metal-1 GND line. Once the transistor sources are identified, then the drain regions are on the other side of the polygate line. These structures are clues to quickly locating markers, reading a layout, and reconstructing the circuit. Designers often reconstruct layouts to transistor schematics when debugging a design.

Remember that a plan view layout represents several mask layers each with a separate function. The layout represents a three-dimensional structure as seen from the top down. The metal-1 drawing has a single mask as do the contacts, vias, well, polygate, and drain and source diffusion regions.

11.2. Layout Approach: Boolean Equations, Transistor Schematic, and Stick Diagrams

The layout process begins with a Boolean statement specifying an operation. The Boolean equation is transformed to a schematic with sized transistors and then to a layout. We will show how to lay out a circuit manually using Microsoft PowerPoint. PowerPoint is a universal drawing tool that is simple to use, and it will force manual design rule checking.

Our manual process will use an intermediate tool between the transistor schematic and layout called a *stick diagram*. A stick diagram is a simple shorthand layout description of the schematic showing transistor location, power lines, signal lines, and contacts. The stick diagram has a spatial property but does not include element sizes and design rules.

The transfer of a stick diagram to a real layout simply adds design rules and structural dimensions. A layout is a stick diagram with dimensions. Examples with an inverter, 2NAND, and 2-NOR gates will illustrate the technique.

An inverter schematic and its stick diagram on the right are shown in Figure 11-3. The stick diagram shows two thick horizontal lines representing the n-diffusion in the bottom line and the p-diffusion on the top line. The vertical line in the middle is the polysilicon gate with input-A. The thin lines are metal-1 and the output signal line is the thin horizontal metal line connected to the output-B. The $\times$ symbol is a contact connecting metal-1 to transistor silicon diffusion regions. A transistor is formed when poly intersects diffusion so the pMOS transistor is at the top intersection, and the nMOS transistor is at the bottom intersection. The source of the pMOS transistor is connected to V_{DD} and the source of the nMOS transistor is connected to GND. The two drain regions are connected to metal-1 contacts and form the output-B.

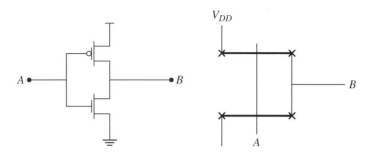

FIGURE 11-3.

Inverter schematic and stick diagram.

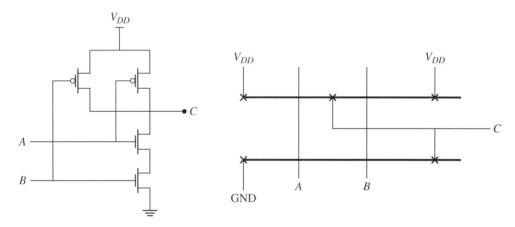

FIGURE 11-4.

2NAND gate schematic and stick diagram.

The 2NAND logic gate schematic and stick diagram in Figure 11-4 have two pMOS transistors shown connected in parallel in the schematic and in the stick diagram. The poly input-A and input-B are shown. The signal output is taken from the shared pMOS and nMOS drain regions, and the sources are connected at the ends of the poly line. We could put the V_{DD} line in the middle of the two pMOS transistors and take the signals from the ends of the poly regions. However, signal metal has a shorter length when taken from the middle thus reducing signal capacitance. Capacitance on the V_{DD} and GND rails is good since it helps buffer transients on the power rails.

The two nMOS transistors are connected in series in the schematic. They are also in series in the stick diagram. The GND contact is at one end of the bottom diffusion line and the signal output at C at the other end. Notice how the nMOS transistors in series share a common n-diffusion region for the drain of one transistor and the source of the other. This is a standard practice for series stransistors.

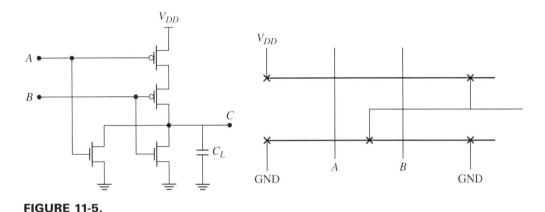

FIGURE 11-5.

2NOR gate schematic and stick diagram.

The 2NOR logic gate schematic and stick diagram have two *p*MOS transistors connected in series (Figure 11-5). They share a common *p*-diffusion region. The VDD contact is at one end of the top diffusion line, and the signal output at C at the other. The two *n*MOS transistors are connected in parallel. The signal output is taken from the shared drain regions, and the sources are connected to GND at the ends of the poly line. Again, we could put the GND line in the middle of the two *n*MOS transistors and take the signals from the poly ends, but signal metal is shorter in Figure 11-5, thus reducing signal capacitance.

11.3. Laying out a Circuit with PowerPoint

Special computer-aided design (CAD) tools exist to perform the layout artwork. Tanner Tools (L-Edit, Ureyama), MAGIC (University of California–Berkeley), Mentor Graphics, and Cadence are four CAD tools found in industry and universities to create mask patterns from transistor schematics. These tools have a stored design rule error-checking feature that frees engineers from the tedium of measuring all geometric spaces between structures. While these tools are efficient, some have a significant learning time in an undergraduate class. In addition, there is no uniform tool that all universities accept, and these CAD tools can disappear in time.

We therefore found a simple expedient using Microsoft Office PowerPoint as a layout tool. The instruction is rapid, its distribution is universal, and it has the capability to implement simple logic gate color-coded layout. It has no design rule checker, but that is a learning advantage to understand exactly what DRs imply. As circuits became more complex, manual design rules checking became numerically impossible, so we must use special CAD tools.

Next we describe the mechanics of drawing the layout for a CMOS inverter. The process uses the transistor schematic and its W/L values, a stick diagram, and the conversion of the stick diagram to a physical plan view of the circuit. Color-coded rectangles represent the *n*-well, *p*-diff, *n*-diff, polygate, metal-1, and contacts. We will show how to reuse standard layouts of CMOS inverters, NAND gates, and NOR gates to build more complex gates without having to redraw each standard logic gate.

11.4. Design Rules and Minimum Layout Spacing

A sequence of partial layouts of the inverter components will show how design rules and transistor dimensions are folded into the final layout. We will layout the *p*MOS transistor first, the *n*MOS transistor second, then assemble both transistors to a common polysilicon gate, and finally add the poly contact and final metal-1ines. The inverter will obey the design rules listed in Table 11-1.

Table 11-1 lists a set of CMOS design rules for a 2 μm technology in single level metal and *n*-well. The word *active* refers to the *p*- or *n*-diffusion regions. Although 2 μm is a much larger gate dimension than typically used today, the 2 μm DRs have integer numbers making the drawings a little easier to check. The principles are the same as for smaller nm technologies, and integers are easier to deal with in the manual layout exercises later in the chapter.

TABLE 11-1 2 μm Design Rules (DR)

DR #	DR Description	DR
1	*n*-well enclosure of *p*-type active	5 μm
2	*n*-well space to *p*-type active	3 μm
3	*n*-well space to *n*-type active	5 μm
4	*n*-well width	5 μm
5	*n*-well space to *n*-well	5 μm
6	active width	3 μm
7	active space to same type active	3 μm
8	active space to opposite type active.	5 μm
9	polysilicon width	2 μm
10	polysilicon space to polysilicon	2 μm
11	polysilicon space to active	1 μm
12	gate space to gate	4 μm
13	polysilicon endcap length	2 μm
14	contact width	2 μm
15	contact space to contact	4 μm
16	active enclosure of contact	1 μm
17	polysilicon enclosure of contact	2 μm
18	polysilicon contact space to active	3 μm
19	active contact space to polysilicon	2 μm
20	metal-1 width	2 μm
21	metal-1 space to metal-1	3 μm
22	metal-1 enclosure of contact	1 μm

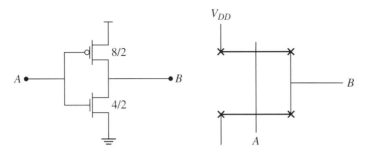

FIGURE 11-6.

Inverter sizes and stick diagram.

Design and manufacturing engineers must read layouts and convert them to transistors and their logic gates. Figure 11-6 shows an inverter stick diagram that will be converted to a layout with minimum design rule spacing in an *n*-well technology. The inverter transistor sizes are shown in the schematic.

11.5. Laying out a CMOS Inverter

11.5.1. *p*MOS Transistor Layout

How do we scale the 2 μm or 8 μm to the PowerPoint page? We could set 1 μm = 1 inch of PowerPoint, but that would make the layout larger than the paper size. But a scaling of 1 μm = 0.1 inch on PowerPoint will work.

There are three ways to size and check dimensions in PowerPoint.

1. PowerPoint will let you specify the exact dimensions of a rectangular box. Click on the box to select it and then click on "format" at the top of the monitor screen. Click on *autoshape* at the bottom of the menu and select *size*. You can type in exact dimensions of the box. This works for diff, polygates, contacts, metal, or any arbitrary shape.
2. Use the PowerPoint "snap to grid" feature so that all dimensions are referenced to a small grid cell. Let 1 μm equal one-grid step. Then the *p*-diff height is 8 μm or 8 grid steps. If your PowerPoint doesn't have a "show grid" feature then this method is tedious.
3. Another method makes calibration squares the size of the design rules. For example, make four squares of dimension 1 μm, 2 μm, 3 μm, and 5 μm. Let 1 μm = 0.1 inch on PowerPoint. These calibrated squares can be positioned anywhere on the layout with the mouse to check a dimension (DR). In practice this is an easy technique for simple circuits.

A *p*MOS transistor is drawn in Figure 11-7 with *n*-well, *p*-diff, polygate, contacts, and metal-1 to V_{DD}. The spacings and dimensions are drawn to minimal size layout. The *p*-diff rectangle is drawn first where its vertical dimension is the *p*MOS transistor channel width

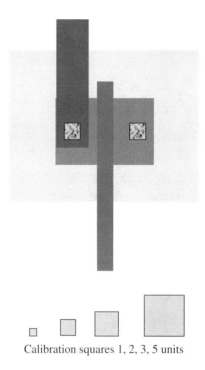

Calibration squares 1, 2, 3, 5 units

FIGURE 11-7.

*p*MOS transistor layout with color code poly (red), *p*-diff (brown), *n*-well (yellow), metal-1 (blue), and contact (metal grain gray). $W/L = 8\ \mu\text{m}/2\ \mu\text{m}$. The colors refer to images on the inside front cover.

of 8 μm. The final *p*MOS transistor layout has two contacts and a metal-1 connection to V_{DD}. In addition to an open PowerPoint, follow file to draw these structures as they are described using the four calibration squares to ensure DR adherence.

Procedure for laying out a *p*MOS transistor (Figure 11-7):

1. Draw the *p*-diff active region that is 8 μm high and about twice that in width. We will adjust the active region width later. Color-code the *p*-diff *brown* as in the figure using the fill command.
2. Insert a contact on the *p*-diff left side. The contact is $2 \times 2\ \mu$m (DR-14). DR-16 states that the active *p*-diff active region must enclose or overlap the contact by at least 1 μm. Place the contact in the vertical middle of the *p*-diff and 1 μm from the left edge. Use the calibration squares to check your DRs. Make sure your "snap to grid" command is off; otherwise you can't smoothly control the calibration box position.
3. DR-19 states that the polygate must be at least 2 μm from the contact. Draw a vertical poly line that is 2 μm in horizontal dimension (DR-9), and place the poly top edge 2 μm above the top of the *p*-diff (endcap, DR-13). Color-code the poly a *red* color. Place the poly line 2 μm to the right of the source contact edge. Run the poly bottom edge an arbitrary distance below the lower edge of the *p*-diff.

4. Place the drain contact 2 μm to the right of the poly (DR-19) and extend the *p*-diff to overlap the contact by 1 μm (DR-16). Color-code the contact using the designated *metal grain gray* fill pattern.

5. Draw a vertical metal-1 line from the *p*-MOS source contact arbitrarily about 6 μm higher than the *p*-diff. This will later connect to a metal-1 VDD line. The metal-1 must overlap (enclose) the contacts by 1 μm (DR-22).

6. Draw an *n*-well (*yellow*) around the *p*-diff. It must extend beyond the *p*-diff edges by 5 μm (DR-1).

Your figure will be clearer if you do not put a line border around the rectangles. However, a line around the contacts may add clarity. Be aware that color coding is not standard, and that the code may differ with specific layout tools. That is not an important issue.

11.5.2. Revisiting the Design Rules of the *p*MOS Transistor Layout

A layout sequence will reinforce these important points about DRs just described. Figure 11-8 shows three stages in the evolution of the *p*MOS transistor. The first picture on the left shows the contact placed in the *p*-diff, the second picture shows the addition of the poly line and its DRs, and the third picture shows the addition of metal-1 to the source contact. Relevant DR spacings are indicated.

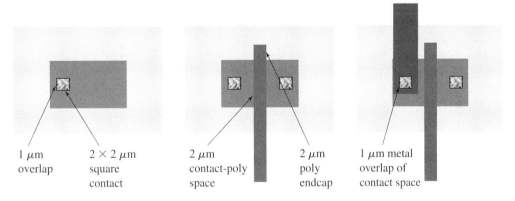

| 1 μm overlap | 2 × 2 μm square contact | 2 μm contact-poly space | 2 μm poly endcap | 1 μm metal overlap of contact space |

FIGURE 11-8.

Layout stages for *p*-MOS transistor illustrating design rules. The colors refer to images on the inside front cover.

11.5.3. *n*MOS Transistor Layout

Figure 11-9 shows an *n*MOS transistor layout taken from stick diagram in Figure 11-6. The polygate line (red), the *n*-diff (green), and source metal to GND (blue) are drawn.

The procedure for laying out an *n*MOS transistor (Figure 11-9) is as follows:

1. Draw the *n*-diff rectangle with a vertical height of 4 μm, which is the transistor channel width. Color code the *n*-diff region green and make it 4 μm (0.4 inch) wide.

Calibration squares 1, 2, 3, 5 units

FIGURE 11-9.

nMOS transistor layout with S-D contacts, polygate, and GND metal-1 line. The $W/L = 4\,\mu$m$/2\,\mu$m. The colors refer to images on the inside front cover.

2. The source contact is drawn on the left edge obeying the same DRs and size as the pMOS transistor. The contact is $2 \times 2\,\mu$m (DR-14).
3. Draw the $2\,\mu$m wide polygate line extending beyond the bottom of the n-diff by $2\,\mu$m (DR-13).
4. Add the $2\,\mu$m contact $2\,\mu$m to the right of the poly line (DR-14, 19).
5. Draw the metal-1 overlapping the source contact by $1\,\mu$m (DR-22). Extend the metal down and draw a horizontal GND line. GND and VDD are typically wider than the minimum width signal lines because current is higher in power lines. The VDD and GND lines are drawn with an arbitrary width for these examples.

11.5.4. Merging Transistors to a Common Polygate

The pMOS and nMOS drain diffusion regions have the same horizontal dimension and line up (Figure 11-10). This makes it easier to drop the metal-1 line connecting the two drain contacts. This partial layout is made by lining up the polygate lines of the previous two layouts for the pMOS and nMOS transistors. Delete one of the poly lines and extend the other to get one continuous connection poly for both transistors. The n-well must be at least $5\,\mu$m from the n-diff active region (DR-3). The metal-1 signal line leading to node-B is thinned to reduce capacitance. The current density in this line is small compared with VDD and GND lines that are arbitrarily widened in this example.

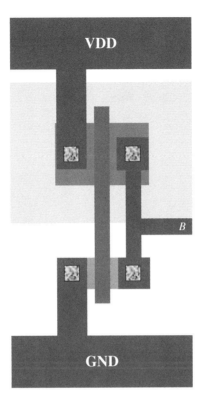

FIGURE 11-10.

*n*MOS and *p*MOS transistors layout with merged polygate line, output node (B), and VDD and GND metal-1 lines. The colors refer to images on the inside front cover.

11.6. Completed CMOS Inverter Drawn to Design Rule Minimum Dimensions

The polygate contact is added next (Figure 11-11). DR-17 states that there must be 2 μm of poly enclosure space around a poly contact. The metal must have 1 μm of space outside the contact (DR-22). The poly near the contact must have a separation of 1 μm to a neighboring active (*p*-diff) region (DR-11), and it actually has 2 μm in the layout.

The poly line input-A entered from the left side of the layout. A different style extends the vertical polygate lines beyond the edges of the VDD and GND lines. It is often more efficient to run signal routing lines in the channel region above and below the logic gate. Poly contacts can be made with either layout style depending on design connections. Both styles are useful.

You should go down the DR list in Table 11-1 and check spacings with the calibration boxes. If a DR violation escapes the layout process, then the DR violation can cause the IC to fail or be seriously impaired. The DR violation must be corrected, which means making a new mask and sending the design through another fab cycle. This is expensive

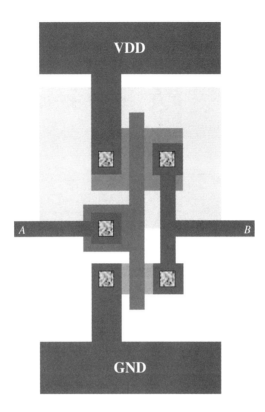

FIGURE 11-11.

Completed inverter layout. The colors refer to images on the inside back cover.

in money for masks and engineering labor and also in delaying time to market. Industry refers to design success as "first silicon, second silicon, third silicon, . . . or cancel project for excessive delays to market."

11.7. Multi-Input Logic Gate Layouts

It is an incremental step from inverter layouts to multi-input logic gates. The major addition is accommodating more polygate input lines, and transistor merging or sharing common diffusion regions. Additional poly input lines require more design rules.

2NAND Gate Layout

The 2NAND schematic and stick diagram are shown as preliminary information to laying out the 2NAND (Figure 11-12). It is easiest to build the 2NAND layout by copying and expanding the previous inverter layout.

Table 11-2 reproduces the 2 μm DRs to reduce back paging when following the multi-input logic gate layout. The 2NAND gate layout will build on the previous inverter layout to reduce the labor.

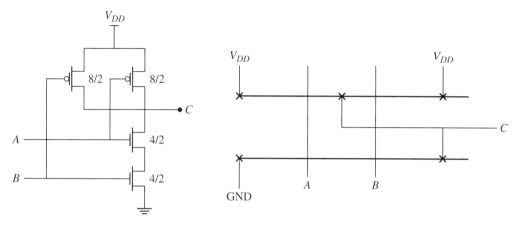

FIGURE 11-12.

2NAND gate schematic and stick diagram.

TABLE 11-2 2 μm Design Rules (DR). Reproduced from Table 11-1

DR #	DR Description	DR
1	n-well enclosure of p-type active	5 μm
2	n-well space to p-type active	3 μm
3	n-well space to n-type active	5 μm
4	n-well width	5 μm
5	n-well space to n-well	5 μm
6	active width	3 μm
7	active space to same type active	3 μm
8	active space to opposite type active.....	5 μm
9	polysilicon width	2 μm
10	polysilicon space to polysilicon	2 μm
11	polysilicon space to active	1 μm
12	gate space to gate	4 μm
13	polysilicon endcap length	2 μm
14	contact width	2 μm
15	contact space to contact	4 μm
16	active enclosure of contact	1 μm
17	polysilicon enclosure of contact	2 μm
18	polysilicon contact space to active	3 μm
19	active contact space to polysilicon	2 μm
20	metal-1 width	2 μm
21	metal-1 space to metal-1	3 μm
22	metal-1 enclosure of contact	1 μm

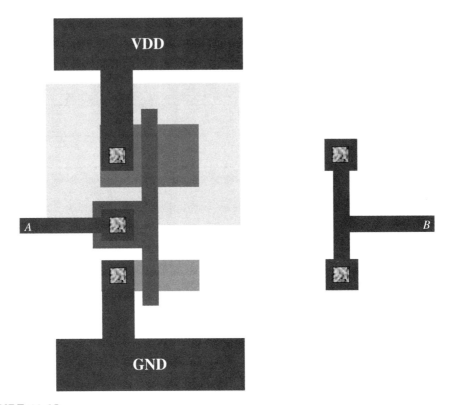

FIGURE 11-13.

Inverter with drain output and right side metal and contacts moved to the right and out of the way. The colors refer to images on the inside back cover.

Figure 11-13 shows the inverter layout in which the drain output metal-1 and drain contacts are selected and dragged out of the way. The gate layout is for the transistors of Input-A. The p-diff and n-diff regions must be widened to accommodate the second poly input line and the contact in the middle p-diff region. The poly-to-poly spacing is a minimum of 2 μm (DR-10). The actual spacing is 8 μm because of the shared contact of the pMOS transistors.

Figure 11-14 shows the completed 2NAND layout when adding the second poly line, the shared middle p-drains and contacts, and the new metal-1 to VDD, GND, and the output-C. The layout follows the 2NAND stick diagram. The design rules from Table 11-2 were enforced in these new spacings.

Self-Exercise 11-1
Draw the transistor schematic, stick diagram, and layout for a 2NOR gate. Use the design rules of Table 11-2 and keep the power rails height and width the same as previous layouts.

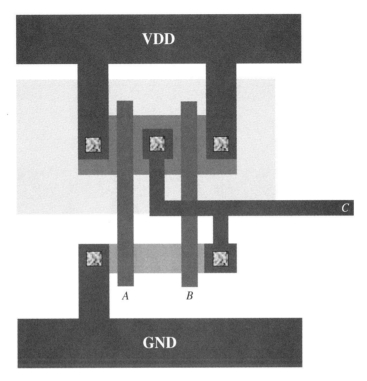

FIGURE 11-14.

Completed 2NAND layout. The colors refer to images on the inside back cover.

Self-Exercise 11-2
If the n-diff region minimum height design rule in Table 11-2 were changed from 4 μm to 3 μm, what would be the impact on the circuit?

11.8. Merging Logic Gate Standard Cell Layouts

The layouts of the inverter, 2NAND, and 2NOR can be used as building blocks to build more complex functions without relaying out each basic gate. In integrated circuit language, these basic gate layouts are called *standard cells*. For example, a 2AND layout can be made from a 2NAND layout in series with an inverter layout (Example 11-1). Since we left the power rails at the same height and width and separation distances, these cells will easily couple. These power rails are said to be at the same pitch. We will leave the transistor sizing intact to make the coupling easier. These transistor W/L ratios are not optimal for achieving maximum transition times for worst-case pull-up or pull-down states, but we express the principle in its simplest form. An example with a 2AND will demonstrate. The poly inputs A and B are shown vertical and are unconnected. Typically, they might be brought out above and below the power rails allowing easy connections in the channel space outside the main layout.

EXAMPLE 11-1

Draw the stick diagram for a 2AND using a 2NAND in series with an inverter. The *p*MOS and *n*MOS *W/L* are 8/2 and 4/2. See inside back cover for color.

First copy and align the layout of the two basic circuits as

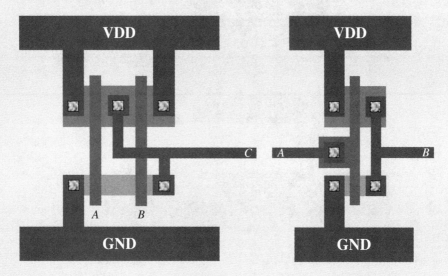

Then join the circuits observing all design rules to get

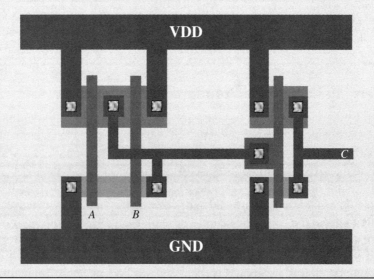

EXAMPLE 11-2

Write the Boolean equation and draw the transistor schematic for this stick diagram. Note GND and VDD lines are drawn.

$$F = \overline{(AB) + (CD)}$$

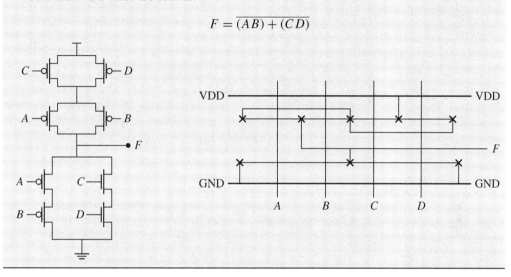

Self-Exercise 11-3
Draw the transistor schematic, stick diagram, and layout for a 2XOR gate. Use the design rules of Table 11-2 and keep the power rails height and width the same as previous layouts.

11.9. More on Layout

The basics of layout are: What is a layout, what are design rules, how do you make a plan view figure representing several masks, and how do you reverse engineer a schematic from the layout? Since inverter, NAND, NOR, and other logic gates are so abundant, it makes sense to layout basic reusable circuits when many millions of logic gates are used. Individual layout of millions of gates would stop the project. For die of all sizes, automated assembly of logic gates and wire interconnections allows a shorter time to market. So the common sense approach carefully lays out individual logic gates and replicates them. This also allows a computer to rapidly assemble the design.

While the standard cells are indispensable for rapid design of moderns ICs, they also have disadvantages. Individual sizing of transistors is awkward or not possible to optimize rise and fall times, or to minimize layout area. Therefore, it is a common practice to hand-craft certain sections of the design that call for high-speed data calculations. An arithmetic logic unit (ALU) might be such an example. Hand crafted layouts are called *custom layouts*. The objective is to minimize circuit area (and load capacitances) and optimally size the transistors for speed. The industry rapidly moved to treating larger macrocells as it does standard logic gate layouts. Actual microprocessor designs are available and are imported as standard cells in a part of a larger layout assembly.

11.10. Layout CAD Tools

Automatic design rule checking is a standard layout feature. Since the x-y coordinates and material type of every rectangle (polygon) are known, a computer can measure distances between different layers or on the same layer. A computer can do logical AND operations between coordinates of different mask layers and detect the presence of a transistor, contact, and metal lines. The AND operation of a particular diffusion layer with the polysilicon layer identifies location and type of a transistor. This computer-based technique is called *circuit extraction*. A complete schematic can be drawn by the computer with these principles.

11.11. Summary

Designers get involved with layout and design rules especially when a violation is found. Designers may want to push a circuit for better performance by compromising design rules, but that is done with caution. The PowerPoint tool used here is educational. In real practice CAD tools make it easier. The CAD tool has many ways to reduce the tedium of manual layout. In addition, manual layout is more prone to error. However, the basic layouts done here should ingrain the role of the geometric shapes to actual transistor construction and increase your knowledge and attention to design rules.

CHAPTER 12

How Chips Are Made

It would appear that we have reached the limits of what it is possible to achieve with computing technology, although one should be careful with such statements, as they tend to sound pretty silly in five years.

John von Neuman, circa 1953

Chip making is a complex sequence of chemical, physical, and photolithographic operations. Engineers in the IC industry should know these basic operations and their language. Designers often must trouble shoot (debug) their designs when chips are returned, and fabrication problems are often the trouble.

The layout exercises in Chapter 11 relate to chip fabrication. A CMOS circuit requires different layout geometries for wells, diffusion regions (drain and source), polysilicon gates, contacts, vias, and metal interconnections. Each circuit level with its unique rectangular shapes has a physical *mask* that is the heart of the photolithographic process in chip fabrication.

Integrated circuit fabrication blends the overall process with several sub-methodologies buried within the process. There are thousands of details that we won't attempt. Our learning objective is a visual image of the major methodologies and how they fit into the overall process. An initial overview of the whole process is followed by a description of the major steps. A concluding section walks through a sample fabrication of a CMOS inverter.

There are no problems or examples in this chapter. Our goal is a qualitative description. There are graduate and advanced undergraduate courses in chip fabrications that more fully develop the subject. Chip manufacturing is often referred to as chip fabrication (or just fab) or sometimes integrated circuit (IC) processing. The plant where this occurs is often referred to as a foundry.

Process operations remained relatively stable for many years. However, about the year 2000, significant changes in materials and equipment occurred as the technology nodes went to 130 nm and below. Significant technology changes are expected as nanometer size technology targets of 22 nm, 15 nm, and 11 nm are achieved.

The challenge is to understand how billions of transistors and several miles of metal interconnect can be put on a silicon chip that can be 1–2 centimeters square. Our goal is assimilation of the larger picture without dwelling on too many details. This gives better understanding of layout, interconnections, and CMOS transistor gates.

12.1. IC Fabrication Overview

A fab process begins with a silicon wafer that is typically 200 or 300 mm in diameter and about 500 μm thick. The wafer is a single crystal material usually with an n- or p-doping. During the subsequent process steps, a single wafer will form a few hundred or thousands of identical circuit die. Each die may contain millions or billions of transistors. A series of repetitive photolithography operations impress images on the wafer that become the circuit wells, diffusion regions (drain-source), thin oxide, polysilicon gates, contacts, vias, and metal layers. SiO_2 and Si_3N_4 are dielectrics that isolate node or line voltages, provide mechanical protection and insulation from airborne corrosive molecules and atoms, and are integral in supporting other constructions. SiO_2 and Si_3N_4 are critical in blocking certain doping atoms from implanting in a wrong region during the fab process.

Die that passed the wafer test are separated by a diamond saw operation and mounted in IC packages and tested again. The packages and test operations are time consuming, complex, and expensive. Some modern, large ICs may use 50% of the manufacturing cost on the test operation. Testing must guarantee that the IC meets the specifications expected by the customer, and as a subset of that the IC must be free of defects that escaped the test process or might become reliability failures in the field.

12.2. Wafer Construction

We are fortunate that silicon material for ICs is plentiful. Silicon is the second most plentiful element making up 27% of the earth's crust. Silicon is mined from beach sand that is SiO_2. Silicon is separated from the SiO_2 sand into aggregates of polysilicon crystals. These clusters of small silicon crystals are poured into a vessel heated above the melting point of silicon (1425°C). The molten Si bath is converted to a large single crystal by lowering a small crystal seed into the molten silicon. The seed grows larger as it is rotated and slowly withdrawn from the melt. p- or n-dopant elements are typically added to the melt to begin the electrical alteration of the crystal. The emerging gray crystalline Si cylinder is called an ingot (Figure 12-1). The process may take one week to one month to pull the crystal. Large diameter ingots can weigh a few hundred pounds.

The ingot is next mechanically rounded to a uniform diameter and sliced with a diamond saw wafers less than 1 mm thick (Figure 12-2). The wafers must be lapped, smoothed, and heated again to recrystalize the saw damage sites. The doped wafers are delivered to the customer as free from crystalline or contaminant defects as possible. This is critical.

FIGURE 12-1.

Ingot photo. (Reproduced by permission of Intel Corp.)

FIGURE 12-2.

An ingot illustrating a diamond saw cutting the ingot into wafers. (Reproduced by permission of Intel Corp.)

12.3. **Front and Back End of Line Fabrication**

Two distinct processing segments bring a chip from the wafer level to a die with all transistors and interconnections. The first is called the front end of Line (FOL) and the second is the back end of Line (BOL). The FOL refers to the construction of the transistors. Transistors are constructed first, and the BOL begins with the metal contact and metal layer operations. Many procedures in FOL and BOL are common and repeated throughout the process.

We will describe eight major operations common to all ICs. Many are repetitive operations.

1. Oxidation of silicon
2. Photolithography
3. Etching
4. Deposition–diffusion, CVD, and ion implantation
5. Cleaning and safety operations
6. Metal fabrication
7. Interlevel dielectric and final passivation
8. Packaging

There are many subcategories in each operation, but our goals are better attained by avoiding excessive details. Fab has several distinctions. Dielectric materials (SiO_2 and Si_3N_4) are formed on the wafer surface, and typically they are etched off in selected regions by a mask. When elements are deposited on the surface, we call this a *deposition*. When these atoms (As, B, P) are inserted beneath the surface we call this *implantation*. Sometimes the terms merge as when we lay down O_2 on a SiO_2 surface and then use diffusion at elevated temperature to drive O_2 beneath the surface to react with unexposed Si.

12.4. FOL Fab Techniques

12.4.1. Oxidation of Silicon

The first process step typically grows a thin 1 μm dielectric layer of silicon dioxide on the top layer of the Si wafer. SiO_2 has three uses. The first is that SiO_2 is an electrical insulator that separates and physically supports metal interconnections. It also electrically isolates transistors from each other. A second use of SiO_2 is the transistor gate thin oxide. The third use of SiO_2 is its ability to mask the entry of injected atoms into certain areas of the wafer surface. We can block the entry of some dopant or metal atoms with a section of SiO_2. Our fab techniques allow us to grow SiO_2 or selectively remove it during processing.

SiO_2 is formed on the wafer surface by passing oxygen over its surface at 900–1200°C. If the oxygen environment is dry, then the SiO_2 growth is slower but produces a higher quality oxide. The dry method gives more precise thin oxides on the transistor. If the oxygen environment has water vapor, then the SiO_2 growth is faster but produces thicker less precise volumes of SiO_2. The wet oxygen method is used to grow the less precise thicker field oxides that electrically isolate sections of the die. The reactions for dry and wet SiO_2 growth are

$$Si + O_2 \rightarrow SiO_2$$
$$Si + 2H_2O \rightarrow SiO_2 + 2H_2$$

SiO_2 forms at the wafer surface so that subsequent reactions of the Si and O_2 atoms rely on O_2 diffusing through the already formed SiO_2 layer to reach the pure Si atoms at the bottom of the SiO_2 boundary. The volume density of SiO_2 is larger than Si, so the SiO_2 bulges up from the initial Si surface.

In summary, the initial processing step grows a thin SiO_2 layer of about 1 μm over the whole wafer surface. The next step will remove certain areas of the SiO_2 film exposing the underlying pure Si to injected dopant materials that form p- or n-doped wells. The Si regions beneath the 1 μm SiO_2 layer are protected from these dopants by the masking properties of SiO_2 to these injected atoms.

12.4.2. Photolithography

Wikipedia, the Internet-based free encyclopedia (http://en.wikipedia.org), offers a good description of photolithography.

> "Photolithography (also optical lithography) is a process used in micro fabrication to selectively remove parts of a thin film (or the bulk of a substrate). It uses light to transfer a geometric pattern from a photomask to a light-sensitive chemical (photoresist, or simply "resist") on the substrate. A series of chemical treatments then engraves the exposure pattern into the material underneath the photoresist. In a complex integrated circuit, a wafer will go through the photolithographic cycle up to 50 times.
>
> Photolithography resembles the conventional lithography used in printing, and shares some fundamental principles with photography. Photolithography allows precise control over the shape and size of the objects it creates, and because it can create patterns over an entire surface simultaneously. Its main disadvantages are that it requires a flat substrate to start with, it is not very effective at creating shapes that are not flat, and it requires extremely clean operating conditions."

Figure 12-3 shows an oblique magnification of a modern IC with five levels of metal. The upper metals were removed leaving the bottom two metal layers. The vertical vias and contacts are visible. The contacts have a long rectangular shape. The transistor locations can be found by locating the gate and diffusion (drain-source) contacts. The wider metal line on the lower right is a GND line and the upper wider line is VDD. The transistor diffusion, well, and polysilicon gate regions are not visible. Photolithography describes how these fine-resolution physical geometries are accomplished. These exact structures are made using photolithography.

Photolithography has similarities to normal photography. IC fab takes the basic principle of photography and extends the camera resolution to tiny nanometer sizes for circuits. Metal line widths can be resolved to as little as 22 nm and smaller. This is smaller than the smallest wavelength of visible light at about 350 nm, so these images cannot be viewed with an optical microscope.

The photolithography process projects a sharp image onto a die through a mask. For example, the image may define n-wells or p-wells, drain or source regions, thin oxide, polysilicon gates, contacts and vias, and all of the upper level metal interconnects. We saw these shapes in the circuit layouts of Chapter 11. The photolithography process makes these geometric definitions one mask at a time. If we define a p-well region and implant a boron

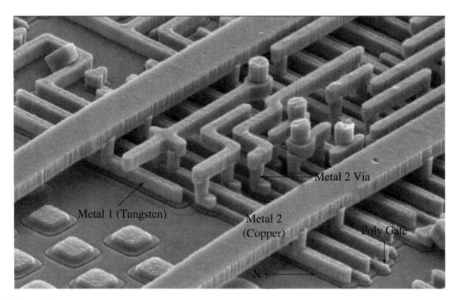

FIGURE 12-3.

Lower metal levels of a microprocessor showing contacts to gate, drain, and source. (Reproduced by permission of Chipworks Inc.)

dopant, then we don't want boron to implant in other areas of the die. The simultaneous targeting of the well for boron implant and protection for the non-well areas requires a *photoresist* material.

There are two types of photoresist: *positive* and *negative*. Positive photoresist breaks down under ultraviolet (UV) light while negative photoresist hardens. The weakened bonds of the photoresist can be chemically removed. Positive photoresist is better for making the small nanometer dimensions of modern circuits, so it will be used in our descriptions. The mask is made of glass or quartz using chrome for the opaque areas. Photoresist is applied by placing a drop on the wafer and spinning the wafer at high speed to achieve a thin 1 μm film.

The photolithography process uses UV radiation through a mask onto the die photoresist and is repeated several times in subsequent circuit construction. *n*- and *p*-wells, thick and thin oxides, drain and source diffusions, contacts and vias, and all upper-level metals are constructed using this basic photolithography process. Only the instruments that deposit the specific atoms change.

12.4.3. Etching

The end result of a photolithographic step often requires removal of material such as SiO_2 or Si_3N_4 dielectrics. This is typically done with a *dry etch* that uses high-energy halogen gases to attack the wafer surface. These source elements are often Cl_2 or Fl_2 or compounds containing them. These halogen elements are introduced into an RF chamber containing an exposed, electrically grounded wafer. Low chamber pressure and application of an RF field of about 13.56 MHz converts the source materials to highly reactive free radicals. The free radicals etch the exposed target material on the wafer.

The dry etch can be *anisotropic* meaning that it etches with a strong directionality. Straight geometric etch definitions are made in the material under attack. This etch property contrasts with the *wet etch* method that uses strong acid or base chemicals. Wet etches are *isotropic* meaning that the etch moves in all directions. This is unacceptable in the small geometry features of today's circuits, so dry etching is typically preferred.

12.4.4. Deposition and Implantation

After the photolithographic step, another step places atoms into regions so that doped transistors can be formed. The popular deposition methods are *diffusion, ion implantation,* and *chemical vapor deposition.* Diffusion is used where large area, non-dimensionally critical depositions are needed such as growing a thick SiO_2 surface over the whole wafer. Ion implantation accelerates charged atoms (ions) into the wafer solid surface, and can implant very fine dimensions. Chemical vapor deposition (CVD) deposits thin film layers on the wafer surface by a chemical reaction.

CVD uses a reaction of multiple molecules and atoms above the wafer surface at elevated temperature. The exact temperature depends on the reactant chemicals but is typically between 600°C and 900°C. CVD gives excellent uniformity of thin films, and its relatively lower processing temperatures reduce the diffusion of atoms previously introduced into the wafer. For example, precisely positioned shallow drain and source regions will undergo significant diffusion of the dopant atoms if the temperature is 1000°C or higher. This will alter the desired electrical properties of the final transistor. Process engineers refer to a *thermal budget* that is a temperature-time budget maintained throughout the process to keep atoms from significant displacement by diffusion. CVD has a lower process temperature than a diffusion process.

An example of CVD of silicon nitride deposition is a CVD 800°C reaction using silane (SiH_4) and ammonia (NH_3)

$$3\,SiH_4 + 4NH_3 \rightarrow Si_3N_4 + 12\,H_2$$

Si_3N_4 forms on the surface and grows vertically.

Another silane example is CVD 650°C deposition of pure silicon to form an epitaxial layer.

$$SiH_4 \rightarrow Si + 2\,H_2$$

Epitaxial layers are thin, doped silicon crystals grown on top of the substrate. Epitaxial layers can protect the circuit against a damage mechanism called latchup that is a runaway current condition that can severely damage the circuit.

CVD is flexible allowing surface depositions of several molecule types and atoms. Table 12-1 lists a few among many capabilities. CVD can deposit metals.

TABLE 12-1 CVD thin films

Deposited Film	
Silicon (epitaxial)	Si
Polysilicon	Si
Silicon dioxide	SiO_2
Silicon nitride	Si_3N_4

Ion implantation converts atoms in a chamber to an ionic state and then accelerates those ions into a silicon substrate. The ion impact energy and mass determine the average depth, and the flux magnitude (ions/(cm$^2 \cdot$ s)) determines the amount of material implanted. Ion implantation is dominant in modern ICs for drain and source doping, gate polysilicon doping, threshold adjustments, and making shallow wells in small geometry modern ICs. Ion implantation allows better depth precision and control of doping concentration of the implants. The impurity concentration forms a Gaussian distribution at the targeted depth. Importantly, ion implantation allows good impurity depth control at a low temperature.

Ion implantation is the preferred technique to implant impurities in modern nm technologies. It typically uses SiO_2 or Si_3N_4 as implant masks. A side effect is the crystalline damage caused by the high-energy atoms. The ions scatter in the Si crystal disrupting the ordered array of atoms. A recovery technique uses rapid thermal anneal (RTA). RTA uses high wattage lamps to rapidly heat the wafer to about 1000°C in about 20 seconds, holding the temperature for about 1–2 minutes, then allowing rapid cooling. These fast anneals reduce the high temperature and diffusion exposure of the concentration profiles of previously implanted atoms.

Diffusion was the dominant method in forming deep well (>0.5 μm) and drain-source junctions. Modern nm technologies require accurate well junction depths placed at <300 nm and drain-source junctions <200 nm. But the isotropic diffusion property and lack of precise control led to the ion-implant technique. Diffusion theory is important since it is relevant to the undesirable rearrangement of ion-implanted junctions by subsequent high temperature operations. Diffusion dynamics in a solid is modeled by Fick's laws, which is a topic beyond this course. Unintended diffusion is ever present during the fab process, and it must be controlled. Since diffusion deposition occurs from a high surface concentration the depth–concentration profile falls off with distance from the surface in a half Gaussian distribution.

Figure 12-4 illustrates these first operations.

12.5. Cleaning and Safety Operations

An intense cleaning operation occurs prior and after each processing operation. Residual particles from a previous step would kill the functionality of a circuit. Room cleanliness is measured in particles per cubic foot for particles <0.5 μm. The required cleanliness in

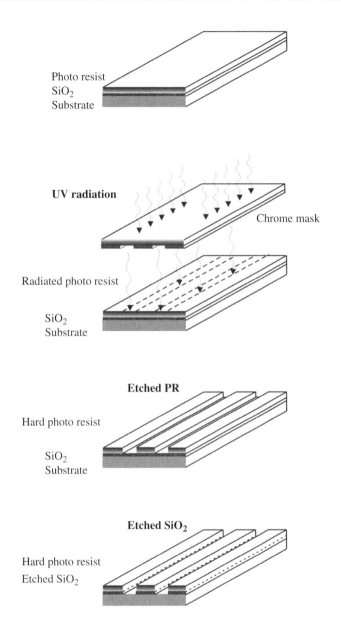

FIGURE 12-4.

SiO_2 deposition on substrate, photoresist layering, UV radiation through the mask, and etching of softened photoresist.

modern fabs are Class 1, Class 10, and Class 100 meaning that the filtered air has less than 1, 10, or 100 particles per cubic foot. Typically, unfiltered air has on the order of several million particles per cubic foot. Humans, dust, and moving machinery are major sources of particles. Humans are now excluded from major processing steps being replaced by robotic equipment. Cleaning solutions include very pure deionized water, organic solvent, and acids or bases.

A fab uses many chemicals that are hazardous to humans. Silane and arsenic are highly toxic, while sulfuric acid, hydrofluoric acid, NaOH, and high temperatures are extreme hazards. Safety is an issue any time that a person is potentially in contact with these chemicals.

Yield is defined as the fraction of good die surviving the fab process. Yield and fab cleanliness are tightly linked. IC fabrication has an unusual property that good die yield gets better as die size gets smaller. Imagine five randomly placed killer particles that land on a wafer with 50 die. Five die fail and the yield is $Y = (50 - 5)/50 = 90\%$. If the die size is shrunk and the wafer now has 100 die, then the same five killer particles fail five die and the yield is $Y = (100 - 5)/100 = 95\%$. This is true to a first order and is a major reason personal computer chips are the same cost or cheaper than they were 15 years ago.

12.6. Transistor Fabrication

Transistors require a doping in substrate or wells, polysilicon gates, and drain and source diffusion regions. Figure 12-5 shows an initial step of a twin well construction. The p- and n-wells are implanted on top of the silicon substrate. Electrical isolation is done using a sheet of SiO_2 in a technique called *shallow trench isolation* (STI). The middle figure shows the thin SiO_2 layer over each transistor channel area. The bottom figure shows the polysilicon laid down on the thin SiO_2. The poly width of the p-channel gate is drawn wider than the n-channel gate to illustrate typical poly geometry. The drains and sources are implanted with either As^- or B^+ in the p- and n-channel transistors after the poly is deposited. The poly provides a mask for the implanted ions so that they are located only in the designated diffusion regions. BOL will place contacts in the gate, source, and drain regions.

12.7. Back End of Line BOL Fab Techniques

The back end of line process constructs the metal contacts, vias, and layers of the flat line interconnects. The BOL also inserts the dielectric that electrically separates and mechanically supports these metals. A third BOL function places a barrier metal silicon compound (Ta Si_2, $TiSi_2$) at the surface of the drain, source, and polysilicon gate regions to reduce resistance. This compound is called a *silicide*, and its lower resistance increases chip performance (speed).

The dominant trend in metal technology today is copper for the upper metal vias and flat lines, and tungsten for the contacts and first level metal. Copper has propagation delay and economic advantages over the more traditional aluminum technology, but Al is still active for many products. We will first describe the metal *sputtering* deposition technique that is useful for Cu, Al, W, barrier metals, and metal alloys.

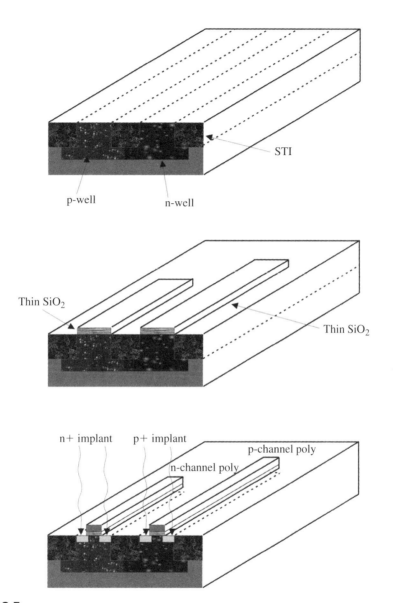

FIGURE 12-5.

Transistor construction preceded by STI, thin oxide deposition, poly gate deposition, and n^+ and p^+ ion implant using poly gates as masks for implant.

12.7.1. Sputtering

Sputtering is a low-temperature, low-pressure technique that deposits metal atoms on the surface of an exposed wafer. Argon (Ar) gas is placed in a low-pressure chamber containing a target wafer (anode) and a source block of the metal that is to be deposited (cathode).

A high DC voltage is placed across the source and target material that ionizes the Ar to Ar^+ allowing the ions to accelerate into the cathode source material at high velocity. The impact ejects the source metal atoms, and these neutral atoms are then deposited on the wafer. While no heat is applied during this operation, the metal absorption on the wafer can heat the wafer to $>300°C$.

Sputtering can deposit most metals and some dielectrics (SiO_2). Cu technology uses sputtering to deposit tungsten as contacts and first level metal flat lines, and to seed layers of Cu for flat interconnect lines. If the source material is an alloy, then the alloy composition is deposited on the wafer. Sputtering has good target definition for small, sharp features.

12.7.2. Dual Damascene

Cu technology evolved a more economic way to deposit metal. Its strategy combines vias and flat line construction in one process step for a given metal layer. This is known as the *dual damascene* process. It is much cheaper than using two steps to make the vias and then the flat line above.

Cu dual damascene first etches the via holes and flat line trenches. Via etching is challenging since the hole diameter is small, on the order of 22–100 nm. The ratio of hole depth to diameter (aspect ratio) is on the order of 2–5 making etching and subsequent post-etch cleaning difficult. Many IC designs have billions of vias connecting up to ten levels of metal making vias a primary defect site.

The vias and trenches are patterned with photoresist, radiated, developed, and dry etched with a plasma followed by a plasma cleaning to eliminate the residue ash from the etch. The via etch is typically done first followed by a separate etch of the flat line trenches. The next operation deposits a thin barrier metal (TaN) on the sides and bottoms of the vias and trenches. The barrier metal is necessary to contain copper from diffusing through the dielectric and poisoning the transistors. The barrier metal layer tries to maintain a thickness on the order of 10% of the metal width. Barrier resistivity can be $200\times$ that of Cu. The thinner the barrier, the lower the total via resistance, but the higher the barrier electrical resistance.

After the barrier metal is deposited, a thin seed of Cu is deposited onto the barrier metal. Next, the wafer is immersed in a Cu electroplate bath to fill the vias and trench volumes. The electroplate operation ends with an overfill of Cu covering the whole top of the wafer. Since we only want Cu covering the trenches and vias, the overfill must be removed. This is done by polishing off the overfill to make a smooth, planarized surface. This complex operation is called *chemical mechanical planarization* (CMP). Initially, a chemical reaction softens the surface, and that is followed by a polishing pad abrasion using a slurry of fine, hard particles to remove the Cu surface. It is the combined vertical force on the pad, the slurry, and the horizontal pad motion that removes surface atoms. The Cu is polished so that the trenches are separated from each other by a dielectric. Subsequent photolithography steps require CMP to make a very planar surface so that the mask-photoresist exposures can focus the UV in one plane.

TABLE 12-2 Material Relative Dielectric

Dielectric	ε_r
SiO_2	3.9
SiO_2– Fl	3.5
Aerogel	< 2.2
Air (vacuum)	1.0

12.7.3. Interlevel Dielectric and Final Passivation

Metal layers that build up need a dielectric filler to electrically isolate the wires and to provide mechanical stability. This dielectric is referred to as the interlevel dielectric (ILD). The dielectric constant of the ILD affects the coupling capacitance and therefore should be kept as low as possible. The traditional SiO_2 has an $\varepsilon_r = 3.9$ while fluorinated SiO_2 has about $\varepsilon_r = 3.5$. Table 12-2 shows the dielectric constant of a few materials considered as ε_r is driven lower. Aerogel dielectrics provide lower ε_r but have challenging mechanical properties particularly in the thermal cycling environment of the fab. Air (vacuum) is the ultimate dielectric and is being explored as of this publication.

12.8. Fabricating a CMOS Inverter

12.8.1. Front End of Line Operation

The techniques and tools described thus far are coordinated in a complex fab process to make a whole circuit. This section will describe how these individual processes come together to make a twin well CMOS inverter. A twin well has a p-well and an n-well.

We will start with a bare single crystal wafer that is p-doped during the ingot growth. A more lightly p-doped *epitaxial layer* is grown over the whole wafer surface by CVD. The epi layer will serve as the p-substrate for the nMOS, and its higher resistivity gives protection against the latchup failure mechanism mentioned previously. Next, a thin SiO_2 layer is grown over the whole wafer on top of the epi layer. This thin SiO_2 layer will later become the transistor thin oxide. A thin sacrificial nitride of Si_3N_4 is then grown by CVD over the thin SiO_2 and epi layers and then patterned. The Si_3N_4 acts as a sacrificial buffer meaning it will not be part of the inverter construction, but temporarily serves to protect the transistor thin oxide from an etch that comes next.

The first phase of the photoresist-mask-radiate-resist develop and etch operation exposes areas that will be dry-etched to allow SiO_2 thick field oxide pockets. SiO_2 grown in these pockets provide electrical isolation. These SiO_2 structures include the thick field oxides and SiO_2 shallow trench isolation (STI) regions that protect the p-channel transistors from the n-channel transistors. The whole surface is then smoothed and planarized with a CMP process that removes the sacrificial nitride pattern. The top layer now has three SiO_2 regions: the thick field oxide, the STI, and the transistor thin oxide.

A thick Si_3N_4 layer is then grown over the whole wafer. The subsequent n-well mask and photolithography etch operation defines the regions to be implanted with arsenic (As) to form the n-wells. That region has the transistor thin oxide exposed at the surface. Arsenic atoms are implanted at a depth just below the Si–SiO_2 interface surface followed by a boron (B) implant that adjusts the p-MOS transistor threshold voltage.

The next photoresist-mask-radiate-resist develop operation exposes regions upon which the polysilicon gates are grown. The Si_3N_4 sacrificial layer is removed by CMP followed by a fresh layering of Si_3N_4 necessary as a pre-step to implanting the p-wells. The p-well mask and photolithography etch operations define the regions to be implanted with boron to define the p-wells. A subsurface boron implant forms the n-wells, and this is followed by a different dosage and energy level of B implant to adjust the n-transistor threshold voltage. A CMP operation removes the sacrificial nitride and planarizes the surface in readiness for the next step, which is the construction of the polysilicon gates.

The polysilicon gates are constructed using another Si_3N_4 layer repeating the photoresist-mask-radiate-resist develop operation to expose regions upon which polysilicon is grown. Polysilicon is a good blocker of ion implanted As and B that will form the drain and source regions of the nMOS and pMOS transistors. Although the polysilicon blocks most of the ions being implanted, it still acquires enough of these ions to become n-doped for the nMOSFET gates and p-doped for the pMOSFET gates. This doping is desired since it reduces the resistivity of the gate and subsequent signal RC time constant. The thin oxide not protected by the poly gates is dry etched removed, and an insulating layer of SiO_2 is grown over the whole wafer. The transistor operation is complete. The BOL operation begins with the metal contacts that are described next.

12.8.2. Back End of Line Operation

The resistance between the transistor silicon and the contacts can be reduced by depositing a barrier metal that forms a surface compound with silicon. When these silicides are layered on the transistor diffusion and gate regions, they significantly lower the resistance.

Transistor contact holes are defined by the resist-mask-radiate-resist develop operation to expose regions upon which dielectric holes will be etched. The holes for the gate, drain, and source contacts are dry etched and cleaned followed by a tungsten (W) sputtering into these holes. The first level of metal in copper systems is made of W. Tungsten contacts and first level metal provides a physical separation between the copper on the higher metal levels and the transistors. This protection is needed since copper readily diffuses in SiO_2 and Si, and Cu will poison the transistors. Although W resistivity is much higher than Cu, the signal lines at the first metal level are kept short so that RC delay is not a serious factor.

The first-level metal W trenches are defined and etched followed by a W sputtering. Subsequent metal vias and layers are made of copper. The dual damascene technique builds up these higher-level metal layers. A CMP step is necessary to preserve planarization for each subsequent focused mask radiation step. After all metal levels are constructed, the final step layers a protecting insulation over the top of the whole wafer. This is typically Si_3N_4 that mechanically protects the circuit as well as gives protection against humidity and any other airborne contaminants to the circuit.

12.9. Die Packaging

The completed die are typically mounted in a mechanical package that comes in many configurations. Figure 12-6 shows four of the package types. The package variables might include the number of IC pins, typically from 12 to over 600, the heat generated by the IC and its thermal conductivity, the protection needed from humidity, and the pin layout geometry. Packages are expensive and may require thermal and stress analysis to ensure long life for the part. Failure analysts report that as much as 90% of failure returns from the field are due to packaging problems.

Figure 12-6(a) shows a Dual-In-Line (DIP). It is an original package still in use, but many die require more pins than supplied by the smaller packages. The progressive pin count increase required different pin geometries. Figures 12-6 (b,c,d) show a pin grid array covering the flat surface of the package, a quad flat package with leads on four sides, and a package with no leads that solders directly to the metal pad on the printed circuit board.

ICs with large pin requirements often use ball grid arrays with tiny solder balls that are melted into contacts on the PCB. The grid arrays allow more freedom in insertion of the

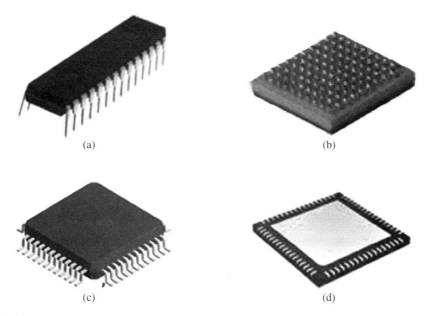

(a)   (b)

(c)   (d)

FIGURE 12-6.

(a) Dual- In-Line Package (DIP), (b) Pin Grid Aray (PGA), (c) Quad Flat Package (QFP), (d) Quad No Leads Package (TQFN). (Reproduced with permission of Silicon Far East Corp.)

VDD, GND rails, and timing signals into the inner area of the die. Small and thin outline packages are light providing an advantage in miniaturized products.

12.10. IC Testing

One of the last processes is the complex IC test operation. Testing is done at the wafer level by small diameter contact probes. Testing is repeated after the packaging. Testing ensures that all specifications claimed by the manufacturer are fulfilled. This includes the operational specifications and pin parameters, and also seeks to guarantee that the IC is free of defects. These test operations are costly and complex with costs going as high as 50% of the IC manufacturing costs.

12.11. Summary

Chip fabrication is a complex process whose basic operations should be known by any engineer dealing with chips. Chip designers, test engineers, reliability engineers, failure analysts, managers, and quality control people are all affected by the successes and failures that go on in fabrication. This chapter only touched on basic processes that hopefully will lead you to take a full course on the subject [1].

Reference

[1] Richard Jaeger, *Introduction to Microelectronic Fabrication, Volume V*, 2nd Ed., Prentice-Hall Pub., 2002.

ANSWERS TO ALL EVEN NUMBERED END OF CHAPTER EXERCISES

Chapter 1

1-2 DeMorgan $F = \overline{(W + X) + (\overline{Y\,Z})}$

1-4

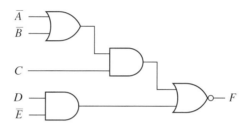

1-6 $F = XYZ$

1-8 $F = X + Y$

1-10 $R_{eq} = R1//R2 + [(R3//R4)]//[(R5//R6) + R7]$

1-12 (a) $V_O = \left[\dfrac{50}{100 + 50}\right] 5\,V = 1.667\,V$

     (b) $V_O = \left[\dfrac{5\,V}{100\,k + 50\,k}\right] 50\,k\Omega = 1.667\,V$

1-14 (a) $V_{O1} = 0.6554\,V$

     (b) $V_{O2} = 1.663\,V$

1-16 $I_{300} = 2\,mA$

    $I_{400} = 1.385\,mA$

    $I_{900} = 0.6154\,mA$

    $I_{2k} = 0.8571\,mA$

    $I_{1.5k} = 1.143\,mA$

1-18 (a) $V_O = 0.692\,V$

     (b) $I_{2M} = 346\,nA$

1-20 $I_{2k} = 72.5\,\mu A$

1-22 $I_{6k} = 925.9\,\mu A$

1-24 $C_{eq} = 60.87\,pF$

    $W_{parallel} = 35.3\,pJ$

1-26 $C_1 = 6.48$ nF
$W = 23.27$ nJ
1-28 $V_O = -4.675$ V
$I_1 = 14.675$ mA
$I_3 = 2.208$ mA
$I_2 = 16.88$ mA
1-30 Depending on assumption, you may get two solutions. Which one is more accurate?
$V_O = 46.59$ mV or 41.84 mV
1-32 $V_D = 380.7$ mV
1-34 $I_D = 2.12$ mA
$V_D = 378.7$ mV
1-36 $I_D = 451.5$ μA
$V_D = 141.0$ mV

Chapter 2

2-2 The fraction is 2.044×10^{-11} carriers/Si atom or about one free electron per 48.9 billion silicon atoms.

2-4

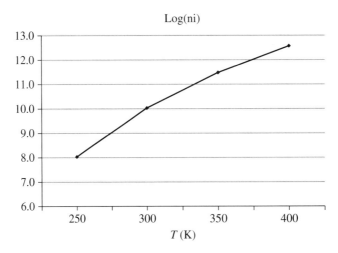

2-6 $n_o = 1.128 \times 10^3$ electrons/cm^3
2-8 $n_o = 5.67 \times 10^5$ cm^{-3}
2-10 $T \approx 344.3$ K
2-12 $p_o = 8.824 \times 10^4$ holes/cm^3
2-14 $N_A = 4.34 \times 10^{16}$ holes/cm^3

2-16 **(a)** $E = 52.1$ V/cm

(b) $d = 384$ μm

2-18 $N_D = 2.315 \times 10^{15}$ atoms/cm^3

2-20 $dp_o/dx = 8.013 \times 10^{14}$ cm^{-1} and $dn_o/dx = 2.404 \times 10^{15}$ cm^{-1}

2-22 $J = 541$ A/cm^2

2-24 **(a)** $V_{bi} = 0.835$ V

(b) $V_{bi} = 0.704$ V

2-26 $N_A = 1.45 \times 10^{15}$ atoms/cm^3

2-28 $I_D = 275$ mA

2-30 $\Delta V = 119.7$ mV

2-32 $Cj = 17.5$ fF

2-34 $C_j = 49.2$ pF

Chapter 3

3-2 **(a)** $I_D = 2.688$ mA

(b) $R_D = 260.4$ Ω

3-4 $V_G = 2.507$ V

3-6 $V_O = 4.161$ V

3-8 $V_O = 0.602$ V

$I_D = 12$ μA

3-10 $R_1 = 2.07$ $k\Omega$

3-12 $R_O = 326.8$ Ω

3-14 **(a)** $V_G < V_D + V_m$, so transistor M1 is saturated and emitting visible light from its drain-channel region.

(b) $I_D = 40.43$ μA $V_{GS} = 1.119$ V $V_{DS} = 3.119$ V

3-16 $I_D = 29.4$ μA and $V_O = 2.94$ V

3-18 $V_O = 1.446$ V and $I_D = 28.92$ μA

3-20 $V_O = 1.730$ V and $I_D = 10.27$ μA

3-22 Transistor is in nonsaturation state and $I_D = 2.88$ mA

3-24 **(a)** $I_D = 392$ μA

(b) Non saturated and $W = 24.3$ nJ

3-26 $V_S = 1.575$ V

$V_D = 0.7125$ V

3-28 $V_G = 100$ mV and $I_D = 180$ μA

3-30 $R_1 = R_2 = 20.83$ $k\Omega$

3-32 **(a)** $V_G = -3.6$ V

(b) $I_{Dp} = 108$ μA

3-34 $R = 466.7 \ k\Omega$
3-36 $I_D = 140 \ \mu A$
3-38 $V_O = 1.26 \ V$

Chapter 4

4-2 **(a)** $RC = 860 \ fs$
  (b) The resistance divides by two so new effective resistance is $R' = R$. The effective capacitance doubles since the area of the metal doubles, therefore the time constant remains the same –no gain in RC.

4-4 $R = 9.74 \ \Omega$
4-6 $R \ (50°C) = 38.46 \ \Omega$
  $R \ (150°C) = 52.0 \ \Omega$
4-8 $T = 154.3°C$
4-10 **(a)** $R_{sheet} = \dfrac{43 \ m\Omega}{\text{square}}$

  (b) $R_{sheet} = \dfrac{22.9 \ m\Omega}{\text{square}}$

4-12 $t = 0.708 \ \mu m$
4-14 $R_{Al} = 143.6 \ m\Omega$
  $R_{Cu} = 1.09 \ \Omega$
4-16 $C_{SiOC} = 20.52 \ fF$
  $C_{Air} = 7.1 \ fF$
4-18 $V_O = V_{in}e^{-\frac{t}{RC}}$
4-20 **(a)** $RC = 2.02 \ ns$
  (b) $\tau = 1.35 \ ns$

4-22 $\dfrac{di}{dt} = 2.605 \times 10^7 \ A/s$
4-24 $n = 125 \ lines$

Chapter 5

5-2 **(a)** $NM_H = 1.7 - 1.6 = 100 \ mV$
  (b) $NM_L = 0.3 - 0.2 = 100 \ mV$
  (c) A $-50 \ mV$ spike voltage wouldn't cause a false logic transition.
  (d) A $-150 \ mV$ spike would put the logic high input at a level not guaranteed to be read correctly. The $-150 \ mV$ spike puts the input to the gate close to the linear transition bias region.

5-4

$NM_L = 0.18$ mV and $V_{IL} = 280$ mV

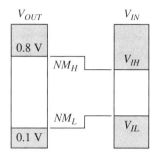

5-6 (a) $\dfrac{W_p}{W_n} = 2.8$

(b) $\dfrac{W_p}{W_n} = 6.3$

5-8 73.4%

5-10 $I_{Dn} = 45.94\ \mu A$

5-12 $V_{tp} = -0.472$ V

5-14 $V_i = 0.947$ V

5-16 102 ps

5-18 34.5%

5-20 $F_{MAX} = 347$ MHz

5-22 (a) $I_D = 42.2\ \mu A$

(b) $I_D = 3.44\ \mu A$

5-24 $n = 7$ inverters

5-26 (a) $C_L = 7.4$ pF

(b) $W_L / W_{in} = 54.6$

Chapter 6

6-2 $I_{DD} = 89\ \mu A$

6-4 $I_{DD} = 89\ \mu A$

6-6 Node-$A = 1$, Node-$B = 0$

6-8 $V_O = 0.333$ V and $V_{in} = 0.833$

6-10 $V_{in} = 0.833$ V and $V_O = 1.333$ V

6-12 (a) $AB\ CDE\ FG = 11 \times 11\,00$

(b) $AB\ CDE\ FG = 11\,000\,0\times$

6-14 (a) $AB\ CD\ EF = 00\ 0\times\ 11$

(b) $AB\ CD\ EF = 00\ 00\ \times 0$

6-16 Node-$A = 0$, Node-$B = 1$

6-18 *Student Answer*

6-20 **(a)** $F = \overline{ABCD + E}$

(b) See sizing in figure.

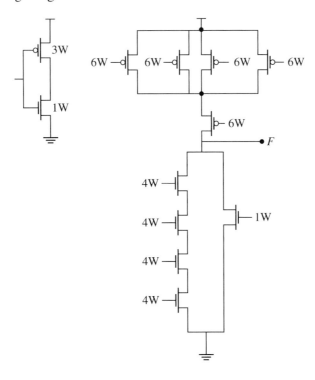

6-22

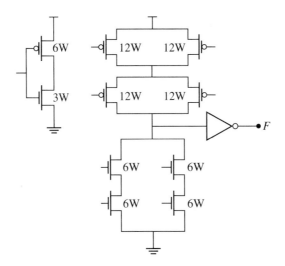

6-24

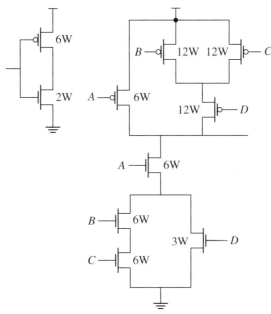

6-26 $V_S = 1.472$ V and $V_{GS} = 0.528$ V

Chapter 7

7-2 $F = \overline{A} + \overline{B} + ABC$

7-4

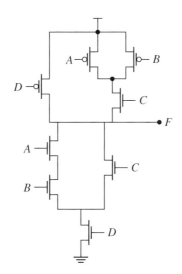

7-6 The circuit is an odd parity exclusive OR gate.
7-8

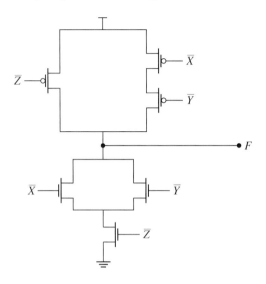

7-10

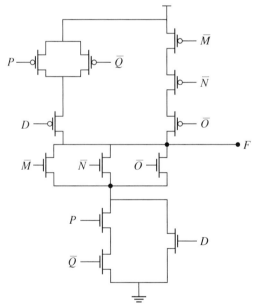

7-12

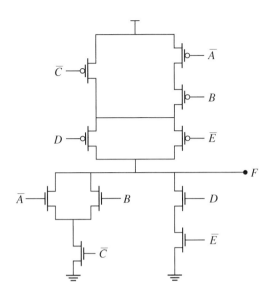

7-14 $F = \overline{\left(\overline{A} + \overline{B}\right) \left(\overline{C} + \overline{D}\right) \overline{E}}$

7-16 $F = B + AC$

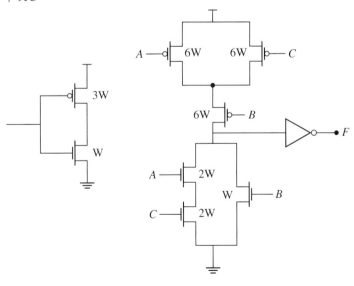

7-18

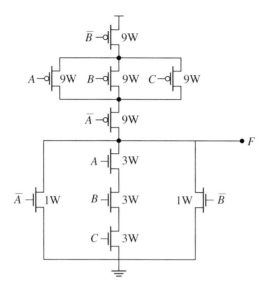

7-20

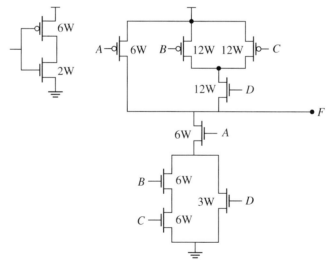

7-22 $V_{final} = 1.0$ V
7-24 $V_f = 1.33$ V
7-26 $V_f = 1.84$ V
7-28 $F = (A + B)CD$

7-30

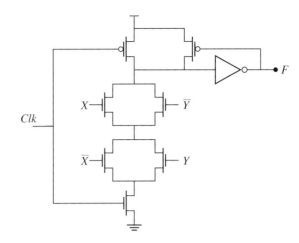

7-32

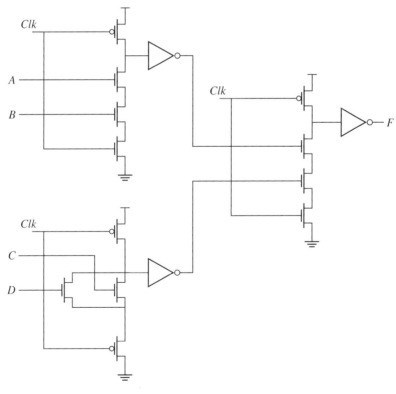

7-34 $V_{OUT} = 1.5$ V

7-36

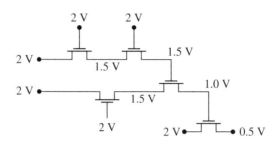

7-38 Node-$D = 0.1875$
Power node-$D = 675$ nW
Node-$F = 0.1094$
Power node-$F = 590.6$ nW

7-40 $\alpha_{0 \to 1} = 0.152$
$\alpha_{1 \to 0} = 0.152$

7-42 $P_{dyn} = 468.75 \ \mu W$

Chapter 8

8-2

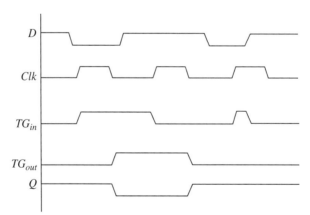

8-4 (a) $t_{su} = 1.2$ ns
$t_{hold} = 0$ ns
$t_{cq} = 1.2$ ns

(b) CW(low) ≥ 1.2 ns
CW(high) ≥ 1.2 ns

(c) No effect

8-6 $I = 1.5$ ns and T-gate $= 1.5$ ns

8-8

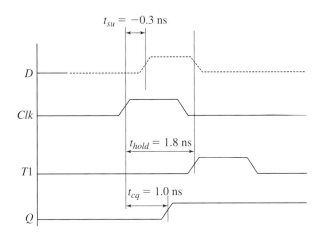

8-10 $t_{logic} \leq 376.7$ ps

8-12 $\delta \geq 180$ ps

8-13 $F_{MAX} = 1.087$ GHz

8-14 $F_{MAX} = \dfrac{1}{780 \text{ ps}} = 1.282$ GHz

8-16 $F_{MAX} = 1.136$ GHz

8-18 $t_{logic} < 204.5$ ps

8-20 $F_{MAX} = 1.24$ GHz

8-22 **(a)** $F_{MAX} = \dfrac{1}{1.905 \text{ ns}} = 525$ MHz

(b) $t_{hold} < 575$ ps

Chapter 9

9-2 $\dfrac{\beta_1}{\beta_5} = \dfrac{\beta_2}{\beta_6} = 2.88$

9-4

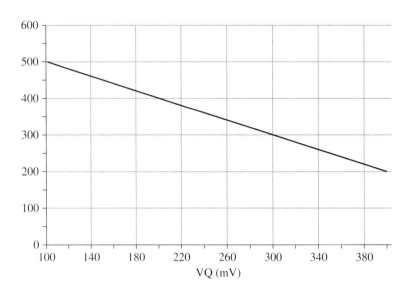

9-6 $V_{tn} = 0.528$ V

9-8 $\left(\dfrac{\beta_4}{\beta_6}\right) = \left(\dfrac{\beta_3}{\beta_5}\right) = 1.24$

9-10

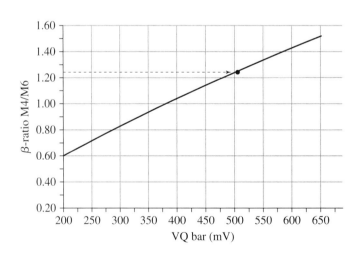

9-12 $W_4 = 0.364 \ \mu$m

INDEX

Note: Page numbers followed by "*f*" and "*t*" indicate figures and tables respectively.

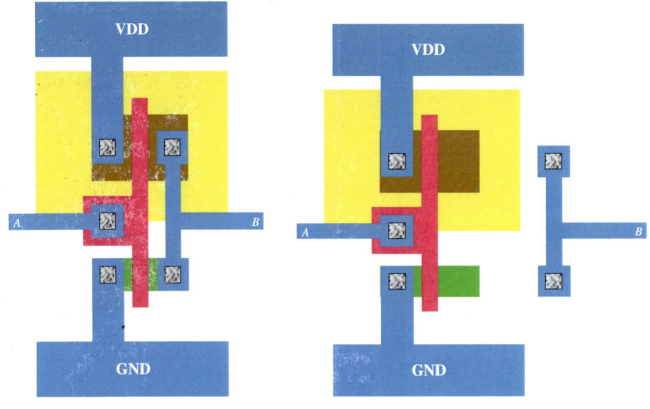

FIGURE 11-11

FIGURE 11-13

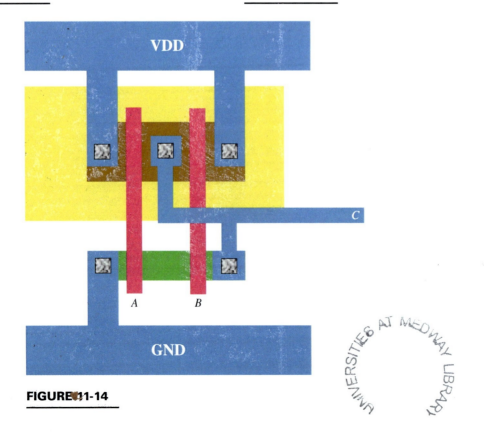

FIGURE 11-14